AF691411

A Handbook of Electronics & Telecommunications Engineering

A Handbook of Electronics & Telecommunications Engineering

Editors
Andris Krupin, Juris Medved
& Rahul Khanna

scitus
academics

A Handbook of Electronics & Telecommunications Engineering
Edited by **Andris Krupin, Juris Medved & Rahul Khanna**

Printed in 2017

ISBN: 978-1-68117-195-1
Library of Congress Control Number: 2015952018

© 2016 by
SCITUS Academics LLC,
616, Corporate Way, Suite 2, 4766,
Valley Cottage, NY 10989

www.scitusacademics.com

This book contains information obtained from highly regarded resources. Copyright for individual articles remains with the authors as indicated. All chapters are distributed under the terms of the Creative Commons Attribution License, which permits unrestricted use, distribution, and reproduction in any medium, provided the original author and source are credited.

Notice

Reasonable efforts have been made to publish reliable data and views articulated in the chapters are those of the individual contributors, and not necessarily those of the editors or publishers. Editors or publishers are not responsible for the accuracy of the information in the published chapters or consequences of their use. The publisher believes no responsibility for any damage or grievance to the persons or property arising out of the use of any materials, instructions, methods or thoughts in the book. The editors and the publisher have attempted to trace the copyright holders of all material reproduced in this publication and apologize to copyright holders if permission has not been obtained. If any copyright holder has not been acknowledged, please write to us so we may rectify.

Preface

Electronics and Telecommunication Engineering is a field that involves complex electronic apparatus, circuits and equipment's that help in executing speedy and efficient telecommunication systems. These engineers design, fabricate, maintain, supervise and manufacture electronic equipment's used in entertainment industry, computer industry, communication and defence. Ever increasing pace of development in electronics, audio and video communications systems and the automation in industry have made an electronic engineer a catalyst for the change of the modern society.

A Handbook of Electronics and Communication Engineering covers the engineering syllabus of several examinations. The Electronics Engineering section gives detail on non-linear and active electrical components which are used to design circuits, chips and devices. It also focuses on implementation of principles, applications and algorithms. Communication Engineering is divided into two parts: Analog and Digital.

Handbook of Electronics and Communication Engineering deals on anextensiveassortment of topics, including transistors, diodes, microprocessors, signals and systems, network theory and microwave engineering. The book highlights important terms and definitions, along with illustrated formulae to make learning easy, with appropriate diagrams, whenever it is appropriate. Anextensive coverage of key points for additional information is also given.

Table of Contents

A Handbook on Electronics and Communications Engineering

1

Electronic Devices & Circuit

CONTENTS

1 Energy Band and Charge Carriers in SC

Electrons are restricted to sets of discrete energy levels within atoms, with large gaps among them where no energy state is available for the electron to occupy. Electrons in solids also are restricted to certain energies and are not allowed at other energies. Difference in the solid, the electron has a range (or band) of available energies. The discrete energy levels of the isolated atom spread into bands of energies in the solid because

- In the solid, the wave functions of electrons in neighbouring atoms overlap, thus, it affects the potential energy term and the boundary conditions in the equation, and different energies are obtained in the solution, and
- An electron is not necessarily localized at a particular atom.

The influence of neighbouring atoms on the energy levels of a particular atom can be treated as a small perturbation, giving rise to shifting and splitting of energy states into energy bands.

Fermi Level

Fermi level is the term used to describe the top of the collection of electron energy levels at absolute zero temperature. So at absolute zero they pack into the lowest available energy states and build up a "Fermi sea" of electron energy states. The Fermi level is the surface of that sea at absolute zero where no electrons will have enough energy to rise above the surface.

Fermi Energies for Metals

The Fermi energy is the maximum energy occupied by an electron at 0K. By the Pauli Exclusion Principle, we know that the electrons will fill all available energy levels, and the top of that "Fermi Sea" of electrons is called the Fermi energy or Fermi level. The conduction electron population for a metal is calculated by multiplying the density of conduction electron states r(E) times the Fermi function f(E). The number of conduction electrons per unit volume per unit energy is

$$\frac{dn}{dE} = \rho(E) f_{FD}(E) = \frac{8\sqrt{2}\pi m^{3/2}}{h^3}\sqrt{E}\frac{1}{e^{(E-E_F)kT}+1}$$

The total population of conduction electrons per unit volume can be obtained by integrating this expression

$$n = \int_0^\infty \rho(E) f_{FD}\, dE = \frac{8\sqrt{2}\pi m^{3/2}}{h^3}\int_0^\infty \frac{\sqrt{E}}{e^{(E-EF)/kT}+1} dE$$

At 0K the top of the electron energy distribution is defined as EF so the integral becomes

$$n = \frac{8\sqrt{2}\pi m^{3/2}}{h^3}\int_0^{E_F}\sqrt{E}dE = \frac{8\sqrt{2}\pi m^{3/2}}{h^3}\left(\frac{2}{3}E_F^{3/2}\right)$$

This expresses the conduction electron density n in terms of the Fermi energy EF. We can also turn this around and express the Fermi energy in terms of the free electron density.

$$E_F = \left(\frac{(hc)^2}{8mc^2}\right)\left(\frac{2}{\pi}\right)^{2/3} n^{2/3}$$

Mass Action Law

This law states the relation between concentrations of minority and majority carriers at a constant temperature. It states that product of concentration of majority and minority carrier is constant at fixed temperature.

According to the law

$$n * p = n_i^2 \text{ at cons tan t temperature}$$

- n_i^2 is the intrinsic concentration given by $n_i^2 = KT^3 e^{\frac{-Eg}{KT}}$ and is constant at a temperature
- n is the concentration of electrons and p is the concentration of holes. So in p-type material, p is concentration of majority carriers and n is concentration of minority carriers. Similarly in n-type, n is concentration of majority carriers and p is concentration of minority carriers.

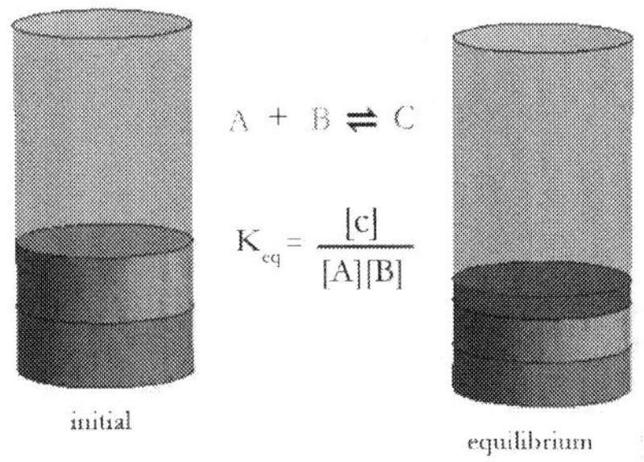

According to the law, if we increase the doping level of the material i.e. concentration of majority carriers, then concentration of minority carriers would decrease.

The mass action law is usually taught in high school chemistry, so we know what we want to find: We look at some chemical reaction, e.g.

$n_0 p_0 = n_i^2$ Where, $n_0 \rightarrow$ concentration of electron in conduction band

$p_0 \rightarrow$ Concentration of hole in valence band

$n_i \rightarrow$ Intrinsic concentration at given temperature

Intrinsic concentration is given by

$n_i = \sqrt{V_c N_v} e - Eg / 2kT$ Where, $E_i \rightarrow$ band gap

n_i Can also be given as

$n_i^2 = A_0 T^3 e^{-Eg/2kT}$ Where, $A_0 \rightarrow$ is a constant

Concentration of electron in conduction band can also be given as

$n_0 = n_i e^{(EF-Ei)/kT}$ Where, $E_i \rightarrow$ intrinsic level lies near the middle of band gap

Concentration of holes in valence band can also be given as

$p_0 = n_i e^{(Ei-EF)/kT}$

Intrinsic concentration depends on temperature. As temperature increases the intrinsic concentration increases as $T^{3/2}$.

Space Charge Neutrality

If the material is to remain electrostatically neutral, the sum of positive charges must balance the sum of negative charges i.e.

$p_0 + N^+_d = n_0 + N^-_a$

Where $N^+_d \rightarrow$ concentration of donor atoms

$N^-_a \rightarrow$ Concentration of acceptor atoms

$p_0 \rightarrow$ Concentration of holes

$n_0 \rightarrow$ Concentration of electrons

Mobility of Charge Carrier

It is drift velocity per unit electric field.

It defines how fast the charge travels from one place to other and is given by

$\mu = \frac{V_d}{E}$ Where $V_d \rightarrow$ drift velocity, $E \rightarrow$ electric field

	Ge	Si
e^- mobility	$3800 cm^2/V sec$	$1300 cm^2/V sec$
hole mobility	$1800 cm^2/V sec$	$500 cm^2/V sec$

Electron's mobility is always greater then whole mobility in a given material. Hence electron can travel faster so contribute more current for same electric field than hole as explained by quantum mechanical physics.

Mobility of charge carrier decreases with temperature and varies as:

$\mu \propto T^{-m}$ Where, m is a constant

$m = 1.66$ for e^- and 2.33 for hole for Ge

$m = 2.5$ for e^- and 2.7 for hole for Si

Mobility also varies with electric field applied as

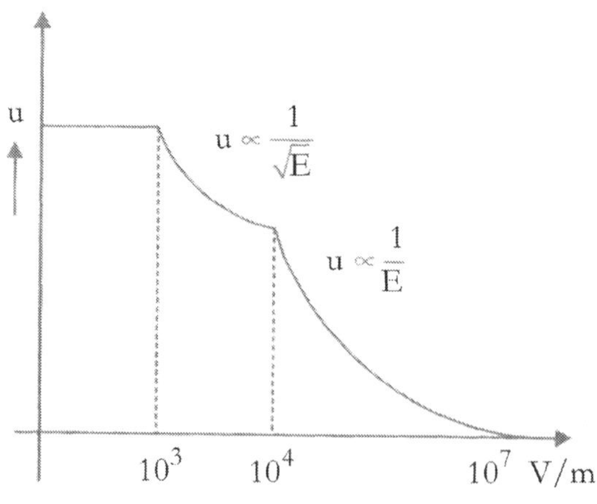

At smaller electric field mobility is constant. At very high electric field product of mobility and electric field becomes constant and is equal to saturation value of drift velocity. There are two types of scattering mechanism that influence electrons and whole mobility are lattice scattering and impurity scattering.

The variations of mobility in accordance with scattering is as

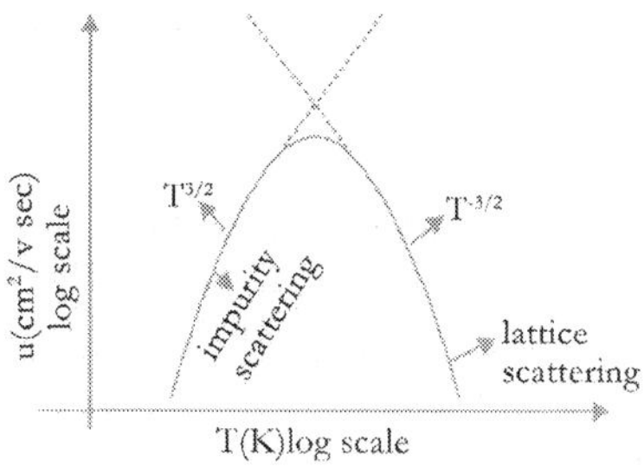

Overall mobility is given by

$$\frac{1}{\mu}=\frac{1}{\mu_1}+\frac{1}{\mu_2}+\frac{1}{\mu_3}+\ldots\ldots\ldots\ldots\ldots\ldots$$

Where, $\mu \rightarrow$ overall mobility

$\mu_1, \mu_2, \mu_3 \rightarrow$ mobility corresponding to different scattering mechanism

Energy Gap(Eg)

Energy gap depends on temperature and interatomic spacing. Variation of energy gap with temperature is as

$$E_g = E_{go} - \beta_0 T \quad \text{ev}$$

Where, $E_{go} \rightarrow$ energy gap at 0K, $\beta_0 \rightarrow$ material constant,

$\beta_0 = 2.2\times10^{-4}\,\text{eV / K for Ge} = 3.6\times10^{-4}\,\text{eV / K for Si}$

$E_g \rightarrow$ energy gap at temperature T(K)

Ego-0.785V(Ge),1.21V(Si) Eg300-0.72eV(Ge),1.1V(Si)

Ego-xxx(GaAs) Eg300-1.47eV(GaAs)

Half Effect

It states that if a speciman(metal or semiconductor) carrying a current I is placed in transverse magnetic field B an electric field is induced in a direction perpendicular to both I and B.

Value of hall voltage is given by

$V_H = Ed$ Where, $E \rightarrow$ electric field induced

$d \rightarrow$ separation between upper and lower surface of specimen or height of speciman

$$V_H = \frac{BI}{\rho W}$$

$W \rightarrow$ width of specimen

$B \rightarrow$ applied magnetic field

$I \rightarrow$ current flowing in specimen

$\rho \rightarrow$ charge density

$\frac{1}{\rho}$ is called Hall coefficient R_H i.e. $R_H = \frac{1}{\rho}$

VH can also be given as $V_H = R_H \frac{BI}{W}$

By hall experiment, mobility is given by

$\mu = \frac{8}{3\pi} \sigma R_H$ where, $\mu \rightarrow$ mobility of charge carriers

Vconductivity of material

Hall effect is utilized in determining whether a given material is metal, n-type SC or p-type SC. For metal, value of VH is lesser compare to SC. Hall voltage is +ve for metal or n-type SC. Hall voltage is –ve for p-type SC. Hall voltage is zero for intrinsic SC. It can be used in finding mobility of charge carrier, concentration of charge carrier, and types of semiconductor. It is utilized in Hall effect multiplier. In metal, R_H increaseswith temperature. In pure SC, R_H decrease with temperature. In extrinsic SC, R_H increases with temperature.

Conductivity: It is the reciprocal of electrical resistivity, and measures a material's ability to conduct an electric current.

$$\sigma = q.n\mu_n + qp\mu_p$$

Where, n→concentration e-s in conduction band

p→ concentration of holes in valence band

μ_n →mobility of e-s

μ_p →mobility of holes

Conductivity of pure SC is given by $\sigma = qn_i\left[\mu_n + \mu_p\right]$

Where, ni→intrinsic concentration

Conductivity of pure SC increases with temperature.

Conductivity of pure SC at 0 K is zero.

Conductivity of extrinsic SC is given by

$\sigma = qn_d\mu_n$ Where, Nd →donor concentration

$\sigma = qn_a\mu_p$ Na→acceptor concentration

Conductivity of extrinsic SC decrease with increase in temperature above normal temperature. Conductivity OF extrinsic SC initially increases when, temperature rises from 0 K. Conductivity increases with increase in doping temperature, Conductivity of extrinsic SC at 0K is zero. At curie temperature conductivity becomes equal to intrinsic conductivity.

Fermi Level in Intrinsic and Extrinsic SC

- Fermi level E_F in intrinsic SC is given by

$$E_F = \frac{E_c + E_v}{2} - kT\ln\left(\frac{N_c}{N_v}\right)$$

Where, E_C →lowest energy level of condition band

E_V →highest energy level of valence band

N_C →density of states in conduction band

N_V→density of states in valence band

k→Boltzmann's constant

If $N_V \cong N_C$ then Fermi level lies in the middle of energy gap. At 0 K Fermi level lies in the middle of energy gap. Fermi level in pure SC depends on temperature. As temperature increase Fermi level moves away from the middle of band gap.

Fermi level in n-type SC is given by $$E_F = E_c - kT\ln\left(\frac{N_C}{N_D}\right)$$

Where, N_C→density of states in conduction band. N_D→donor concentration

In n-type SC Fermi level depends on both temperature and donor concentration. At OK Fermi level coincides with E_C. As temperature increases Fermi level moves towards the middle of band gap. As donor concentration increases Fermi level moves towards the EC. At NC=ND. EP coincides with EC and this is the saturation value of EF, EF cannot be greater than this value. Normally Fermi level lies close to EC.

Fermi level in p-type SC is given by $$E_F = E_V - kT\ln\left(\frac{N_V}{N_A}\right)$$

Where, NV→density of states in valence band, NA→concentration of acceptors. In p-type SC Fermi level EF at 0K coincides with EV.

In p-type SC Fermi levels lies close to valence band. In p-type SC Fermi level moves away from valence band as temperature increases. In p-type SC Fermi level moves towards valance band as NA increases and attains a saturation value of EV at NA=NV. Further increase in NA will not change EF.

Shift in the position of Fermi level with respect to intrinsic Fermi level in n-type SC is given by

$$\text{Shift} = kT\ln\left(\frac{n}{n_i}\right) \simeq KT\ln\left(\frac{N_D}{n_i}\right)$$

Shift in position of Fermi level in p-type SC with respect intrinsic Fermi level is given by

$$\text{Shift} = kT\ln\left(\frac{p}{n_i}\right) \cong KT\ln\left(\frac{N_A}{n_i}\right)$$

In a material at equilibrium there is no discontinuity in Fermi level. More generally we can say that Fermi level at equilibrium must be constant throughout the material i.e.

$$\frac{dE_F}{dx} = 0$$

2 Excess Carriers in SC

Excess carriers, essential for device operation, are created by optical excitation, electron bombardment, or injected across a forward-biased p-n junction. These excess carriers can dominate the conduction process in semiconductor materials.

Let us assume that excess electron and hole concentration is created in a p-type SC at t=0 and initial concentration of excess electron and holes is Δn and Δp, then concentration of excess electrons at any other time t is given by

$\delta(t) = \Delta n e^{-1/\tau_n}$ Where, $\tau_n \rightarrow$ recombination life time or minority carrier life time.

$\Delta n \rightarrow$ Concentration at t=0

$\tau_n = (\alpha_r p_0)^{-1}$ Where, $\alpha_r \rightarrow$ constant of proportionality for recombination

P0→concentration of holes

A more general expression for carrier life time is given by

$$\tau_n = \frac{1}{\alpha r(n_0 + p_0)}$$

This expression is valid for n-type or p-type material if the injection level is low.

Steady State Carrier Generation (Quasi Fermi Level)

If we optically generate excess charge carrier in a SC by falling light on it

then optical generation rate is given by $g_{op} = \frac{\delta n}{\tau_n}$.

Where, $\delta n \rightarrow$ equilibrium access concentration electrons

$$\tau_n = \frac{1}{\alpha_r(n_0 + p_0)}$$

$$\delta n = \delta p$$

$\delta p \rightarrow$ excess hole concentration

$g_{op} \rightarrow$ optical generation rate

When excess carrier is generated then there exist different Fermi levels F_n and F_p for electron and hole called quasi Fermi levels.

Carrier concentration (resulting) is given by

$$n = n_i e^{(F_n - E_i)/kT}$$

$$p = n_i^{e(E_i - F_p)/kT}$$

Where, $E_i \rightarrow$ int rinsic fermi level

$F_n \rightarrow$ fermi level for e_s^-

$F_k \rightarrow$ fermi level for holes

$n \rightarrow$ net concentration of e_s^-

$p \rightarrow$ net concentartion of holes

Diffusion Process

Diffusion is the natural result of the random motion of the individual molecules. Diffusion happens due to concentration gradient. Diffusion current density is given by

$$J_n(diff) = qD_n \frac{dn(x)}{dx}$$

Where, $J_n(diff) \rightarrow$ diffusion current density due to electrons

$D_n \rightarrow$ electron diffusion coefficient

$$J_p(diff) = -qD_p \frac{dp(x)}{dx}$$

Jp(diff)→diffusion current density due to holes

$D_p \rightarrow$ hole diffusion coefficient

N(x)→ profile of e^-_s with respect to x

P(x)→profile of holes with respect to x

Overall current density is given by

$$J_n(X) = q\mu_n E(X)n(X) + qD_n \frac{dn(x)}{dx}$$

$$J_p(x) = q\mu_p(x)E(x) - qD_p \frac{dp(x)}{dx}$$

$$J(x) = J_n(X) + J_p(X)$$

Where, J(x)→overall current density

$J_n(x) \rightarrow$ current density due to e^-_s

$J_p(x) \rightarrow$current density due to holes

E(x)→ electric field applied

Einstein Relation

This gives the relation between the diffusion coefficient, mobility and temperature and given as

$$\frac{D_n}{\mu_n} = \frac{D_p}{\mu_p} = \frac{kT}{q}$$

Where, $\frac{kT}{q} = V_T \rightarrow$ thermal voltage Continuity Equation

This equation is given by.

$$\frac{\partial \delta p}{\partial t} = -\frac{1}{q}\frac{\partial J_p}{\partial x} - \frac{\delta p}{\tau_p}$$

$$\frac{\partial \delta n}{\partial t} = \frac{1}{q}\frac{\partial J_n}{\partial x} - \frac{\delta n}{\tau x}$$

where, $\delta p, \delta n \rightarrow$ excess carrier concentration

$\tau_n, \tau_p \rightarrow$ carrier life time for e_s^- and holes respectively

Another form of this equation is given by

$$\frac{\partial \delta n}{\partial t} = D_n \frac{\partial^2 \delta n}{\partial x_2} - \frac{\delta n}{\tau_n}$$

$$\frac{\partial \delta p}{\partial t} = D_p \frac{\partial^2 \delta p}{\partial x^2} - \frac{\delta p}{\tau_p}$$

Diffusion Length

Diffusion length for electron is given by

$L_n = \sqrt{D_n \tau_n}$ Where, $L_n \rightarrow$ diffusion length for electron

$D_n \rightarrow$ diffusion coefficient for electron

$\tau_n \rightarrow$ carrier life time for electron

Diffusion length for holes is given by

$L_p = \sqrt{D_p \tau_p}$ Where, $L_p \rightarrow$ diffusion length for holes

$D_p \rightarrow$ holes diffusion coefficient

$\tau_p \rightarrow$ carrier life time for holes

If excess carrier concentration Δn and Δp are introduced at x=0 in n-type material then at any other x excess carrier concentration will be

$$\delta p(x) = \Delta p e^{-x/L_p}$$

Where, $\Delta p \rightarrow$ excess holes concentration at x=0, $\delta p(x) \rightarrow$ excess holes concentration at x=0, $L_p \rightarrow$ diffusion length for hole

Field Intensity in a Uniform SC

Field intensity in a uniform SC is constant throughout the material, When doping is suddenly introduced into SC the conductivity first decreases, then increases with doping.

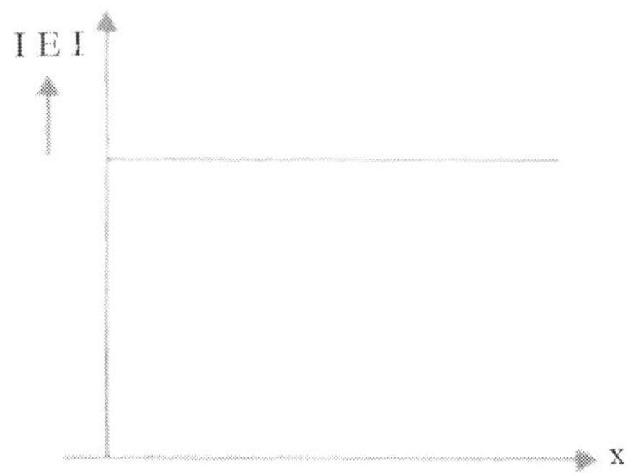

Direct Band gap SC

Most of falling e^-_s will be directly releasing the energy in the form of light and hence called direct bandgap SC. During e^-_s falling from conduction band to valence band the energy of e^- changes (both K.E. and P.E.). Momentum of e^-_s will remain same. Direction of e^-_s will remain same. Carrier life time is comparatively less.

Indirect Band gap Material

Most of falling e^-_s will be directly releasing the energy in form of heat. Momentum of e^-_s will change. Direction of e^-_s will change. Larger carrier life time.

Direct bandgap material having higher carrier life time is used for fabrication of laser. When minority carrier is injected into extrinsic SC they flow inside SC due to diffusion in absence of electric field.

3 Junction Diode

A semiconductor rectifying device in which the barrier between the two regions of opposite conductivity (n-type and p-type) type produces the rectification. All solar cells are junction diodes.

Contact Potential or Built-in Potential (V_0)

$$V_0 = \frac{kT}{q} \ln\left[\frac{NaNd}{n_i^2}\right]$$

where $V_0 \rightarrow$ contact potential

k = Boltzmann's constant (1.38×10^{-23} J/k)

$q \rightarrow$ Electron charge (1.6×10^{-19} coulomb)

$T \rightarrow$ Temperature in Kelvin of p-n junction

$N_a \rightarrow$ Concentration of acceptors (/cm^3) on p-side

$N_d \rightarrow$ Concentration of donors (/cm^3) on n-side

$n_i \rightarrow$ Intrinsic concentration (/cm^3) at given temperature

Ratio of Concentration of Holes or Electrons on the Two Sides of p-n junction

$$\frac{p_p}{p_n} = \frac{n_n}{n_p} = e^{qV_0/kT}$$

where, $p_p \rightarrow$ concentration of holes$(/cm^3)$ on $p-$side

$p_n \rightarrow$ concentration of holes $(/cm^3)$ on $n-$side

$n_n \rightarrow$ Concentration of electrons $(/cm^3)$ on n-side

$n_p \rightarrow$ Concentration of electrons $(/cm^3)$ on p-side

$V_0 \rightarrow$ Contact potential

$q \rightarrow$ Electron charge (1.6×10^{-19} coulomb)

$K \rightarrow$ Boltzmann's constant (1.38×10^{-23}J/k)

$T \rightarrow$ Temperature of p-n junction diode in Kelvin

In case of unbiased p-n junction diode electric field is maximum at the junction and decrease on the two sides of junction and zero outside the space charge region.

- Maximum electric field is given by

$$E_0 = -\frac{q}{\in}.N_d \times_{n0} = -\frac{q}{\in} Na \times_{p0}$$

where, $q \rightarrow$ electron charge$(1.6\times10^{-19}$coulomb)

$N_d \rightarrow$ Concentration of donors $(/cm^3)$ on n-side

$x_{n0} \rightarrow$ Width of depletion region on n-side

$N_a \rightarrow$ Concentration of acceptors $(/cm^3)$ on p-side

$x_{p0} \rightarrow$ Width of depletion region on p-side

$\in \rightarrow$ Permittivity of material from which diode is made

$-ve \rightarrow$ Sign is due to the fact that this electric field is from n to p.

- Contact potential in the terms of maximum electric field is given by

$V_0 = -\frac{1}{2}E_0 W$ Where, $V_0 \rightarrow$ contact potential

$E_0 \rightarrow$ Maximum electric field at junction

$W \rightarrow$ Width of depletion region

- Width of depletion region in unbiased condition is given by

$W = \left[\frac{2 \in V_0}{q}\left(\frac{N_a + N_d}{N_a N_d}\right)\right]^{1/2}$ Where, $\in \rightarrow$ permittivity of material used for formation p − n junction diode

$W = \left[\frac{2 \in V_0}{q}\left(\frac{1}{N_a} + \frac{1}{N_d}\right)\right]^{1/2}$ $V_0 \rightarrow$ Contact potential

$q \rightarrow$ Electron charge $\left(1.6 \times 10^{-19} \text{coulomb}\right)$

$N_a \rightarrow$ Concentration of acceptors (/cm^3) on p-side

$N_d \rightarrow$ Concentration of donors (/cm^3) on n-side

- If we reverse bias the diode by voltage V then in formulae of depletion width V_0 is replaced by $|V_0+V|$
- It is clear that depletion width increase with reserve and decreases with forward biased.
- Depletion width

$W \propto \frac{1}{\sqrt{\text{concentration}}}$ Concentration $\rightarrow$ concentration of donors or acceptors on n − side or p − side respectively

Where, $x_{n0} \rightarrow$ width of depletion on n − side

$x_{n0} N_d = x_{p0} N_a$ $\qquad x_{p0} \rightarrow$ width of depletion region on p − side

Also $x_{n0} + x_{p0} = W$

$N_d \rightarrow$ concentration of donors on n − side

$N_a \rightarrow$ concentration of acceptors on p − side

Forward current

$I_f \rightarrow$ forward current

$I_f = I_0[e^{-qV/kT\eta} - 1]$ Where, $I_0 \rightarrow$ reverse saturation current

V→Forward biased voltage
k→Boltzmann's constant (1.38×10^{-23}J/K)
T→Temperature in Kelvin
q→Electron charge (1.6×10^{-23}coulomb)

- Another form of these formulae is

Where, l_0 → reserve saturation or leakage current

$$l_f = l_0[e^{Vd/\eta VT} - 1]$$

V_d → forward biased voltage

η → Recombination factor (material constant)

=1 for Ge

=2 for Si

VT→thermal voltage

$$V_T = \frac{kT}{q} = 0.0259 \text{ Vat room temperature}$$

VT=

l_f → is majority carrier current

l_f → is diffusion current

l_o → drift current

Dynamic or Incremental Resistance of Diode

$$r = \frac{\eta V_T}{l_f} = \frac{dV}{dl}$$

Where, η → - recombination factor

= 1 for Ge

= 2 for Si

I_f -> forward current

V_T -4 thermal voltages

- Static Resistance

$$R = \frac{V}{1}$$

Static resistance is always greater than A.C. or dynamic resistance. ∝

- In any type of p-n junction

$$B_v \text{ or } V_{Br} \propto \frac{1}{\text{dopping}}$$

Where B_v or V_{Br} -) Breakdown voltage

Reverse saturation current (I_0) doubles for every 10°C rise in temperature.

$$I_{0(T2)} = I_{0(T1)} \times 2^{(T_2 - T_1)/10}$$

Where, $I_0(_{T2}) \rightarrow$ reverse saturation current at temperature T_2

$I_{0(T1)}$ -4 reverse saturation current at temperature T_1

Here, $T_2 > T_1$.

I_0 — only depends on temperature and is minority carrier current.

I_f (forward diode current) is independent of temperature.

Forward voltage across diode decrease with the temperature.

For 1°C rise in temperature it decreases by 2.5 mV

$$\frac{dV}{dT} = -2.5 \text{ mV}/^\circ\text{C}$$

Junction capacitance (Cj)

$$C_j = \frac{A \in}{W}$$

Where, A -> area of cross-section of diode

W -> width of depletion region

$\in \rightarrow$ Permittivity of material C

$$C_j \propto \frac{1}{W}$$

Depletion layer in a p-n junction behave as a parallel plate capacitance and capacitance is given by

$$C_j = \left|\frac{dQ}{d(V_0 - V)}\right| = \frac{A}{2}\left[\frac{2q\in}{(V_0 - V)}\frac{N_a N_d}{N_a + N_d}\right]^{1/2}$$

Where, V→ is applied voltage, A → area of cross-section, → other has usual meanings

$$C_j \propto \frac{1}{(V_0 - V)^{1/2}}$$

C_j is voltage variable capacitance.

In case of reverse bias transition capacitance C_T is same as junction capacitance.

In reverse biased diode transition capacitance.

$$C_T \propto V^{-n}$$

Where, V → applied reserve voltage

n→is a constant and its value is

n = 1/2 for step graded diode (p^+n or $p\text{-}n^+$)

= 1/3 for linear graded diode ($p^+ n^+$)

= 1/2.5 for diffused p-n junction diode

Diffusion Capacitance or Storage Capacitance (C_D)

$C_D > C_T$ always

$C_D \propto \sqrt{\text{dopping}}$

$C_D \propto \text{lf}$

$C_D = \tau.g$

Where, τ → mean life time of minority carriers on either side or time constant of diode

g→dynamic conductance of diode

$$g = \frac{l_f}{\eta V_T}$$

So $$C_D = \frac{\tau l_f}{\eta V_T}$$

Diffusion capacitance of diode when it is operated with sinusoidal signal

(CD') is $C_D' = \frac{1}{2}\tau.g$ at low frequency

At high frequency

$$C_D' = \left(\frac{\tau}{2\omega}\right)^{1/2} \cdot g$$

$\omega \rightarrow$ Angular frequency

$\tau \rightarrow$ Time constant of diode

$g \rightarrow$ Dynamic conductance of diode

Photo Conductors

When light fall on semiconductors its conductivity increases and this property is called photo conductive effect. This is also called photo resistive effect. The minimum photon energy required for intrinsic excitation (from VB to CB) is energy gap E_g. Extrinsic excitation (from donor level to CB or from acceptor level t CB) requires lesser energy than energy gap.

Photo conductive equation is given by

$$E_g = hv = \frac{hc}{\lambda}$$

Where, $h \rightarrow$ Plank's constant

$c \rightarrow$ speed of light in free space

$\lambda \rightarrow$ Wavelength of light incident on SC

Wave length of light incident can be given as

$$\lambda = \frac{1.24}{E_g(eV)} \mu m$$

LED (Light Emitting Diode)

It works on the principle of Electro-luminescence. It is fabricated using direct band gap material. It will emit light when properly excited. It emits light due to a large number of recombination at junction. Colour of light depends on concentration of dopant. It is always operated under forward biased. With a forward current of 20 mA, LED gives out the maximum intense of light. In reverse biased it works like a normal diode.

Power dissipation in LED is of the order of mW. Response time is in μ sec. operating life is 100,000 + hrs. Cutin voltage is 1.3 V. It is used in remote control, as a display device and in designing of opt couplers.

LCD (Liquid Crystal Display)

Operating principle in dynamic scattering of light. Power dissipation is of the order of μW. Response time is in msec. operating life is 50,000+hrs. It is used as display device. Material used is liquid crystal material.

LDR (Light Dependent Resistor)

Also called photo resistor or light activated resistor (LAR). Principle of operation is photo resistive effect. Range of resistance is 5Ω to 0.75MΩ. Dark resistance of LDR is 0.75 MΩ. It is used in opto-couplers. Materials used for fabrication are CdS, Se.

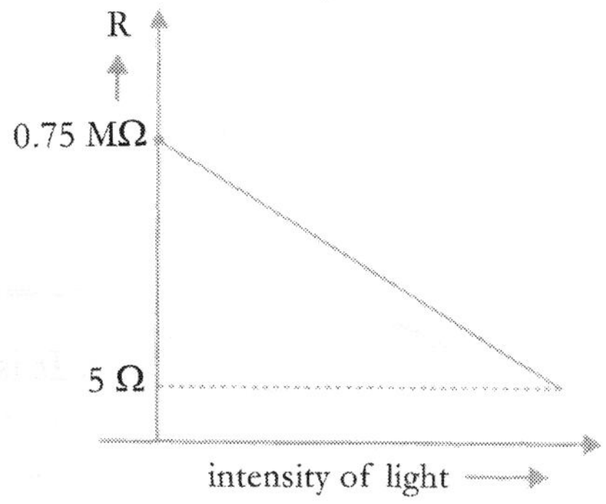

LDR characteristics

Photo Diode

Principle of operation is photo conductive effect. Photo sensitive material used is CdS, Se, ZnS. It is also called light operated switch. Ge-photo diode responds to visible light while Si-photo diode responds to infrared light. Photo sensitive coating is pWrovided at junction only. Compare to normal diode photo diode has larger depletion width obtained from lower level of doping. It is always operated under reverse biased condition. Compare to normal diode it is 10 times fasters, 100 times higher sensitive but power handling capacity is low. Magnitude of photo current increases with increase in intensity of light falling at junction. Current in photo diode is given by

$$I = I_s + I_0\left[1 - e^{V/\eta V_T}\right]$$

Where, l_S→short circuit current of photo diode

l_0 →reserve saturation current

V→voltage applied

V_T→thermal voltage

Photo current flows from n to p. Photo current are a minority carrier current. It does not provide gain. Photo current is a diffusion current It is used in remote control sensor, in designing of optocouplers and read audio track recorded on motion picture film. When photo diode is forward biased it behaves as a normal diode a effect of light on current is zero.

Avalanche Photo Diode (APD)

Basically a photo diode along with Avalanche effect. APD can handle large amount of power compare to PD (Photo diode). Its response time is smaller than PD. APD is widely used in fiber optic communication because of its hi sensitivity.

Photo Transistors

Its principle of operation is photo conductive effect. Coating of photo conductive material is done at collector base junction. It is basically a light operated switch. n-p-n photo transistor is faster than p-n-p photo transistor.

Soler Cell

Its principle of operation is photo voltaic effect. Terminal of voltage of solar cell cannot exceed the barrier potential diode that is why an array solar cells is used to achieve higher voltage. We can measure terminal voltage of solar cell using voltmeter. Popularly used solar cells are Se cells, Ni-cd cells, PbS cell. Ni-Cd cells are rechargable cells used in satellites. It is used in automatic traffic signal lightening. Generally operated under open circuit condition. It can be operated in forward biased condition and has cutin volta equal to zero.

Opto Couplers

These are optically coupled but electrically isolated. Optocouplers are faster than conventional devices. Widely used in industrial application where very good do isolation belt than transformers is required.

Tunnel Diode

It is fastest switch. Its response time is of the order of psec. It is a p+ n+ diode having doping level of $1:10^3$. Worked on the principle of tunneling effect.

It has very narrow depletion layer 100 A to 200 A. It is used as linear device as well as negative resistance device. Best material is Ga. As having highest swing. It is used in designing microwave oscillators, as a relaxation oscillator, in designing of pulse and switching circuits, and as parametric amplifier.

PIN Diode

It is p^+ - I - n^+ diode. If I is replaced by p-type then called p π n diode and if I is replaced by n-type then called p γ n diode. In PIN diode light doped intrinsic SC is sand witched between highly doped p and n. It has low response time because of high resistivity of I-region. It is two terminals, three layers, having single junction device. It is always operated under reverse biased condition. When whole I - region is covered by depletion layer then it is called swept out condition. In PIN diode if whole I region is not swept out then signal loss will occur. It is used in handling microwave power, as microwave mixer, as a duplexer, in designing of transmit receive switch in designing of anti-transmit-receive switch.

Laser

It stands for light amplification by stimulated emission of radiation. It is source of coherent light. It is fabricated with direct band gap material having larger carrier life time. Emission in laser is both spontaneous and stimulated. Population inversion occurs in laser. Lasers are highly directional.

4 Bipolar Junction Transistor and Field Effect Transistor

BJT (Bipolar Junction Transistor)

Bipolar junction transistor is invented by William Schokley, Bartain and Bardeen. In it current flows due to both majorities as well minority caries that is who called bipolar device. It is a current controlled device (CCD). Its input impedance is low and o/p impedance is high. Noisy device due to presence of minority carriers. Thermal stability is lesser because of leakage current or revers saturation current. Doping in emitter is highest and in base it is lowest. Area of collector is highest and that of base is lowest. Transistor action takes place in base region. If both the junction (JE and JC) are in forward biased then transistor is in saturation region. If JE is forward biased and JC is reverse biased then transistor is in active region. If JE is reverse biased JC is forward biased then transistor is in reverse active region. If JE and JC both are reverse biased then it is in cut-off region. If in saturation region collector junction (JC) voltage is greater than emitter junction (JE) voltage then it is said to be reverse saturation region. Transistor can work as an amplifier in active region and can work as a switch operated in saturation and cut-off regions. A transistor represented by two diode connected back to back cannot work as transistor as there is no bonding between base and collector and also the base-transit time is large. In transistor all the major currents are diffusion currents. Base current is also called recombination current. I_{co} is a drift current. For proper transistor action width of base must be less than the diffusion length. Collector current in transistor is given by

$$I_c = \beta I_B + (\beta + 1) I_{CO}$$

This equation is valid in active region.

$\beta \rightarrow$ current gain of common emitter.

Power dissipation in a transistor is given by

$$P_T = |I_C| \times |V_{CE}|$$

Where, I_c collector current

$V_{CE} \rightarrow$ collector to emitter voltage

Power dissipation in saturation and cut-off region is almost zero. Power dissipation is maximum in active region. Emitter cut-off current I_{CB0} is given by $I_{CEO} \cong I_{CO}$ but actually it is slightly greater than I_{CO}. Base cut-off current I_{CE0} is given by $I_{CEO} = (\beta + 1) I_{CBO}$ $I_{CO} \rightarrow$ reverse saturation current across reverse biased collector junction.

Current gain of common base a is given by

$$\alpha = \left| \frac{I_C}{I_E} \right| = \frac{\beta}{\beta + 1}$$

$\beta \rightarrow$ current gain of common collector

β in terms of α is given as $\beta = \dfrac{\alpha}{1-\alpha}$

Current gain of common collector γ is given by $\gamma = \beta + 1 = \dfrac{\alpha}{1-\alpha}$

β for dc analysis is greater than β for ac analysis i.e.

$$\beta_{dc} > \beta_{ac} \text{ or } h_{FE} > h_{fe}$$

- Emitter injunction efficiency γ_E is given by

$$\gamma_E = \frac{i_{Ep}}{i_{Ep} + i_{En}}$$ For p-n-p transistor

Where, $i_{Ep} \rightarrow$ emitter current due to holes

$$\gamma_E = \frac{i_{En}}{i_{Ep} + i_{En}}$$ For n-p-n transistor

i_{En} →emitter current due to electrons

- Base transport factor B is given by

$$B = \frac{i_C}{i_{Ep}}$$ For p-n-p transistor

$$B = \frac{i_C}{i_{En}}$$ For n-p-n transistor

Current gain a in terms of γ_E and B is given as $\alpha = B_{\gamma_E}$. Value of γ_E and B is very close to 1 and in ideal conditions it is 1. In a diffusion transistor base current is only due to diffusion. In a drift transistor base current is due to both diffusion and drift phenomenon. Recombination agent used to trap charge carriers is Gold. Increase in depletion region width due to increase in reverse biased across collector base junction is called Early effect. Due to early effect effective base width decreases. Due to early effect α increases by little amount but β increases by large amount. The effective base width offers a spread resistance at high frequency. Emitter junction breakdown voltage is smaller than the collector junction breakdown voltage. Emitter junction breakdown is due to zener effect while collector junction breakdown is due to Avalanche effect.

FET

It is a voltage controlled device. It is a high input impedance device. It is a majority carrier device. It is less noisy than BJT. It is thermally stable. It is smaller in size compare to BJT. Leakage currents are zero. It is unipolar device. FET is a symmetrical device where drain and source can be interchange while BJT is asymmetrical device. In JFET channel is wedge shaped. Process of varying channel width by variation of Vps is called channel width modulation. In n-channel JFET channel width is narrow near drain. MOSFET is a voltage controlled capacitor (VCC). Input impedance of MOSFET is greater than JFET. MOSFET is basically a capacitor. Internal power consumption in MOSFET is very low. MOSEFTs are less noisy than FET. MOSFETS are faster than JFETs. BJT is faster than FET devices but its power consumption is large compare to FET devices.

In saturation region current through FET is given by

$$I_{DS} = I_{DSS}\left(1-\frac{V_{GS}}{V_p}\right)^2$$

Where, I_{DSS}→saturation current at $V_{GS} = 0$

V_p→ pinch - off voltage.

I_{DS}→ saturation current at any V_{GS}.

Amplification factor p. is given by $\mu = r_d \cdot g_m$

Where, r_d→ drain resistance

g_m→trans conductance

Tran's conductance g_m is given by

$$g_m = -\frac{2I_{DSS}}{V_p}\left[1-\frac{V_{GS}}{V_p}\right] \text{ or } g_m = \frac{2}{|V_p|}\sqrt{I_{DS}I_{DSS}}$$

General equation for gate to source voltage is

$$V_{GS} = V_p\left[1-\sqrt{\frac{I_{DS}}{I_{DSS}}}\right]$$

In case of n-channel Mosfet

If $V_{GS} > V_{Th}$ → Transistor is ON

If $V_{GS} < V_{Th}$ →Transistor is OFF

Where, V_{Th},→ threshold voltage

If transistor is ON then it will be either in saturation or in triode (linear) region.

If $V_{DS} \geq (V_{GS} - V_{Th})$ saturation region

If $V_{DS} < (V_{GS} - V_{Th})$ triode region

In saturation region current is given by

$$I_{DS} = \frac{\mu C_{OX}}{2}\frac{W}{L}\left(V_{GS} - V_{Th}\right)^2$$

Where, $\mu \rightarrow$ mobility of charge carriers

$C_{OX} \rightarrow$ capacitance formed by metal and SC

$W \rightarrow$ width, $L \rightarrow$ length

In triode region current I_D is given by

$$I_D = \frac{\mu C_{OX} W}{2}\left[\left(V_{GS} - V_{Th}\right)V_{DS} - \frac{1}{2}V_{DS}^2\right]$$

Where, $V_{Ds} \rightarrow$ drain to source voltage

In p-channel MOSFET

If $V_{SG} > |V_{Tp}|$ Transistor is ON

$V_{SG} < |V_{Tp}|$ Transistor is OFF

Where $V_{Tp} \rightarrow$ threshold voltage

If transistor is on then

If $V_{SD} \geq \left(V_{SG} + V_{Tp}\right)$ saturation region

If $V_{SD} < \left(V_{SG} + V_{Tp}\right)$ triode (active) region

In saturation region current I_{DS} is given by

$$I_{DS} = \frac{\mu C_{OX} W}{2L}\left(V_{SG} + V_{Tp}\right)^2$$

In triode region, current I_D is given by

$$I_D = \frac{\mu C_{OX} W}{L}\left[\left(V_{SG} + V_{Tp}\right)V_{SD} - \frac{1}{2}V_{SD}^2\right]$$

In triode region FET works as voltage variable resistor (VVR', in saturation region it is used as amplifier.

For depletion MOSFET channel is diffused and current flows even if V is zero.

It operates in two mode depletion mode and enhancement mode For V_{GS} less than threshold voltage it works in depletion mode and for V_{GS} greater than threshold voltage it works in enhancement mode.

A Handbook on Electronics and Communications Engineering

2

Analog Electronics

CONTENTS

1 Junction Diode

A p–n junction diode is made of a crystal of semiconductor, usually silicon, but germanium and gallium arsenide are also used. Impurities are added to it to create a region on one side that contains negative charge carriers (electrons), called n-type semiconductor, and a region on the other side that contains positive charge carriers (holes), called p-type semiconductor.

- Contact potential or built-in potential (V_0)

$$V_0 = \frac{kT}{q} \text{in}\left[\frac{N_a N_d}{n^2}\right]$$

Where, V_0 → Contact potential

k→Boltzmann's const. (1.38×10^{-23}J/K)

q →4 Electron charge (1.6×10^{-19} Coulomb)

T→Temperature in Kelvin of p-n junction

N_a →Concentration of acceptors (/cm^3) on p-side.

N_d→Concentration of donors (/cm^3) on n-side

n_i→Intrinsic concentration (/cm^3) at given temperature.

- Ratio of concentration of holes or electrons on the two sides of junction

$$\frac{p_p}{p_n} = \frac{n_n}{n_p} = e^{qv_0/kT}$$

Where,

p_p → Concentration of holes (/cm^3) p-side

p_n →Concentration of holes (/cm^3) n-side

n_n → Concentration of electrons (/cm^3) on n-side

n_p → Concentration of electrons (/cm^3) on p-side

V_o →Contact potential

q →Electron charge (1.6 x 10^{-19} coulomb)

k→ Boltzmann's constant (1.38 x 10^{-23} J/K)

T→Temperature of p-n junction diode in Kelvin.

- In case of unbiased p-n junction diode electric field is maximum at t junction and decreases on the two sides of junction and zero outside the space charge region.

Maximum electric field is given by

$$E_0 = -\frac{q}{\in}.N_d X_{no} = \frac{q}{\in} N_a X_{po}$$

Where,

q → Electron charge (1.6 x 10^{-19} coulomb)

N_d→ Concentraction donors (/cm^3) on n-side

X_{no} →Width of depletion region on n-side

N_a→ Concentration of acceptors (/cm^3) on p-side.

EPermitivity of material from which diode is made.

-ve sign is due to the fact that this electric field is from n to p.

- Contact potential in the terms of maximum electric field is given by

$$V_0 = -\frac{1}{2}E_0W$$

Where,

V_0 → Contact potential

E_0 → Maximum electric field at Junction

W → Width of deplection region.

- Width of depletion region in unbiased condition is given by

$$W = \left[\frac{2 \in V_0}{q}\left(\frac{N_a + N_d}{N_a N_d}\right)\right]^{1/2}$$

$$W = \left[\frac{2 \in V_0}{q}\left(\frac{1}{N_a} + \frac{1}{N_d}\right)\right]^{1/2}$$

Where.

E → Permitivity of material used for formation p-n junction diode

V_0 → Contact potential

q → Electron charge (1.6 x 10^{-19} coulomb)

N_a → Concentration of acceptors (/cm^3) on p-side

N_d → Concentration of donors (/cm^3) on n-side

- If we reverse biased the diode by voltage V then in formulae of depletion width V_0 is replaced by (V_0 + V).
- It is clear that depletion width increases with reverse and decreases with forward biased.
- Ideal zener diode has zero resistance (dynamic resistance).
- Ripple factor (γ)

$$\gamma = \frac{\text{rms value of A.C. conponent}}{\text{DC value}}$$

$$\gamma = \sqrt{\frac{V_{rms}^2}{V_{dc}^2} - 1}$$

- Form factor (F)

$$F = \frac{\text{rms value}}{\text{DC value}} = \frac{V_{rms}}{V_{dc}}$$

$$\gamma = \sqrt{F^2 - 1}$$

- Crest factor (C)

$$C = \frac{\text{Peak value}}{\text{rms value}}$$

- Ripple factor for half wave rectifier is

$$\gamma = 1.21$$

- Average value of current and voltage of half wave rectifier

$$I_{dc} = \frac{I_m}{\pi} \rightarrow \text{ideal case}$$

$$V_{dc} = \frac{V_m}{\pi} \rightarrow \text{ideal case}$$

Where,

$I_{de} \rightarrow$ average value of current

$V_{dc} \rightarrow$ average value of voltage

$I_m \rightarrow$ maximum value of current

$V_m \rightarrow$ maximum value of voltage

Where

$$I_m = \frac{V_m}{R_L}, R_L \rightarrow \text{load resistance}$$

In practical case

$$I_{dc} = \frac{I'_m}{\pi}$$

Where

$$I'_m = \frac{V_m}{R_s + R_f + R_L}$$

$R_s \rightarrow$ Coil resistance

$R_f \rightarrow$ Diode forward resistance

$R_L \rightarrow$ Load resistance.

$$V_{dc} = \frac{V'_m}{\pi}$$

Where

$$V'_m = I'_m R_L$$

- RMS value of current and voltage

$$I_{rms} = \frac{I_m}{2}, \text{ ideal case for half wave rectifier}$$

$$V_{rms} = \frac{V_m}{2}, \text{ ideal case for half wave rectifier}$$

- Form factor (F) for half wave rectifier

$$F = 1.58$$

- Crest factor(C) for half wave rectifier

$$C=2$$

- Efficiency η for half wave rectifier

$$\boxed{\eta = \frac{\text{O / P dc power}}{\text{A.C. i / p power}}} \quad \boxed{\eta = \frac{I_{dc}^2 R_L}{I_{rms}^2 [R_s + R_f + R_L]}}$$

$$\boxed{\eta = \frac{4}{\pi^2}\left[\frac{R_L}{R_s + R_f + R_L}\right]}$$

- Maximum efficiency (when $R_s = R_f = 0$) ideal case

$$\eta = \frac{4}{\pi^2} = 40.6\%$$

- Peak inverse voltage for half wave rectifier PIV = V_m,
- Transformer utilization factor (TUF) for half wave rectifier TUF =0.286
- O/P frequency = i/p frequency for half wave rectifier.
- Conduction angle for half wave rectifier $\theta = \pi$ for ideal case

For practical case

$$\theta = \pi - 2\sin^{-1}\frac{V_\gamma}{V_m}$$

$V\gamma$ = cut-in voltage of diode

V_m = maximum voltage

- Ripple frequency (fr) for half wave rectifier ,

$f_r = f$

$f \rightarrow$ input frequency

2 Full Wave Rectifier (Centre Tap)

Full Wave Rectifier circuit two diodes are now used, one for each half of the cycle. A multiple winding transformer is used whose secondary winding is split equally into two halves with a common centre tapped connection.

- Average value of current and voltage for full wave rectifier

$$I_{dc} = \frac{2I_m}{\pi}, \text{ ideal case}$$

$$V_{dc} = \frac{2V_m}{\pi}, \text{ ideal case}$$

- RMS value of current and voltage for full wave rectifier

$$I_{rms} = \frac{I_m}{\sqrt{2}}, \text{ ideal case}$$

$$V_{rms} = \frac{V_m}{\sqrt{2}}, \text{ ideal case}$$

For practical case I_m is replaced by I'_m

where

and

$$I'_m = \frac{V_m}{\frac{R_s}{2} + R_f + R_L}$$

$$V'_m = I'_m R_L$$

- Ripple factor (γ) =0.48 for full wave rectifier.
- Form factor (F) =1.11 for full wave rectifier.

Crest factor (C) $=\sqrt{2}$ for full wave rectifier.

Efficiency $\eta = \frac{8}{\pi^2} \frac{R_L}{\frac{R_S}{2} + R_f + R_L}$ for full wave rectifier centre tap.

η is maximum when $R_s = R_f = 0 \rightarrow$ ideal case and $\eta_{max} = \frac{8}{\pi^2} = 81.2\%$

- Peak inverse voltage for full wave rectifier (centre tap) is PIV=$2V_m$
- Transformer utilization factor (TUF) TUF = 0.693 for full wave centre tap.

For full wave rectifier (bridge type) in all the formulas $\frac{R_S}{2}$ replaced by R_s and R_f is replaced by $2R_f$.

- PIV=V_m for bridge type
- TUF=0.812 for bridge type

Conduction angle

- $\theta=2\pi$ ideal case

$$\theta = 2\pi - 4\sin^{-1}\frac{V_\gamma}{V_m} \text{ for centre tap.}$$

$$\theta = 2\pi - 4\sin^{-1}\frac{2V_\gamma}{V_m}$$

$V\gamma \rightarrow$ cut in voltage of diode

Filter Circuit

A filter circuit consists of passive circuit elements i.e. inductors, capacitors, resistors and their combination. The filter action depends upon the electrical properties of passive circuit elements.

- To minimize the ripple in o/p, filter circuits are used.

Half wave rectifier with capacitor filter (C) $\rightarrow$ Peak to peak ripple voltage

$$V_r = \frac{I_{dc}}{f.C}$$

where

$I_{dc} = \frac{V_{dc}}{R_L} \rightarrow$ o/p dc current.

$V_{dc} \rightarrow$ o/p average voltage

$f \rightarrow$ input frequency

$C \rightarrow$ capacitor in o/p circuit used for filtering.

$$V_{dc} = V_m - \frac{I_{dc}}{2fC}, \text{ PIV} = 2V\text{m} \text{ with filter.}$$

Full Wave Rectifier with Capacitor Filter Ripple Voltage (V_r)

Ripple Voltage (Vr)

$$V_r = \frac{I_{dc}}{2fC}$$

$I_{dc} = \frac{V_{dc}}{R_L}$ → o/p average value of current.

f →input frequency

C→ capacitor in o/p circuit used for filtering.

R_L= Load resistance

$$V_{dc} = V_m - \frac{I_{dc}}{4fC}$$

PIV=$2V_m$

Ripple factor for half wave rectifier with capacitive filter

$$\gamma = \frac{1}{2\sqrt{3}\,fCR_L}$$

- Maximum current through diode (surge current)

$$I_{D\max} = \frac{T}{T_c} I_{dc}$$

Half wave rectifier

$$I_{D\max} = \frac{T}{2T_c}.I_{dc}$$

For full wave rectifier

T → Time period of input wave

$T_c \rightarrow$ Conduction time of diode per cycle.

- Exact solutions for half wave rectifier and full wave rectifier with C-filter

$$I_{Davg} = I_L \left[1 + \pi\sqrt{\frac{2V_p}{V_r}}\right]$$

$$I_{Dmax} = I_L \left[1 + 2\pi\sqrt{\frac{2V_p}{V_r}}\right]$$

Where

$$I_L = \frac{V_m}{R_L} = \frac{V_p}{R_L}$$

$V_p \rightarrow$ peak value of voltage

$V_r \rightarrow$ Ripple voltage (peak to peak)

$V_r = I_L / fC$

- For full wave

$$I_{Davg} = I_L \left[1 + \pi\sqrt{\frac{V_p}{2V_r}}\right]$$

$$I_{Dmax} = I_L \left[1 + 2\pi\sqrt{\frac{V_p}{2V_r}}\right]$$

Where

$$I_L = \frac{V_m}{R_L} = \frac{V_p}{R_L}$$

$V_p \rightarrow$ peak value of voltage

$V_r \rightarrow$ Ripple voltage (peak to peak)

$V_r = \frac{I_L}{2fC}$

- Conduction angle for both half wave and full wave rectifier is

$$\theta = \omega T_c \cong \sqrt{\frac{2V_r}{V_p}}$$

Where

$V_r = \frac{I_L}{fC}$ for half wave

$V_r = \frac{I_L}{2f.C}$ for full wave.

- Ripple factor for L-filter is given by

$$\gamma = \frac{2}{3\sqrt{2}} \times \frac{1}{\sqrt{1 + \left(\frac{X_L}{R_L}\right)^2}}$$

Where

$X_L = wL$ for half wave

$X_L = 2\omega L$ for full wave.

- Ripple factor for LC filter

$$\gamma = \frac{\sqrt{2}}{3} \frac{X_c}{X_L}$$

Where

For half wave rectifier

$$X_{C1} = \frac{1}{2\omega C_1}$$

$$X_{C2} = \frac{1}{2\omega C_2}$$

$$X_C = \frac{1}{2\omega C}$$

$X_L = 2\omega L$

LC filter's ripple factor is independent of load.

Ripple Factor for π-Filter or CLC Filter

It basically consists of a capacitor filter, followed by LC section. This filter offers a fairly smooth output and is characterized by highly peaked diode currents and poor regulation. As in L section filter the analysis is obtained as follows.

$$\gamma = \sqrt{2}\frac{X_{C1}X_{C2}}{R_L.X_L}$$

Where

$$X_{C1} = \frac{1}{\omega C_1}$$

$$X_{C2} = \frac{1}{\omega C_2}$$

$X_L = \omega L$

Ripple Factor of CRC Filter

In above formula for CLC filter X_L is replaced by R.

In clamper circuit load resistance circuit load resistance R is chosen according to formula:

$$R = \sqrt{R_f R_r}$$

$R_f \rightarrow$ Forward resistance of diode

$R_r \rightarrow$ Reverse biased resistance of diode

Clamping Circuit Theorem

A circuit that places either the positive or negative peak of a signal at a desired D.C level is known as a clamping circuit.

It states that for any i/p waveform the ratio of area under o/p voltage curve in forward direction to that in reverse direction is equal to ratio $\frac{R_f}{R}$ i.e.

$$\frac{A_f}{A_r} = \frac{R_f}{R}$$

Where,

$A_f \rightarrow$ Area in forward direction

$A_r \rightarrow$ Area in reverse direction

$R_f \rightarrow$ Forward resistance of diode

$R \rightarrow$ Load resistance or o/p resistance.

3 Bipolar Junction Transistor

A Bipolar junction transistor is an active semiconductor device formed by two P-N junctions whose function is amplification of an electric current.

Bipolar junction transistor is a current controlled current source.

$I_E = I_C + I_B$,

$V_{CB} + V_{BE} = V_{CE}$

$\alpha = \frac{I_C}{I_E}$, current gain in case of common base amplifier.

$\beta = \frac{I_C}{I_B}$ current gain in case of common emitter amplifier

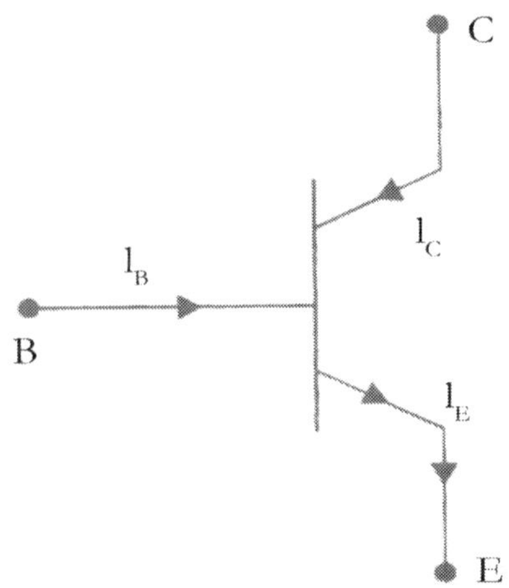

where

$I_c \rightarrow$ Collector current,

$I_B \rightarrow$ Base current,

$I_E \rightarrow$ Emitter current

- Forced current gain

$$\beta_f = \frac{I_{CS}}{I_B}$$

where

$I_{CS} \rightarrow$ Collector current when transistor is in saturation

$$\gamma = -\frac{I_E}{I_B}$$

$$\gamma = (\beta + 1) = \frac{1}{(1-\alpha)}$$

Where

$\gamma \rightarrow$ current gain in common collector transistor.

Relation between α and β

$$\boxed{\beta = \left(g\frac{\alpha}{(1-\alpha)}\right)} \quad \boxed{\alpha = \left(\frac{\beta}{(1+\beta)}\right)} \quad \boxed{I_C = \beta I_\beta + (\beta+1) I_{CO}}$$

{Valid for active region of operation}

$$I_C = \alpha I_E + I_{CBO}$$

Where

$$I_{CO} \cong I_{CBO}$$

$I_{CBO} \rightarrow$ leakage current or reverse saturation current with emitter open.

I_{CBO} is slightly greater than reverse saturation current.

Region of Operation

	Region	Collector Junction	Emitter Junction
1.	Cut off	Reverse	Reverse
2.	Active	Reverse Biased	Forward Biased
3.	Saturation Region	Forward Biased	Forward Biased
4.	Reverse Active	Forward Biased	Reverse Biased

Equivalent circuit of bipolar junction transistor is

$$V_{BE} + V'_{BE} + I_R r_b$$

Where,

$V'_{BE} \rightarrow$ Internal voltage

$V_{BE} \rightarrow$ Net voltage between base and emitter

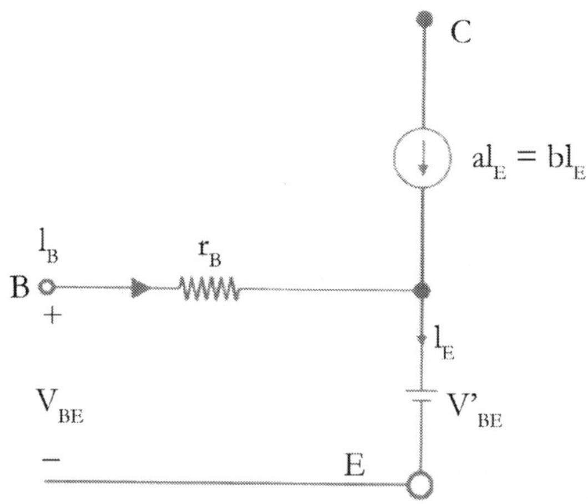

DC Load Lines

DC load line is the locus of all possible operating point at which BJT remains in active region.

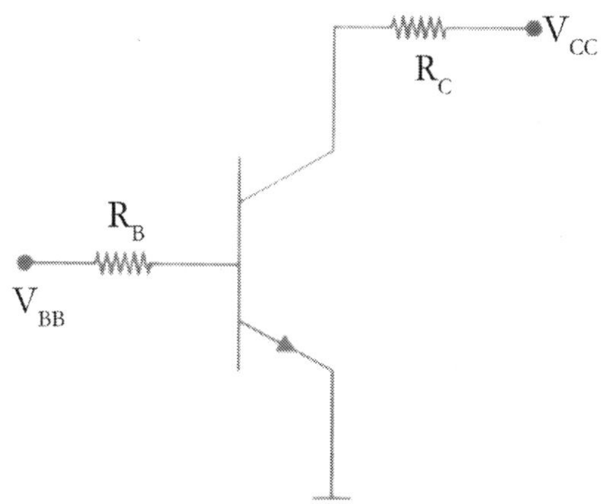

If base current $I_B < \frac{I_{C\,sat}}{\beta}$ then transistor operate in active region.

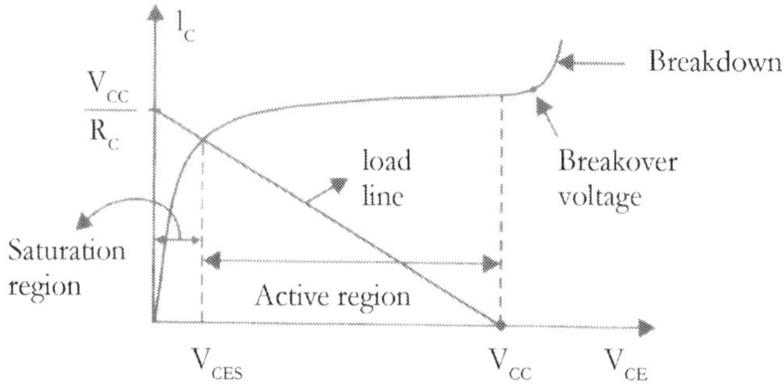

If $I_B < \frac{I_{C\,sat}}{\beta}$ then it operates in saturation region.

Collector current $I_C = \frac{V_{CC} - V_{CE}}{R_C}$ for above figure at saturation $V_{CE}=V_{CES}$

$V_{CE}=V_{CC}$ at cutoff when $I_C=0$

Maximum value of current

$$I_{C\max} = \frac{V_{CC}}{R_C}$$

[Taking VCE = 0 at saturation]

This is the ideal case.

- Power dissipated in a transistor is

$$PD=ICVCE$$

- Power dissipation is maximum in active region and minimum in cutoff region and saturation region.

$$I_{CEO} = (\beta + 1) I_{CBO}$$

Where

$I_{CEO} \rightarrow$ Collector to emitter current when base is open circuited.

$I_{CBO} \rightarrow$ Collector to base current with emitter open.

4 Transistor Biasing Circuits

Transistor Biasing is the process of setting a transistors DC operating voltage or current conditions to the correct level so that any AC input signal can be amplified correctly by the transistor.

Biasing is about stabilizing I_C and V_{CE} so as to ensure that transistor remains in active region for entire range of input signal.

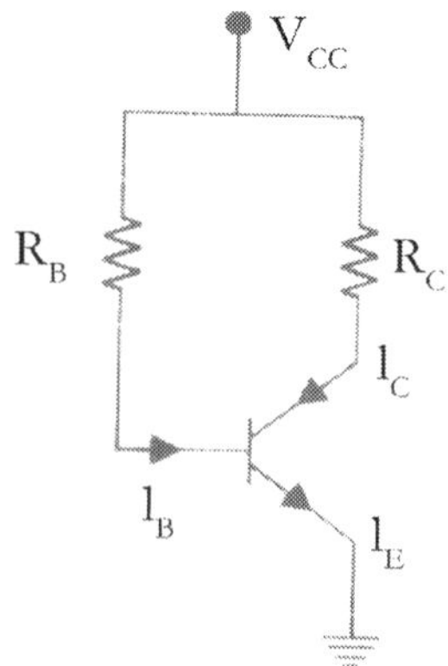

Base Register Biasing

$$I_C = \frac{V_{CC} - V_{CE}}{R_C}$$

$$I_B = \frac{V_{CC} - V_{CB}}{R_B}$$

Emitter Feedback Bias

Collector current I_C

$$I_B = \frac{V_{CC} - V_{CB}}{R_E + R_C}$$

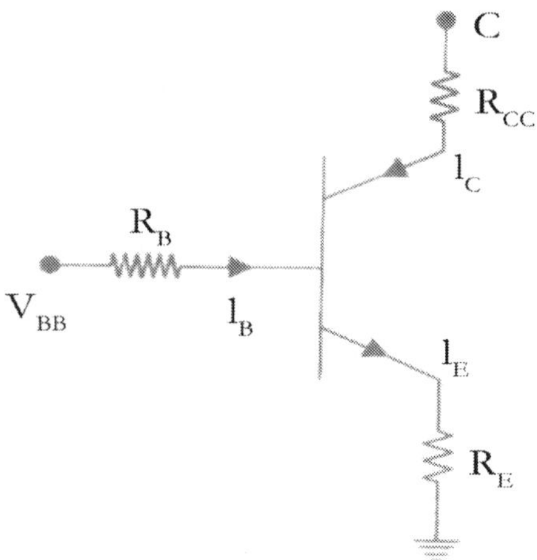

Collector-Feedback Bias (self-bias)

$$I_C \cong \frac{V_{CC} - V_{CB}}{R_C}$$

Or

$$I_C = \left[\frac{V_{CC} - V_{CE}}{R_C}\right] \times \left(\frac{\beta}{\beta + 1}\right)$$

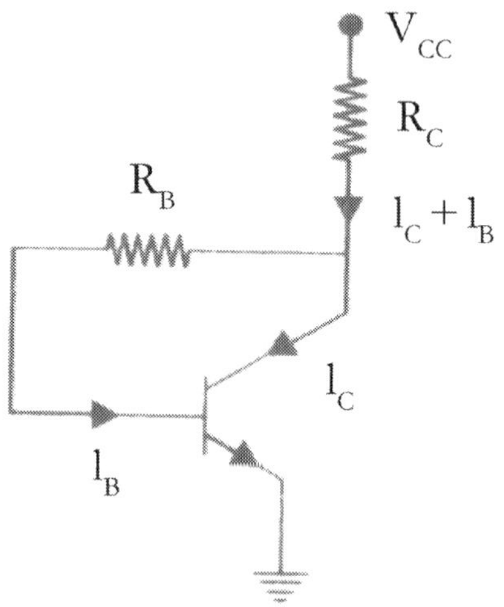

Voltage Divider Vias [universal bias]

Widely used in linear circuits and is as Equivalent circuit will be→ (thevenine equivalent)

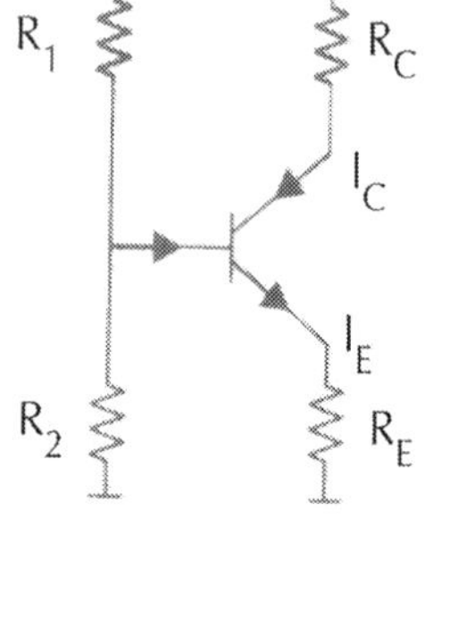

$$I_E = \frac{V_{th} - V_{BE}}{R_E + \left(\frac{R_{th}}{\beta + 1}\right)}$$

Where

$$V_{th} = \frac{R_2}{R_1 + R_2} V_{CC}$$

$$R_{th} = R_1 \parallel R_2$$

$$I_C = \frac{V_{CC} - V_{CE}}{R_C + R_E}$$

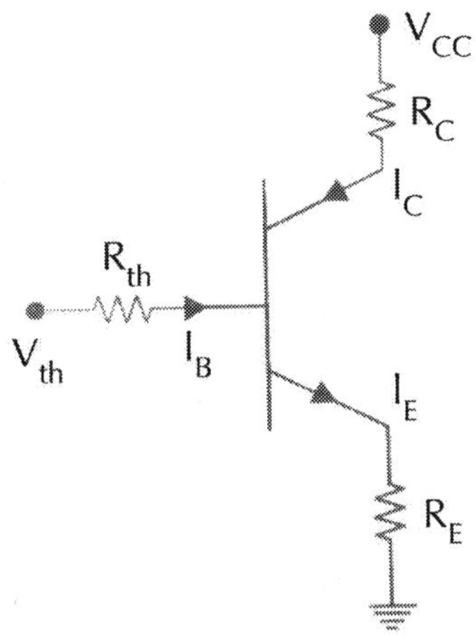

Stabilization

Stabilization is about making the Q-point independent of changes in temperature and changes in transistor parameters.

- If I_{CO}, V_{BE} and β changes simultaneously then net change in I_C

$$\Delta I_C = \frac{\partial I_C}{\partial I_{CO}} \Delta I_{CO} + \frac{\partial I_C}{\partial V_{BE}} \times \Delta V_{BE} + \frac{\partial I_C}{\partial \beta} \Delta \beta$$

$$\frac{\partial I_C}{\partial I_{CO}} = S \rightarrow \text{Current stability factor.}$$

$$\frac{\partial I_C}{\partial V_{BE}} = S' \rightarrow \text{Voltage stability factor}$$

$$\frac{\partial I_C}{\partial B} = S_\beta \text{ or } S'' \rightarrow \text{Amplification stability factor}$$

- Out of three stability factor S is most significant reason being, it S is within tolerable limit then other S' and S" are guaranteed to remain within tolerable limit.

$S_{Ideal} = 1$

Practically S should be less than 20.

For voltage divider bias circuit, current stability factor

$$S = \frac{(\beta+1)(R_{th} + R_E)}{R_{th} + (\beta+1)R_E}$$

R_E must be large for lesser value of S, but it also decreases the gain.

- Alternate Evaluation of S

$$S = \frac{(\beta+1)}{1-\beta\frac{\partial I_B}{I_C}}$$

- Condition for effective stabilization

$$S = \frac{(\beta+1)(R_E + R_{th})}{R_{th} + (\beta+1)R_E}$$

[For voltage divider circuit]

$$S = \frac{(\beta+1)}{1 + \frac{\beta}{\left(1 + \frac{R_{th}}{R_E}\right)}}$$

If

$$\frac{R_{th}}{R_E} << 1$$

Then S $\rightarrow 1$

For base register biasing

S= (β+1) $\rightarrow$ very large $\rightarrow$ highly unstable.

- Thermal resistance (θ)

$$\theta = \frac{T_j - T_A}{P_D} \ (^{o}C\ /\text{watt or K/watt})$$

$T_j \rightarrow$ Junction temperature (collector junction)

$T_A \rightarrow$ Ambient temperature in Kelvin.

$P_D \rightarrow$ Power dissipated across collector junction

- A transistor will be thermally stable if

$$\frac{\partial P_C}{\partial T_j} \leq \frac{\partial P_D}{\partial T_j} \leq \frac{1}{\theta}$$

$\frac{\partial P_C}{\partial T_j} \rightarrow$ Rate at which heat is released.

$\frac{\partial P_D}{\partial T_j} \rightarrow$ Rate at which heat is dissipated.

- For thermal stability

$$\frac{V_{CC}}{2} > V_{CE}$$

5 Amplifier Circuit with Bipolar Transistor

- AC emitter resistance

$$r_e = \frac{\alpha V_T}{I_C} = \frac{V_T}{I_E} = \frac{\alpha}{g_m}$$

Where

α= Current gain

V_T= Thermal voltage

I_C= Collector current

I_E=Emitter current

g_m=Trans conductance of transistor

- AC Beta

$$\beta_{ac} = \frac{\Delta i_C}{\Delta i_B}$$

Normally

$$\beta_{dc} > \beta_{ac}$$

- Common emitter amplifier

$$V_{out} = -i_c R_c$$

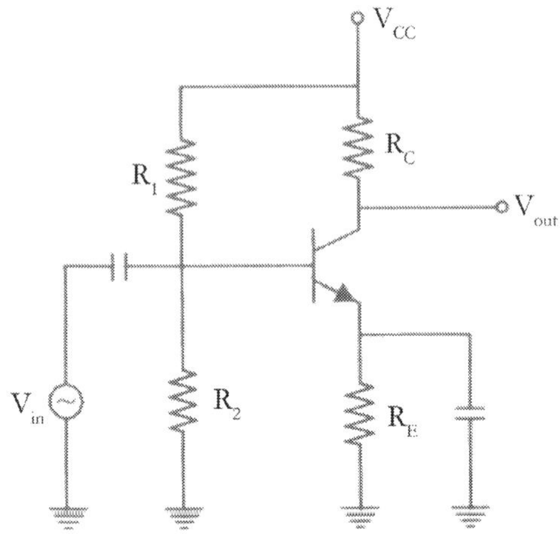

- Voltage gain

$$\frac{V_{out}}{V_{in}} = A_v = -\frac{R_c}{r_e} \times \left(\frac{\beta+1}{\beta}\right)$$

- Input impedance

$$Z_{in} = R_1 \,||\, R_2 \,||\, (\beta+1) r_e$$

- Input impedance seen from base

$$Z_{in(base)} = (\beta+1) r_e$$

- Output impedance

$$Z_{out} = R_c$$

Common Collector Amplifier

The common collector amplifier output is taken from the emitter resistor, is useful as an impedance matching device since its input impedance is much higher than its output impedance. It is also termed a "buffer" for this reason and is used in digital circuits with basic gates.

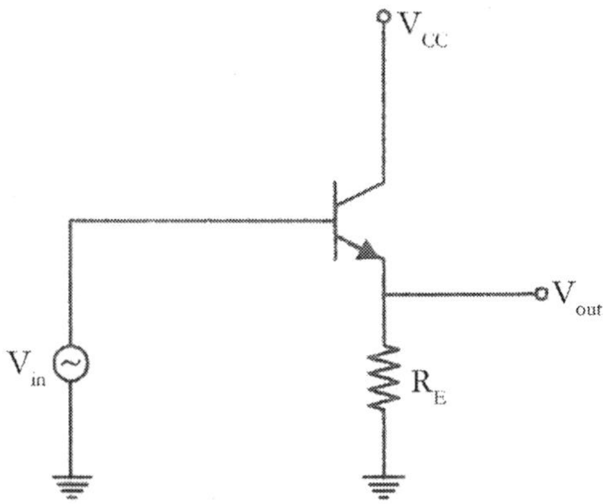

Voltage Gain

The common collector amplifier is characterized by a voltage gain of essentially one since the output is constrained at a voltage about 0.6 volts below the base which is the signal input. It is useful for impedance matching.

$$\text{Voltage gain, } A_v = \frac{V_{out}}{V_{in}} = \frac{R_E}{R_E + r_e} \simeq 1$$

Input Impedance

$$Z_{in} = R_1 \parallel R_2 + (\beta + 1)(r_e + R_E)$$

Input Impedance at base

$$Z_{in(base)} = \beta(r_e + R_E)$$

Output Impedance

$$Z_0 = r_e + \frac{R_S \,||\, R_1 \,||\, R_2}{\beta}$$

if R_s=0

Then Z_0=r_e

With $R_{s\neq}0$, voltage gain

$$A_v = \frac{R_E}{\left(R_E + r_e + \frac{R_S \,||\, R_1 \,||\, R_2}{\beta}\right)}$$

Darlington Amplifier

A current amplifier consisting essentially of two separate transistors and often mounted in single transistor housing. A current amplifier that utilizes a push-pull transistor.

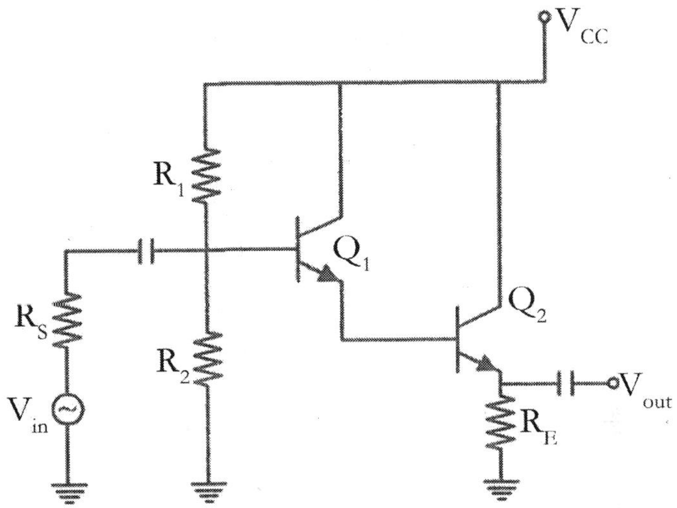

Input Impedance of Second Stage

$$\boxed{Z_{in(2)} = \beta_2 R_E} \quad \beta_2 \rightarrow \text{Current gain of } Q_2$$

Input Impedance of First Stage

$$\boxed{Z_{in(1)} = \beta_1 \beta_2 R_E} \quad \beta_1 \rightarrow \text{Current gain of first transistor } Q_1$$

AC Thevenine Impedance at Input

$$Z_{in} = R_S \parallel R_1 \parallel R_2$$

Output Impedance of First Stage

$$Z_{out(1)} = r_{e_1} + \frac{r_{th}}{\beta_1}$$

Where, $r_{e1} \rightarrow$ Emmiter resistance of Q_1

$r_{th} \rightarrow R_1 \| R_2$

Output Impedance of Second Stage

$$Z_{out(2)} - r_{e_2} + \frac{r_{e_1} + \frac{r_{th}}{\beta_1}}{\beta_2}$$

r_{e_2} › Emitter resistance of Q_2 (A.C.)

Common Base Amplifier

A common base amplifier is one of three basic single-stage bipolar junction transistor (BJT) amplifier configuration, typically used as a current buffer or voltage amplifier. In this configuration, the emitter terminal of the transistor serves as the input, the collector the output, and the base is common and connected to ground.

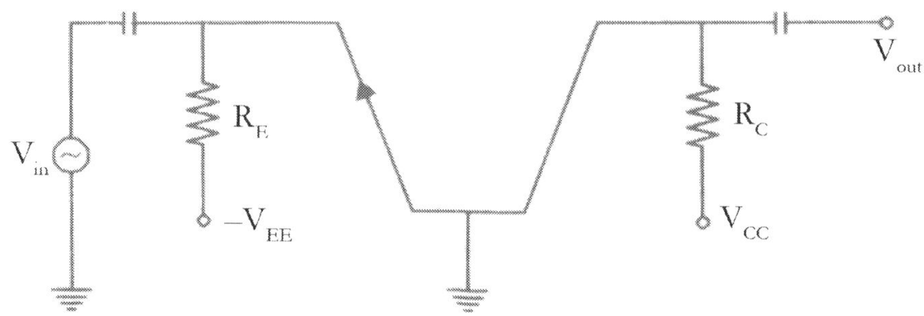

Input Impedance

$$\boxed{Z_{th} = r_e} \quad r_e \rightarrow \text{Emitter resistance}$$

Voltage gain A_V, $A_V = \dfrac{R_C}{r_e}$

Output Impedance

$$Z_{out} = R_C$$

Note:

- Input impedance of common collector amplifier is highest among CE, CC and CB.
- Output impedance of common collector amplifier is lowest among CE, CC and CB.
- Darlington pair is used where high input impedance is required.
- Darlington pair offers very high current gain. Also called current booster or current multiplier.
- Voltage gain of CB is highest and current gain of CC is highest.
- CE has moderate voltage gain, current gain and highest power gain.

6 h- Parameter and ϖ Model

H-parameters are one system for characterizing bipolar transistors. The h-parameters of a transistor will give you a good idea what it can do, how to use it effectively in a circuit, and whether it is appropriate for a particular circuit.

The π (pi) model is a popular circuit model used for analyzing the small signal behavior of bipolar junction and field effect transistors.

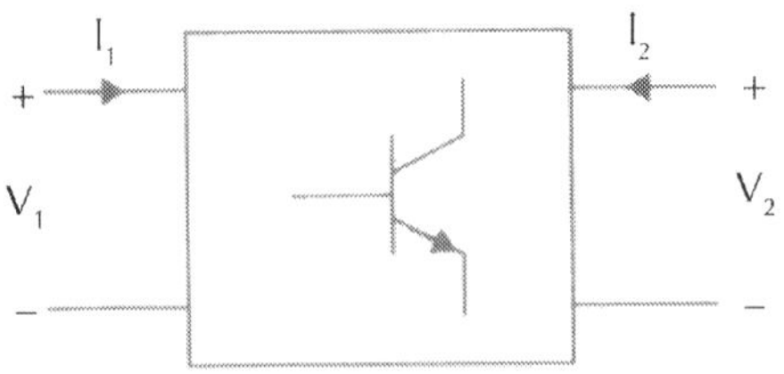

$$\begin{bmatrix} V_1 \\ i_2 \end{bmatrix} = \begin{bmatrix} h_i & h_r \\ h_f & h_0 \end{bmatrix} \begin{bmatrix} i_1 \\ V_2 \end{bmatrix}$$

Where,

$$h_i = \left.\frac{v_1}{i_1}\right|_{v_2=0}$$

$h_i \rightarrow$ Input impedance with output shorted.

$$h_f = \left.\frac{i_2}{i_1}\right|_{at\ v_2=0}$$

$h_f \rightarrow$ Short circuit gain.

$$h_r = \left.\frac{v_2}{v_1}\right|_{at\ i_1=0}$$ hr→ Reserve voltage amplification with input open circuited.

$$h_0 = \left.\frac{i_2}{v_2}\right|_{at\ i_1=0}$$ h_0→O/P admittance with i/p circuited.

For Common Emitter Equivalent Circuit will be

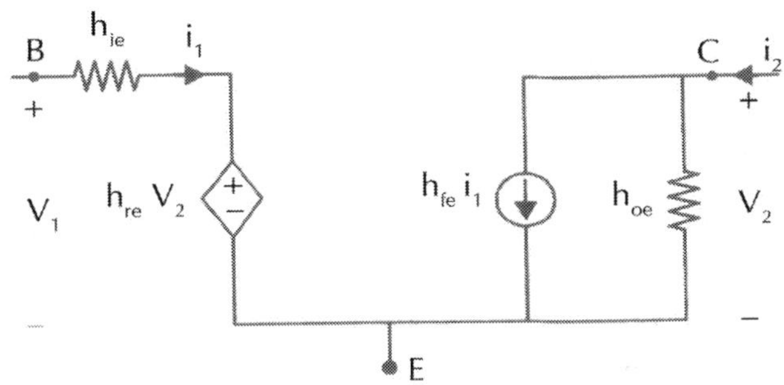

Where h_{ie}, h_{re}, h_{fe} and h_{oe} are h parameters for common emitter configuration.

For Common Collector Equivalent Circuit will be

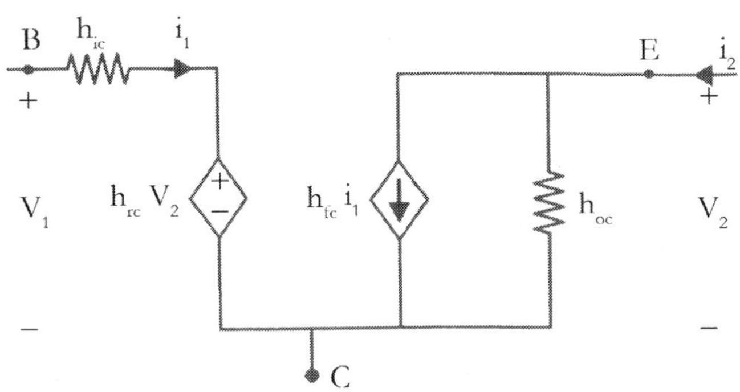

Common collector h-parameters in terms of common emitter h-parameters

$$h_{ic} = h_{ie} \qquad h_{oc} = h_{oe}$$

$$h_{fc} = -(1 + h_{fe}) \qquad h_{rc} = 1$$

Common base h-Parameters in terms of Common Emitter h-Parameters

$$h_{ib} = \frac{h_{ie}}{1 + h_{fe}} \qquad h_{rb} = \left(\frac{h_{ie} h_{oe}}{1 + h_{fe}} - h_{re} \right)$$

$$h_{fb} = -\frac{h_{fe}}{1 + h_{fe}} \qquad h_{ob} = \frac{h_{oe}}{1 + h_{fe}}$$

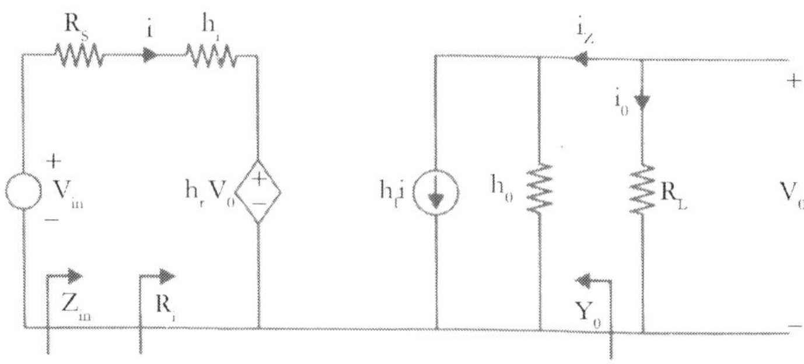

Formula for current gain, voltage gain, input impedance, output Impedance for above equivalent circuit is given by

$$\boxed{A_I = -\frac{h_f}{1 + h_o R_L}}$$

$$\boxed{Z_{in} = R_S + h_i + A_I h_r R_L}$$

$$\boxed{R_I = h_i + A_I h_r R_L}$$

$$\boxed{A_{VS} = \frac{V_0}{V_{in}} = A_v \left(\frac{R_i}{R_i + R_S} \right)}$$

where

$$A_V = A_I \frac{R_L}{R_i}$$

Overall current gain

$$A_I' = \frac{i_O}{i_S} = A_I \frac{R_S}{R_i + R_S}$$

Output admittance

$$Y_0 = h_0 - \frac{h_r h_f}{R_S + h_i}$$

Under the assumption $R_L \rightarrow \infty$ and $V_{in}=0$

ϖ Model or High Frequency Model

When an electronic component handles high frequency signals, it is appropriate to use high frequency models. These models generally contain energy storage elements along with the resistive elements. High frequency model may be small or large signal model depending on the circuit in which the device is used.

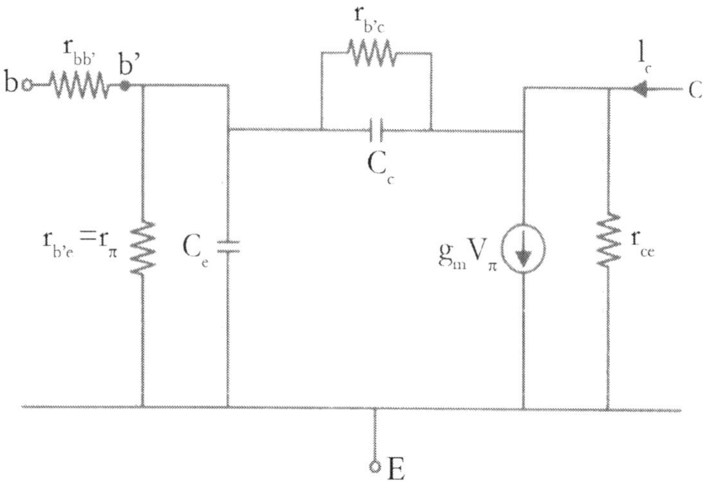

$V_{b'e} = V_\pi$ where, $r_{bb'}$ ohmic base spreading resistance

Note:

- Typical value of $r_{bb'}=100\Omega$
- Typical value $r_{b'e}=r_\pi=1k\Omega$
- Typical value of $r_{ce}=r0=80k\Omega$
- Typical value of $r_{b'c}=r_\mu=1M\Omega$
- Typical value of $C_C= C_\mu=3pf.$
- Typical value of $C_e= C_\pi=100pf.$
- Typical value of $g_m=50mS.$

Where

$C_C \rightarrow$ Collector junction barrier capacitance.

$C_E \rightarrow$ Diffusion capacitance between b' and E.

Relation between h-parameter and Parameter of ϖ-Model

$$h_{ie} = r_{bb'} + r_\pi$$

$$h_{re} = \frac{r_\pi}{r_\mu} \qquad \text{where, } g_0 = \frac{1}{r_0} = \frac{1}{r_{ce}}$$

$$h_{fe} = g_m r_\pi \qquad g_\mu = \frac{1}{r_\mu} = \frac{1}{r_{b'c}}$$

$$h_{oe} = g_0 + (1 + h_{fe}) g_\mu$$

$$g_m = \frac{|I_c|}{V_T} \qquad r_\pi = \frac{h_{fe}}{g_m}$$

$$r_{bb'} = r_b = h_{ie} - r_{b'e} = h_{ie} - r_\pi \qquad r_{b'e} = \frac{r_{b'e}}{h_{re}} = \frac{r_\pi}{h_{re}}$$

Common Emitter Short Circuit Current Gain

$$A_I = \frac{h_{fe}}{\sqrt{1 + \left(\frac{f}{f_\beta}\right)^2}} \qquad \text{where, } f_\beta \rightarrow \text{3dB upper cutoff frequency}$$

And $$f_\beta = \frac{1}{2\pi r_{b'e}(C_e + C_c)}$$

Frequency at which short circuit current gain becomes unity is

$$f_T = f_\beta\sqrt{h_{fe}^2 - 1} \quad \text{Or,} \quad f_T \cong f_\beta h_{fe} \qquad (\because h_{fe} \gg 1)$$

$$f_T = \frac{h_{fe}}{2\pi rb'e(C_e + C_C)} \quad \text{Or,} \quad f_T = \frac{g_m}{2\pi(C_e + C_C)}$$

Current Gain with Resistive Load at the O/P is

$$A_I = \frac{h_{fe}}{\sqrt{1+\left(\frac{f}{f_H}\right)^2}} \quad \text{where } f_H = \frac{1}{2\pi r_{b'e}}[Ce + C_C(1 + g_m R_L)]$$

Note:

Unity gain frequency f_T is highest and f_H is lowest among f_H, f_B and f_T i.e. $f_T > f_B > f_H$

7 Feedback Amplifier

A feedback amplifier is one in which the output signal is sampled and fed back to the input to form an error signal that drives the amplifier.

Feedback in amplifiers gives better performance in several important ways, including:

- Increased stability in the amplification. The gain is less dependent on the parameters of the amplifier elements.
- Feedback reduces distortion in the amplifier.
- The bandwidth of the amplifier is increased.
- It is easier to achieve desired input and output impedances

Diagram of feedback amplifier

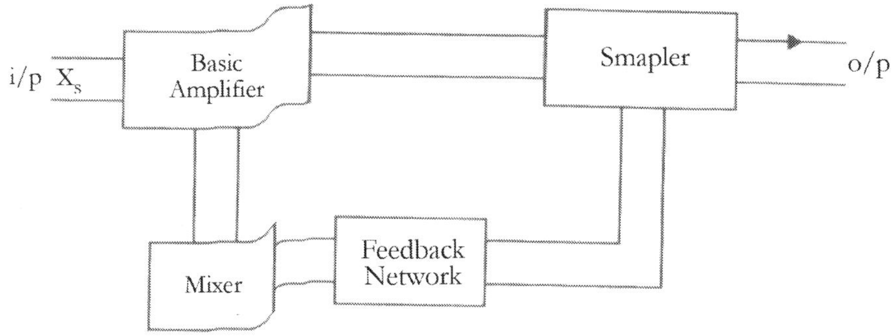

Sampler Circuit	Mixer (Comparator)
1. Voltage Sampler 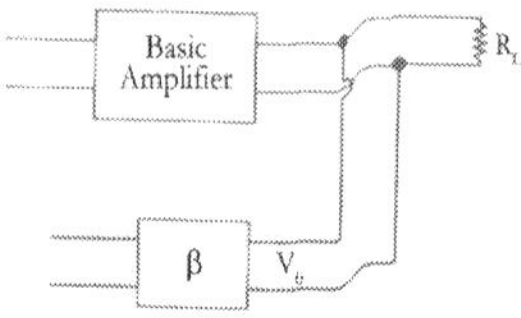	1. Series Mixer (Comparator) 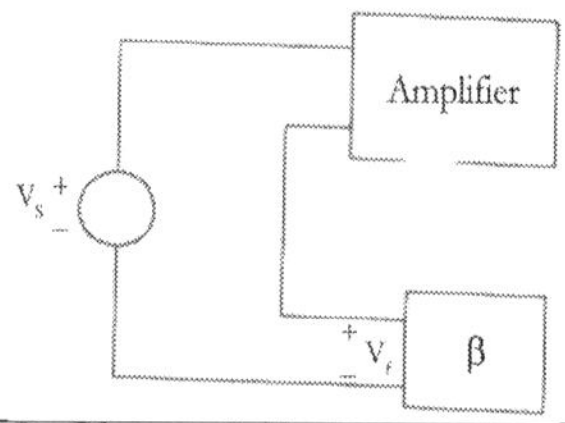
2. Current Sampler 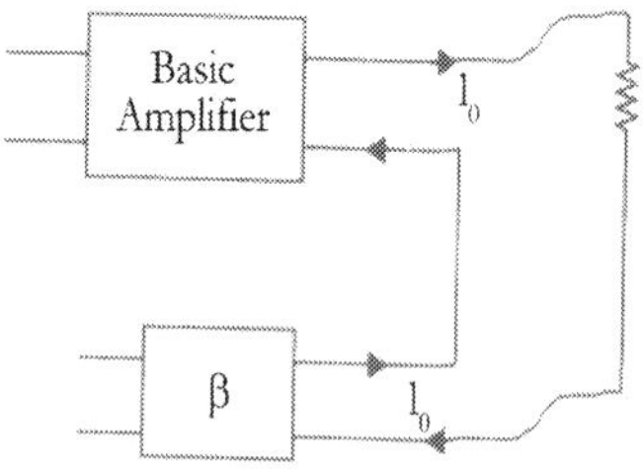	2. Shunt Mixer 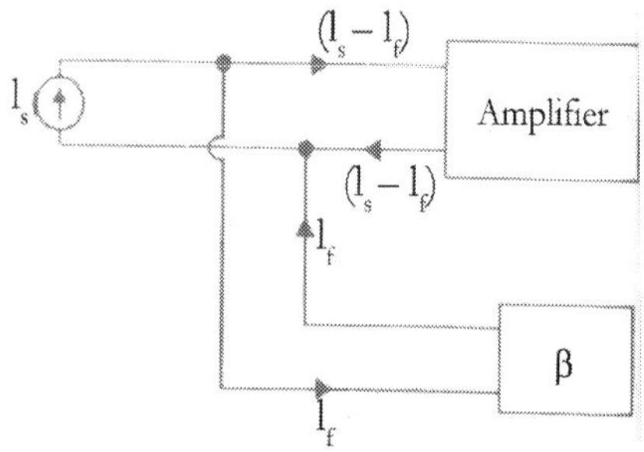

Feedback Topology

Feedback topology are following:

1. Voltage Series

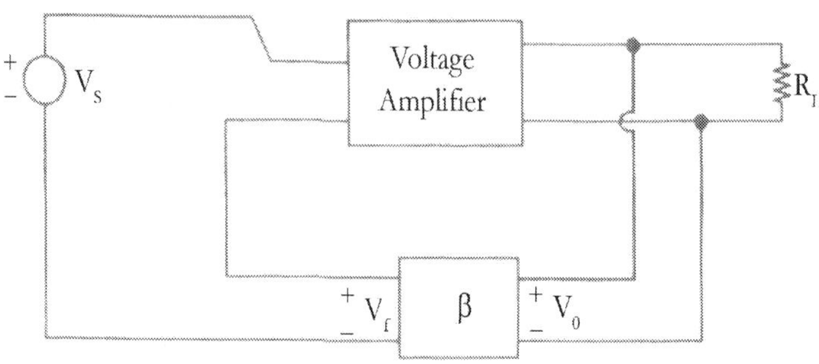

- βis dimensionless
- Voltage series feedbacks stabilize the voltage gain.

2. Current Series

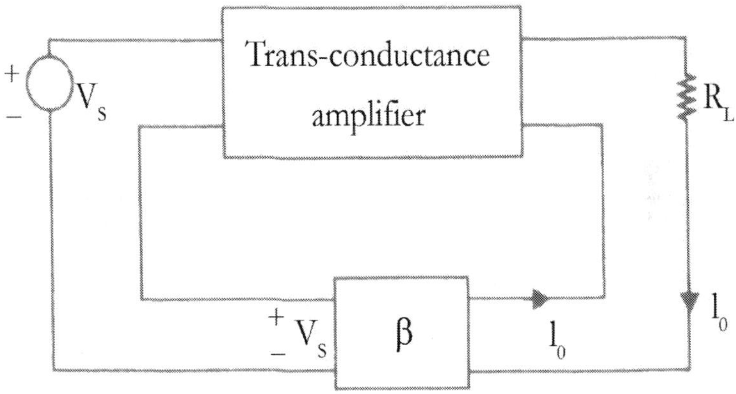

- Unit of β is ohm.
- It stabilizes trance-conductance gain of the amplifier.

3. Voltage Shunt

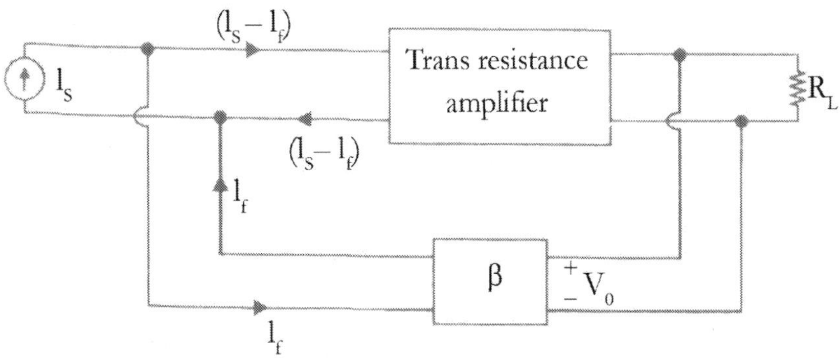

- βhas unit ohm.

- It stabilizes trance-resistance gain of the amplifier.

4. Current Shunt

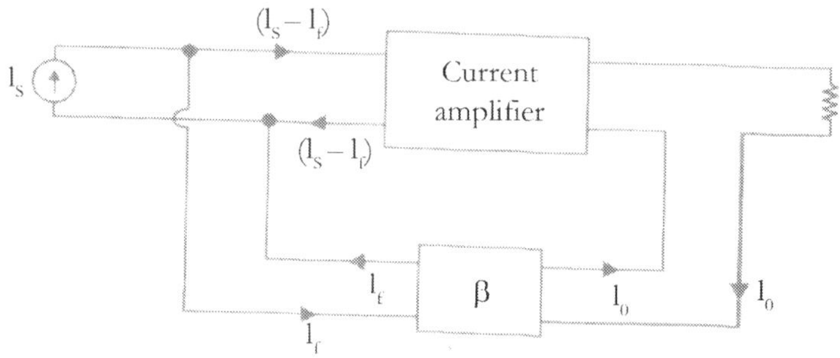

- βhas no unit.
- It stabilizes current gain of the amplifier.

Gain with feedback

$$A_f = \frac{A}{1 + A\beta} \text{for} - \text{ve feedback}$$

$$A_f = \frac{A}{1 - A\beta} \text{for} + \text{ve feedback}$$

Where, A→ Gain without feedback

β → feedback factor

Loop Gain or Return Ratio

loop gain = -Aβ

- Difference between unity and loop gain is called return difference (ID)

$$D=(1+A\beta)$$

- D is also called desensivity.
- The amount of feedback introduce into amplifier is often expressed in dB and is given by

$$N(\text{Fee dback in dB}) = 20 \log\left|\frac{1}{1+A\beta}\right|$$

- In case of –ve feedback, advantages are as

$$\frac{dA}{A} \rightarrow \%\,\text{change without feedback.}$$

$$\frac{dA_f}{A_f} \rightarrow \%\,\text{change with feedback}$$

$$\frac{dA_f}{A_f} = \frac{dA / A}{1+A\beta}$$ i.e. reduction in sensitivity of amplifier.

- Reduction in noise

$$\boxed{N_{of} = \frac{N_O}{1+A\beta}}$$

$N_O \rightarrow$ Noise without feedback
$N_{of} \rightarrow$ Noise with feedback

- Lower cutoff frequency

$$\boxed{N_{lf} = \left(\frac{f_l}{1+A\beta}\right)}$$

$f_{lf} \rightarrow$ Lower 3–dB frequency with feedback
$N_l \rightarrow$ Lower 3–dB frequency without feedback

$$\text{upper cutoff frequency } f_{hf} = f_h(1+A\beta)$$

whre $f_{hf} \rightarrow$ Upper 3 – dB frequency with feedback

$f_h \rightarrow$ Upper 3 – dB frequency without feedback

- For different feedback configurations effect on input and output impedance

	Feedback	**i/p**	**o/p**	**Gain**	**R_i**	**R_o**
1.	Voltage series	Voltage	Voltage	A_V	↑	↓
2.	Current series	Voltage	Current	G_m	↑	↑
3.	Voltage shunt	Current	Voltage	R_m	↓	
4.	Current shunt	Current	Current	A_I	↓	↑

Note:

- N → −ve for –ve feedback. N → +ve for +ve feedback.
- Bandwidth increases with –ve feedback.
- Voltage series feedback is the best arrangement in terms of R_i and R_o.
- Current shunt is the worst for the same.

8 Transistor application in Multi-vibrator

A multivibrator is an electronic circuit used to implement a variety of simple two-state systems such as oscillators, timers and flip-flops. It is characterized by two amplifying devices (transistors, electron tubes or other devices) cross-coupled by resistors or capacitors.

Astable Multi-vibrator

The Astable Multivibrator is another type of cross-coupled transistor switching circuit that has NO stable output states as it changes from one state to the other all the time. The astable circuit consists of two switching transistors, a cross-coupled feedback network, and two time delay capacitors which allow oscillation between the two states with no external trigger signal to produce the change in state.

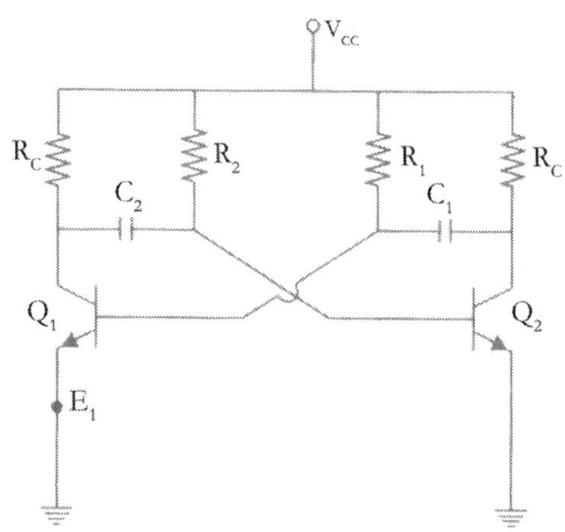

Time period of O/P wave form generated is

$$T = 0.69\left[R_1C_1 + R_2C_2\right]$$

O/P frequency is

$$f = \frac{1}{T} = \frac{1.45}{(R_1C_1 + R_2C_2)}$$

Resolution time for bistable multivibrator is $\tau_s \rightarrow$ smallest allowable interval between trigger. So maximum frequency of operation

$$\boxed{f_{max} = \frac{1}{\tau_s}} \quad [\text{Practical value}]$$

Bistable Multivibrator with Commutating Capacitor

The bistable multivibrator has two absolutely stable states. It will remain in whichever state it happens to be until a trigger pulse causes it to switch to the other state.

$$\boxed{f_{max} = \frac{R_1 + R_2}{2CR_1R_2} = \frac{1}{2CR}}$$

where, $R = R_1 || R_2$
and $\tau_s = CR$

Where,

$R_1 \rightarrow$ Resistance between collector of one transistor and base of other transistor

$R_1 \rightarrow$ Resistance between base and ground of each transistor

$f_{max} \rightarrow$ maximum frequency

In case of monostable multivibrator, width of pulse generated is given by

$$T=0.69\ RC$$

C→Capacitance between collector of one transistor and base of other transistor

R→Resistance between V_{cc} and base of one transistor

Note:

- Astable multivibrator is nothing but square wave generator.
- Monostable multivibrator is used for missing pulse detection, in PWM, PPM.
- Bistable multivibrator is nothing but a flip flop.

9 OPAMP and OPAMP Based Circuit

Op amps are extremely versatile and have become the amplifier of choice for very many applications. The advantages of integration also allow op amps to be included in many application specific integrated circuits (ASICs) where, combined with other circuit elements, a chip can be designed to carry out a specific function, which for example, can vary from a dedicated tone control or a programmable filter network to a complete audio or communications system

Ideal Property of an OPAMP

$$A_{OL} = \infty, R_i = \infty, R_O = 0, \text{Bandwidth} = \infty, \text{Slew rate} = \infty, \text{CMRR} = \infty$$

Where, $A_{OL} \rightarrow$ Open Loop Gain of OPAMP

$R_I \rightarrow$ Input Impedance of OPAMP

$R_O \rightarrow$ Output Impedance of OPAMP

$$\text{CMRR} = \frac{|A_d|}{|A_C|}$$

Where, $A_d \rightarrow$ Differential Gain of OPAMP, i.e. $(v_1\text{-}v_2)$

$V_2 \rightarrow$ Input at –ve terminal

$A_C \rightarrow$ Common mode gain $\frac{V_1 + V_2}{2}$ (usually very small)

Note:

- CMRR = ∞ ideally
- Practically CMRR is very high.

Property of OP-AMP

	Ideal Value	Practical Value
Voltage gain	∞	10^6
i/p resistance	∞	10^6 Ω or 1 MΩ
O/P resistance	0	10 Ω or 100 Ω
B.W.	∞	10^6 Hz or 1 MHz
CMRR	∞	10^6 or 120 dB
Slew Rate	∞	80V/μ sec

Inverting Amplifier

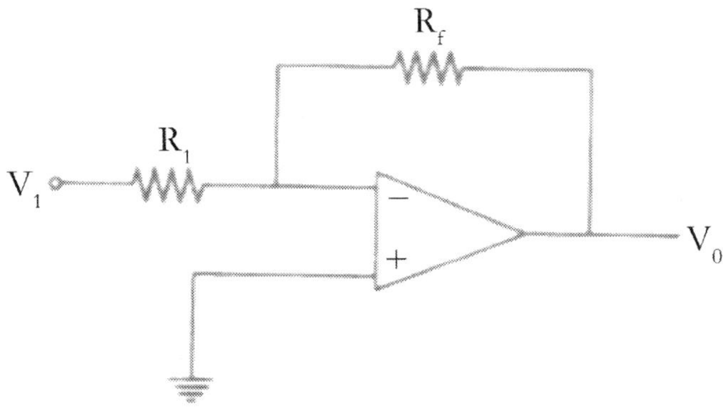

$$A_V = -\frac{R_f}{R_T}$$

Non-Inverting Amplifier

$$A_V = \left(1 + \frac{R_f}{R_1}\right)$$

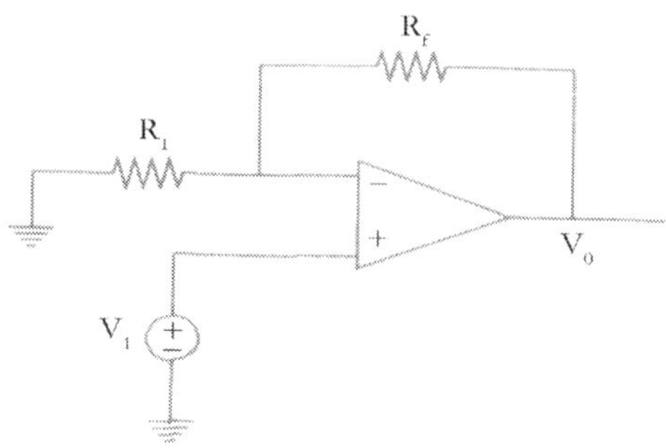

Differential Amplifier

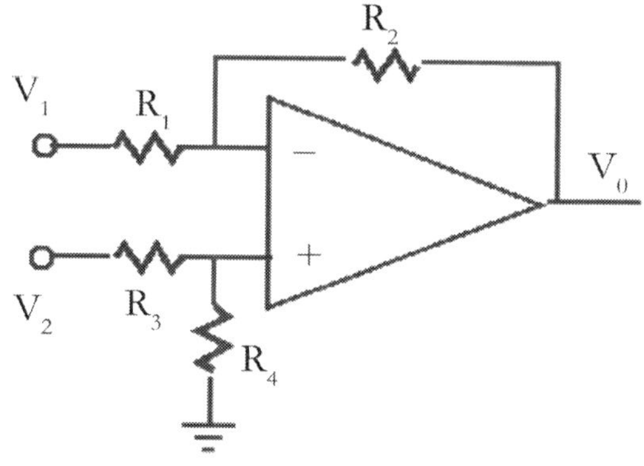

$$V_0 = -\frac{R_2}{R_1}V_1 + \left(\frac{R_4}{R_3 + R_4}\right)\left(1 + \frac{R_2}{R_1}\right)V_2$$

$$\text{if } \frac{R_2}{R_1} = \frac{R_4}{R_3} \text{ then, } V_0 = \frac{R_4}{R_3}\left[V_2 - V_1\right] = \frac{R_2}{R_1}\left(V_2 - V_1\right)$$

In this case when $\frac{R_2}{R_1} = \frac{R_4}{R_3}$

$$A_C = 0 \rightarrow \text{Common mod e gain and CMRR} = \infty$$

Non Linear OP-AMP Circuit

These circuits are commonly known as switching circuits, whose output switches between positive and negative saturation voltage levels.

Integrator

An integrator circuit is a circuit in which the input waveform. Such a circuit is also termed as an integrating amplifier. The circuit is somewhat similar to an OPAMP inverting.

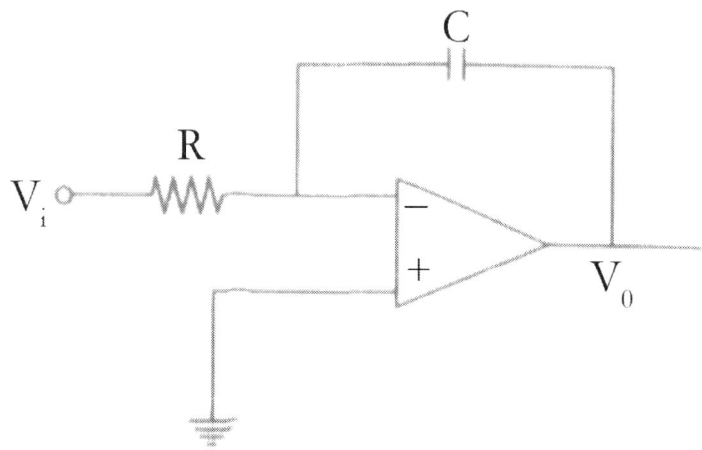

$$V_0 = -\frac{1}{RC}\int V_i dt + V_c(0+)$$

Differentiator

The basic Op-amp Differentiator circuit is the exact opposite to that of the Integrator Amplifier circuit.

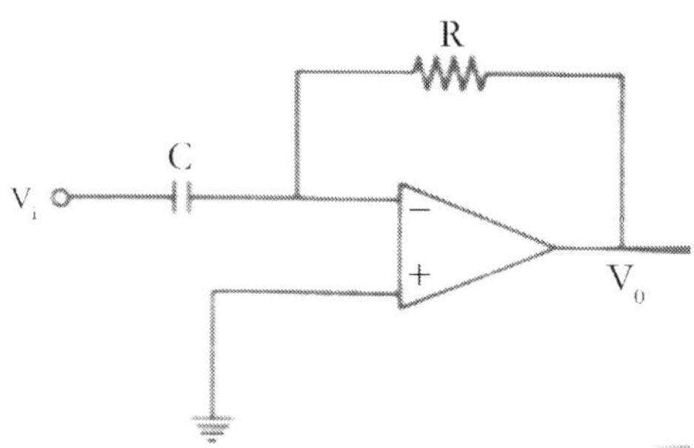

$$V_0 = -R.C\frac{dV_i}{dt}$$

Log Amplifier

Log amplifier finds a lot of application in electronic fields like multiplication or division, signal processing, computerized process control, compression, decompression, RMS value detection etc.

$$V_0 = -\eta V_T \, \text{In} \frac{V_i}{I_0 R}$$

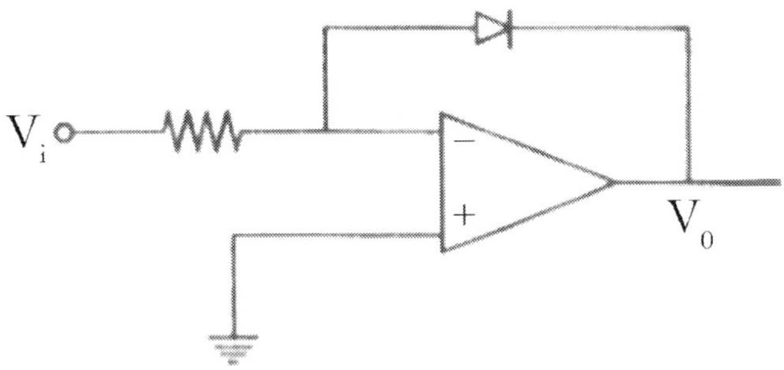

where

$\eta \rightarrow$ Recombination factor

$V_T \rightarrow$ Thermal Voltage

$I_O \rightarrow$ Reserve saturation current of diode

Antilog Amplifier

Antilog amplifier is one which provides output proportional to the antilog i.e. exponential to the input voltage.

$$V_0 = -I_o R \text{ antilog} \frac{V_i}{\eta V_T}$$

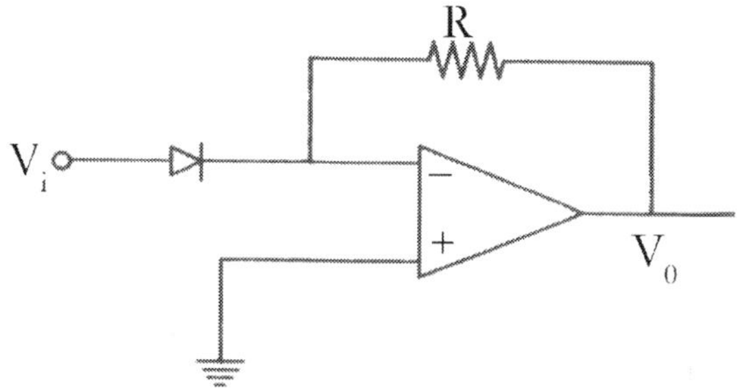

Where

I_O→ Reserve saturation current of diode

η→ Recombination factor

V_T→ Thermal Voltage

Grounded Load

It has been found that a very significant improvement in performance and reduction in circuit-area might be achieved by employing grounded load devices in portions of a system employing large-scale integrated circuits.

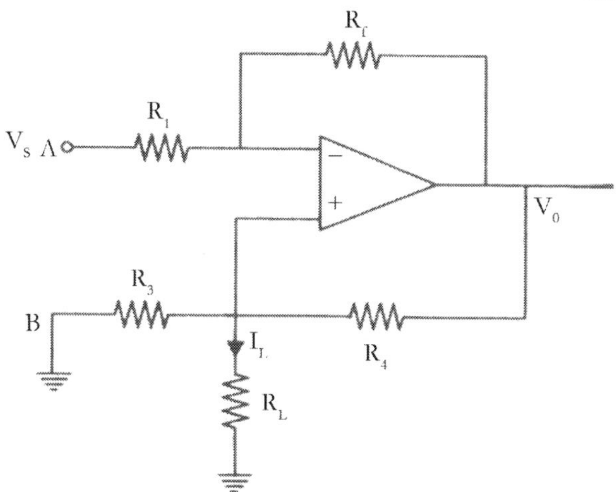

$$\text{if } \frac{R_f}{R_1} = \frac{R_4}{R_3} \text{ then } I_L = \frac{V_S}{R_3}$$

If V_S is applied at point A and B is grounded and

$$\text{if } \frac{R_f}{R_1} = \frac{R_4}{R_3} \text{ then } I_L = \frac{-V_S}{R_3}$$

Square Wave Generator or Astable Multivibrator

This square wave generator is like the Schmitt trigger circuit in that the reference voltage for the comparator action depends on the output voltage.

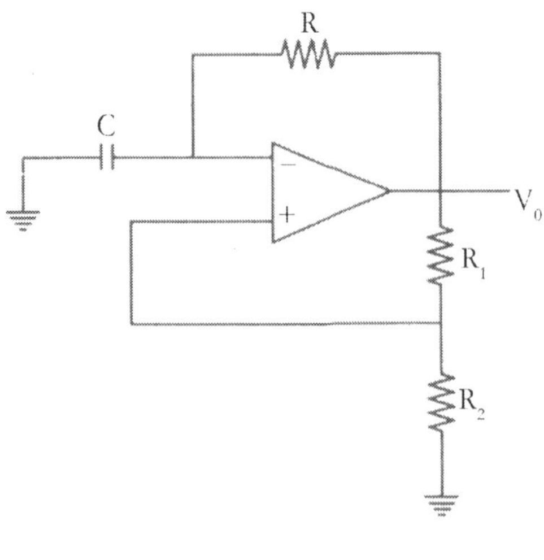

Time period of O/P waveform generated is

T=2RC In3

(For $R_1 = R_2$)

$$T = 2RC\ In\left[\frac{1+B}{1-B}\right]$$

$$\left(\text{Where } B = \frac{R_2}{R_1 + R_2}\right)$$

Monostable Multivibrator

Monostable Multivibrators have only ONE stable state (hence their name: "Mono"), and produce a single output pulse when it is triggered externally.

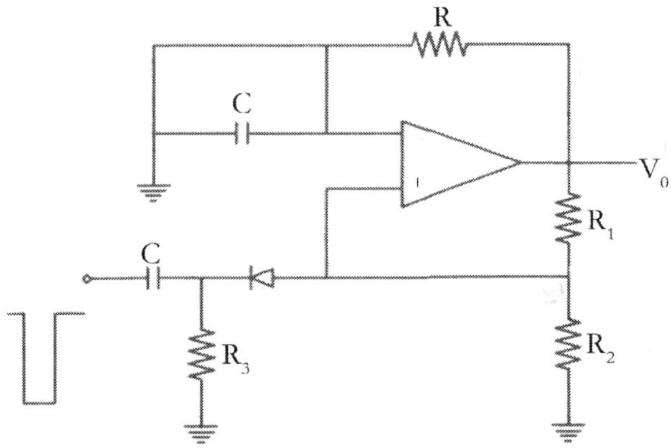

- Width of pulse generated at O/P is

$$T = RC\ In\left(1 + \frac{R_2}{R_1}\right)$$

If $R_2 = R_1$,

Then $T=RC\ In\ 2$

Triangular Wave Generator

The most important feature of a triangular wave is that it has equal rise and fall times while a saw tooth wave has un-equal rise and fall times. The triangular wave includes sampling circuits, thyristor firing circuits, frequency generator circuits, tone generator circuits etc.

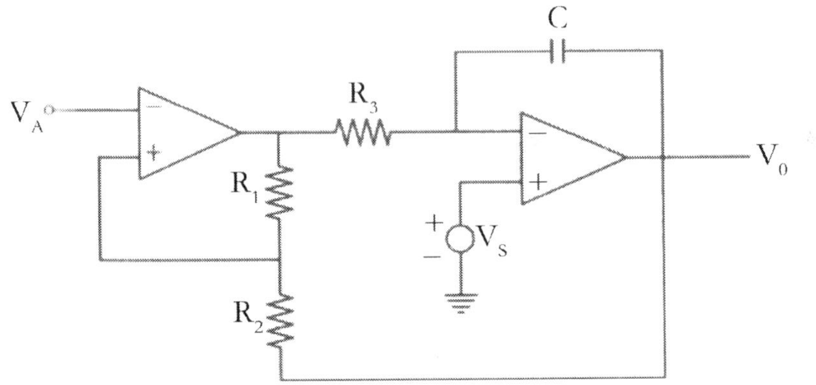

$V_S \rightarrow$ Control the Slop

$V_A \rightarrow$ Control the amplitude of the output

For $V_s=0 \rightarrow$ Symmetric triangular wave is obtained

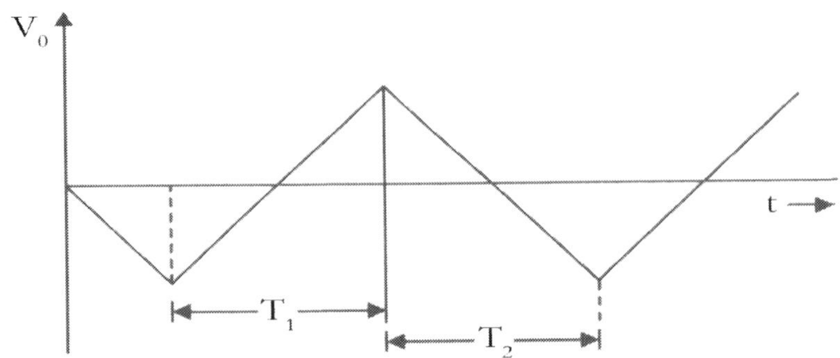

Where

$$T_1 = \frac{2R_2R_3CV_{sat}}{R_1\left(V_{sat} + V_s\right)}$$

$$T_2 = \frac{2R_2R_3CV_{sat}}{R_1\left[V_{sat} - V_s\right]}$$

Bias Current and voltage

An electric current flowing through the base-emitter junction of a transistor and adjusted to set the operating point of the transistor.

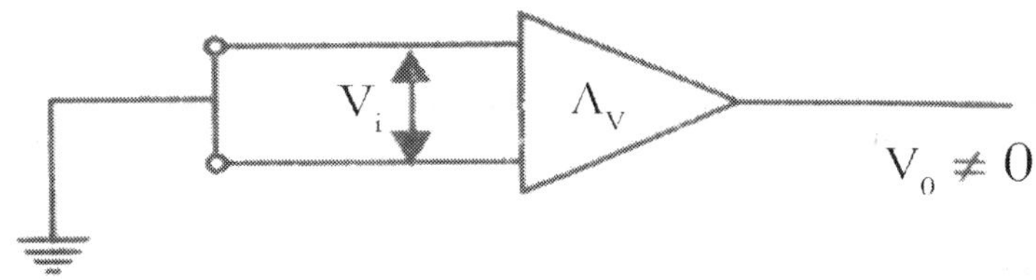

$V_O \neq 0 \rightarrow$ Called offset voltage

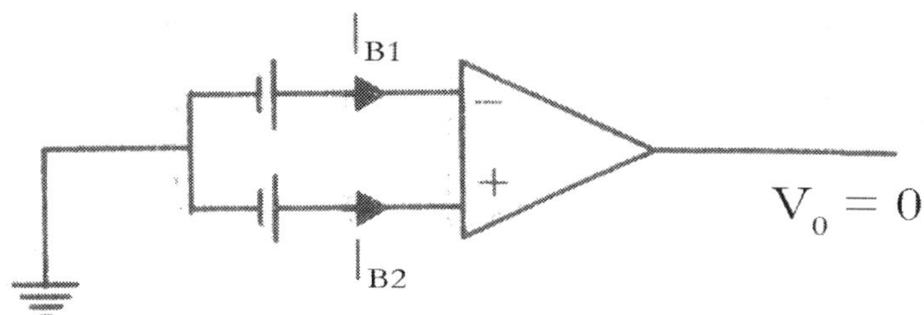

Balanced condition

i/p bias current (I_B) $I_B = \dfrac{I_{B_1} + I_{B_2}}{2}$

input offset current $I_{io} = (I_{B1} - I_{B2})$

input offset voltage is the voltage which must be applied between input terminals to balance amplifier ($V_0 = 0$)

for input $V_m \sin \omega t$

$$f_{max} = \frac{SR}{2\pi A_{CL} V_m}$$

Where, SR→ Slew Rate

- f_{max} → Maximum frequency of operation
- A_{CL} → Closed loop gain of OP-AMP
- In case of monostable multivibrators using 555 timer width of pulses is generated is

T = RC In 3

555 in Astable Multivibrators Mode

An astable multivibrator, often called a free-running multivibrator, is a rectangular-wave generating circuit. Unlike the monostable multivibrator, this circuit does not require any external trigger to change the state of the output, hence the name free-running. An astable multivibrator can be produced by adding resistors and a capacitor to the basic timer IC. The timing during which the output is either high or low is determined by the externally connected two resistors and a capacitor.

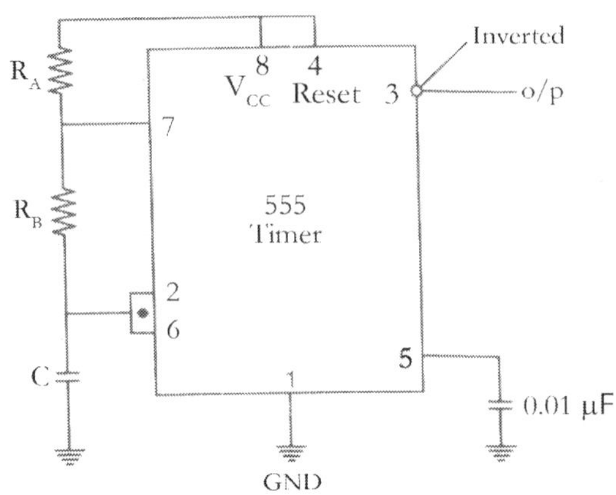

- Time period of O/P wave generated is

$$\boxed{T = 0.69\left[R_A + 2R_B\right]C} \quad \boxed{f = \frac{1.45}{\left(R_A + 2R_B\right)C}}$$

$$\boxed{\text{duty cycle} = \frac{T_{on}}{T} = \frac{R_A + R_B}{R_A + 2R_B}}$$

Note:

- Duty code > 50%

$$\because T_{on} > T_{off} \rightarrow \text{Duty cycle} > 50\%$$

- By placing a diode parallel to R_B and making R_A+R_f → we have square wave of 505 duty cycle at the output.

10 Power Amplifier

The term power amplifier is a relative term with respect to the amount of power delivered to the load and/or provided by the power supply circuit. In general the power amplifier is the last 'amplifier' or actual circuit in a signal chain and is the amplifier stage that requires attention to power efficiency.

AC load Line

The ac load line is a graph that represents all possible combinations of I_C and V_{ce} for a given amplifier. Under normal circumstances, the ac and dc load lines for a given amplifier are not identical. A typical ac and dc load line combination is shown in Figure

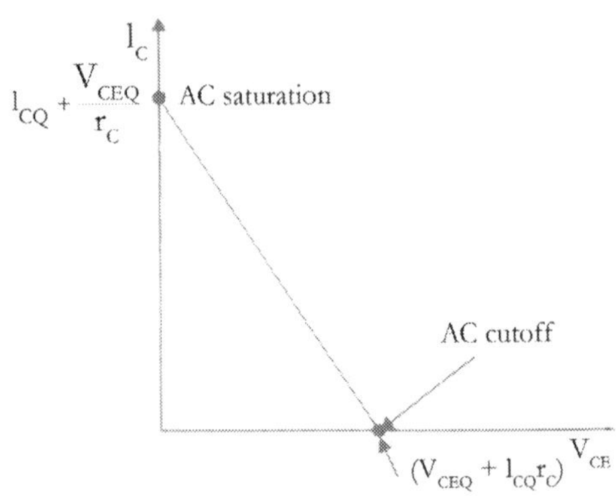

- CE amplifier, where

$I_{CQ} \rightarrow$ Quiescent collector current

$V_{CEQ} \rightarrow$ Quiescent collector-emitter voltage

- AC load resistance seen by collector

$$r_C = R_C \parallel R_L$$

AC saturation current

$$I_{C(sat)} = I_{CQ} + \frac{V_{CEQ}}{r_C}$$

AC cutoff voltage

$$V_{CE(out)} = V_{CEO} + I_{CO} r_C$$

- AC output compliance is the maximum unclipped peak to peak AC voltage that an amplifier can produce.
- AC output compliance of a CE amplifier

$$PP = \min.\left(2I_{CQ} r_C, 2V_{CEQ}\right)$$

Emitter Follower

An emitter follower circuit is widely used in AC amplification circuits. The input and output of the emitter follower is the base and the emitter, respectively, while the collector is at AC zero, therefore this circuit is also called common-collector circuit.

Effective a_C load resistance

$$r_E = R_E \parallel R_L$$

AC saturation current

$$I_{C(sat)} = I_{CO} + \frac{V_{CEQ}}{r_E}$$

AC cutoff voltage

$$V_{CE(cut)} = V_{COQ} + I_{CQ} r_E$$

AC output compliance of an emitter follower

$$PP = \min.\left(2I_{CQ} r_E, 2V_{CEQ}\right)$$

CB Amplifier

- AC load resistance of CB compiler

$$r_C = R_C \parallel R_C$$

- AC load line and ac output compliance of CB amplifier are same as that of CE amplifier

Class A operation

- Transistor operates in active region at all times.
- Unloaded gain of CE amplifier

$$A = -\frac{R_C}{r_e}$$

Where, $r_e \rightarrow$ emitter resistance (AC)

Loaded voltage gain $A_V = -\frac{r_C}{r_e}$ where, $r_C \rightarrow$ AC load resistance

Current gain of the transistor

$$A_i = -\frac{i_C}{i_b}$$

Where,

$A_i \rightarrow$ Current gain

$i_C \rightarrow$ AC collector current

$i_b \rightarrow$ AC base current

- Load Power

$$P_L = \frac{V_L^2}{R_L} = \frac{V_{PP}^2}{8R_L}$$

Where,

$P_L \rightarrow$ AC load power

$V_L \rightarrow$ RMS load voltage

$V_{PP} \rightarrow$ Peak to Peak load voltage

$R_L \rightarrow$ Load resistance

Transistor power dissipation P_D

$$P_D = V_{CEQ} I_{CQ} - i_C V_C$$

Normally $i_C V_C \rightarrow$ is very small so we can neglect it.

Where $P_D \rightarrow$ Power dissipation

$V_{CEQ} \rightarrow$ Quiescent collector – emitter voltage

$I_{CQ} \rightarrow$ Quiescent collector current

Total DC power supplied to an amplifier

$$P_S = V_{CC} I_S$$

Maximum AC load power

$$P_{L(max)} = \frac{PP^2}{8RL}$$

Where, PP = Maximum unclipped value V_{PP}

Note: In most of the circuits you can use the approximation $A_i \cong \beta$

- Power gain $A_P = A_V A_i$
- Stage efficiency

$$\eta = \frac{P_{L(max)}}{P_s} \times 100\%$$

Where, $\eta \rightarrow$ Stage efficiency

$P_S \rightarrow$ DC input power

Class B Operation

- Current flow for half of the cycle of input.
- Q-point is located at cutoff on the dc and ac load line.

$$V_{CEQ} = \frac{V_{CC}}{2}$$

- AC load power of a class B push-pull amplifier

$$P_L = \frac{V_{PP}^2}{8R_L}$$ Where $V_{PP} \rightarrow$ Peak-Peak load voltage

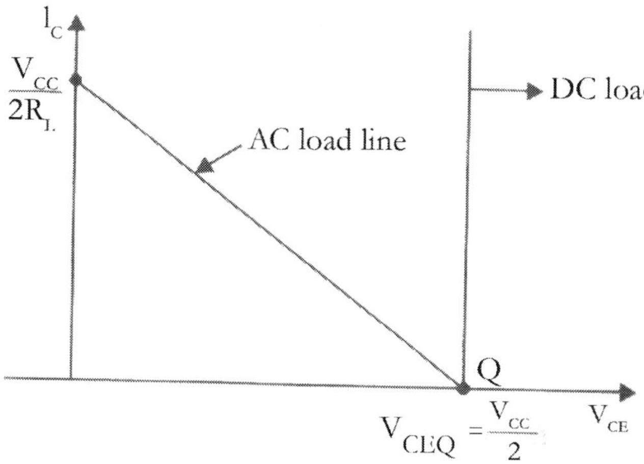

Class C Operation

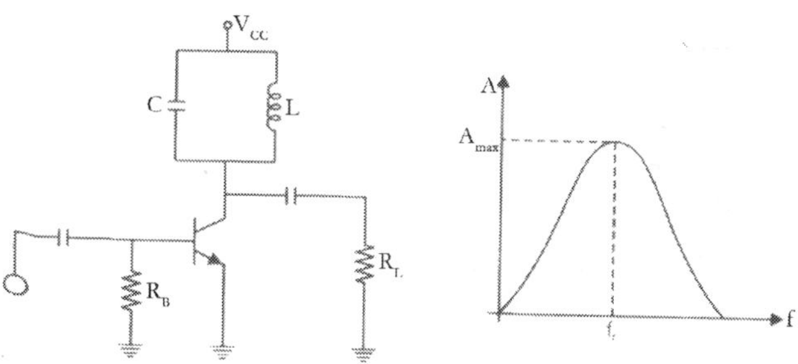

- It is tuned amplifier.

Resonant frequency of tank circuit is given by $f_r = \dfrac{1}{2\pi\sqrt{LC}}$

Where, f_r = Resonant frequency

L→Inductance

C→Capacitance

AC load power for class C amplifier $P_L = \dfrac{V_{PP}^2}{8R_L}$

Where, V_{pp}→Peak to peak load voltage

R_L→Load resistance.

- General equation for drain current →for JEET is

$$I_D = I_{DSS}\left[1 - \frac{V_{GS}}{V_P}\right]^2$$

in saturation region

Where IDSS→Maximum value of current with VGS=0

ID→Current at other values of VGS

VP→Pinch off voltage

VGS→Gate to source voltage.

$$r_d = \left.\frac{\Delta V_{DS}}{\Delta I_{DS}}\right|_{at\ V_{GS}}$$

Drain resistance (ed) is given by

Range is from 100 kΩ to 500 kΩ

$$g_m = \left.\frac{\Delta I_D}{\Delta V_{GS}}\right|_{at\ V_{DS}}$$

Trans conductance (gm) is given by

For JEET range of gm is from 0.1mS to 10mS.

$$\mu = \left.\left(\frac{\Delta V_{DS}}{\Delta V_{GS}}\right)\right|_{I_D}$$

Amplification factor (μ) is given by

Range of μ is from 2.5 to 150.

- Relation between μ, rd and gm is

$$\mu = r_d \cdot g_m$$

Formulae for Transconductance

Transconductance is a property of certain electronic components. Conductance is the reciprocal of resistance; transconductance is the

current variation at the output to the voltage variation at the input. It is written as g_m

- In saturation region

$$g_m = \frac{-2I_{DSS}}{V_P}\left[1-\frac{V_{GS}}{V_P}\right]$$

Or

$$g_m = -\frac{2}{V_P}\sqrt{I_D I_{DSS}}$$

Or

$$g_m = \frac{2}{|V_P|}\sqrt{I_D I_{DSS}}$$

Or

$$g_m = g_{mo}\left[1-\frac{V_{GS}}{V_P}\right]$$

Where, $g_{mo} = -\frac{21_{DSS}}{VP}$

Where, I_{DSS}→Maximum value of saturation current

V_P→Pinch off voltage

V_{GS}→Gate to source voltage

g_{mo}→Maximum value of transconductance

$$V_{GS} = V_P\left[1-\sqrt{\frac{I_D}{I_{DSS}}}\right]$$

Small Signal Low Frequency Equivalent of JFET

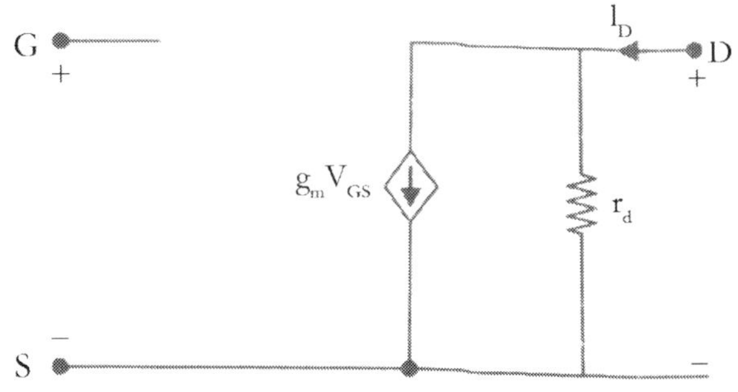

High frequency and small signal equivalent circuit is

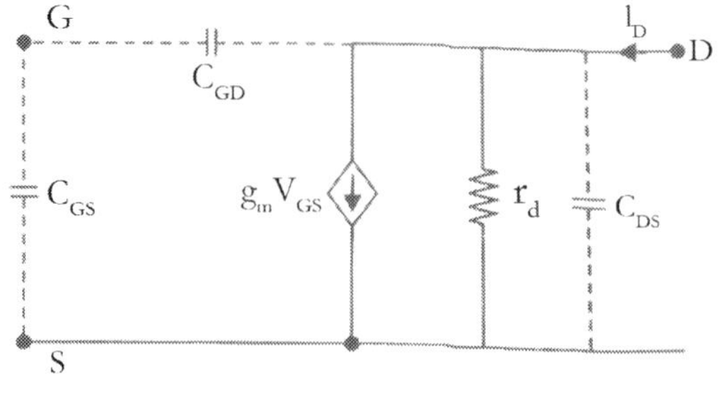

- Net i/p capacitance between gate and source is

$$C_i = C_{GS} + C_{GD}\left[1 - A_v\right]$$

Where C_{GS}→Capacitance between gate and source

C_{GD}→ Capacitance between gate and drain

A_V→Voltage gain of Amplifier

- Net O/P capacitance between drain and source is

$$C_o = C_{DS} + C_{GD}\left[\frac{A_V - 1}{A_V}\right]$$

C_O→O/P capacitance

C_{DS}→ Capacitance between drain and source

C_{GD}→ Capacitance between gate and drain

A_V→Voltage gain of Amplifier

Note:

If two FETs are connected in parallel having transconductance g_{m1} and g_{m2} drain resistance r_{d1} and r_{d2}, amplification factor μ_1 and μ_2 then effective Trans conductance g_m, drain resistance r_d and amplification factor μ is given by

$$g_m = g_{m1} + g_{m2}$$

$$r_d = r_{d_1} \parallel r_{d_2} = \frac{r_{d_1} r_{d_2}}{r_{d_1} + r_{d_2}}$$

$$\mu = \frac{\mu_1 r_{d_2} + \mu_2 r_{d_1}}{r_{d_1} + r_{d_2}}$$

Self-Bias

The establishment of an operating point on the transistor volt-ampere characteristics by means of direct voltages and currents.

$$V_{GS} = -I_D R_S$$

Source resistance

$$V_S = -\frac{V_P}{I_{DSS}}$$

V_P→Pinch off voltage

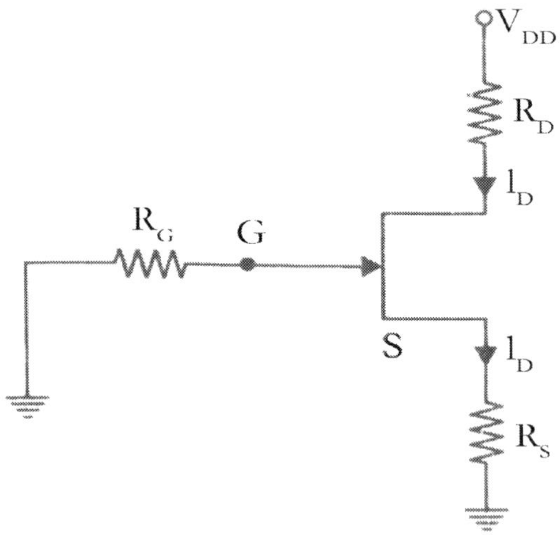

Voltage Divider and Source Bias

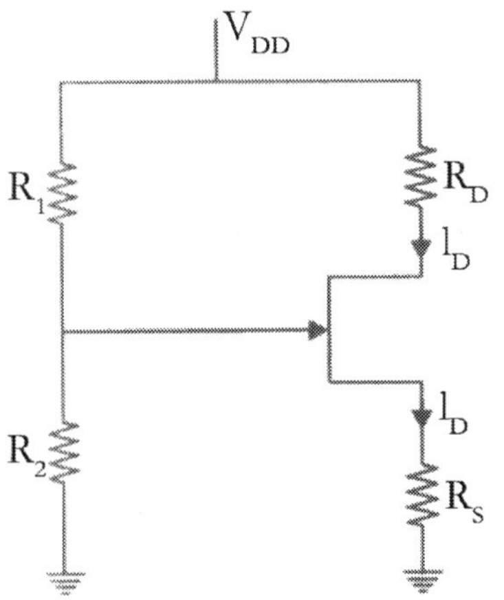

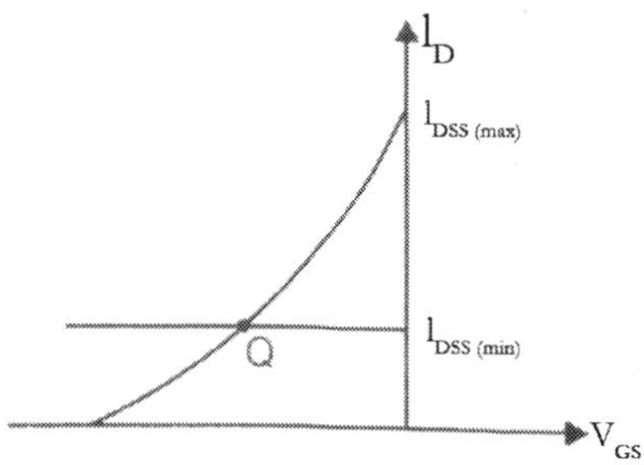

$$\text{Drain current } I_D = \frac{V_{th} - V_{GS}}{R_S}$$

$$\text{Where } I_{th} = \left(\frac{R_2}{R_1 + R_2}\right).V_{DD}$$

Current Source Bias

- Integrated circuits have transistors which are manufactured simultaneously with the same device parameters.
- One common technique is current source biasing, which allows the designer to take advantage of matched devices
- We will begin by looking at some simple current source circuits
- A current source is not a "naturally" occurring device. It can be simulated by a network of transistors and circuit elements.

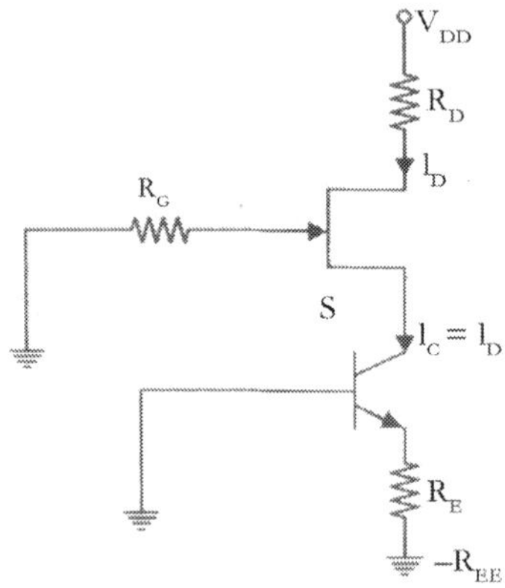

$$I_D = I_C = \frac{V_{EE} - V_{BE}}{R_E}$$

Note:

In Ohmic region, in case of FET

$$r_{DS} = \frac{r_0}{\left(1 - \frac{V_{GS}}{V_P}\right)^2}$$

Where $r_0 \rightarrow$ resistance (drain to source) when $V_{GS}=0$

$r_{DS} \rightarrow$ Resistance at any V_{GS} between drain and source

$V_P \rightarrow$ Pinch off voltage

$V_{GS} \rightarrow$ Gate to source voltage

Enhancement Type MOSFET

The metal–oxide–semiconductor field-effect transistor is a type of transistor used for amplifying or switching electronic signals.

For n-channel Enhancement Type MOSFET

- VGS must be greater than V_t (threshold voltage) for MOSFET to be ON.

i.e. $V_{GS} > V_t \rightarrow$ for ON

if $V_{GS} < V_t \rightarrow$ Cutoff region

- For $V_{GS} > V_t$, there are two region of operation

  If $V_{DS} < (V_{GS} - V_t) \rightarrow$ Triode region of operation

  If $V_{DS} \geq (V_{GS} - V_t) \rightarrow$ Saturation region

- Formulae for current in triode region → (n-channel) is

$$I_D = \frac{\mu_n C_{ox} W}{L}\left[(V_{GS} - V_t)V_{DS} - \frac{1}{2}V_{DS}^2\right]$$

$\mu_n C_{ox} = K'_n \rightarrow$ Process trans conductance

$\frac{W}{L}$ = aspect ratio

$r_{DS} \rightarrow$ drain to source resistance in triode region

We take $\frac{\mu_n C_{ox} W}{L} = k_n$

So $I_D = k_n\left[(V_{GS} - V_t)V_{DS} - \frac{1}{2}V_{DS}^2\right]$

- Formulae for current in saturation region for (n-channel)

$$I_{DS} = \frac{\mu_n C_{ox} W}{2L}(V_{GS} - V_t)^2$$

Where $\frac{\mu_n C_{ox} W}{2L} = k_n$

So $I_{DS} = K_n\left(V_{GS} - V_t\right)^2$

For p-channel Enhancement Type MOSFET

If $V_{SG} > |V_t| \rightarrow$ ON

And $V_{SG} < |V_t| \rightarrow$ OFF (cutoff region)

$V_t \rightarrow$ threshold voltage

If $V_{SG} > |V_t|$

And $V_{SD} < (V_{SG} + V_t) \rightarrow$ triode region

And if $V_{SD} \geq (V_{SG} + V_t) \rightarrow$ saturation region

- Formulae of current in triode region for p-channel

$$I_D = k_n [(V_{SG} + V_t) V_{SD} - \frac{1}{2} V_{SD}^2$$

- Formulae of current in saturation region for p-channel

$$I_{DS} = k_n [V_{SG} + V_t]^2$$

Where $k_n = \frac{\mu_n C_{ox} W}{2L}$

And $k_n = \frac{\mu_n C_{ox} W}{L}$

- For depletion type MOSFET formulas remains same.

g_m for MOSFET is

$g_m = \frac{\mu_n C_{ox} W}{L}\left(V_{GS} - V_t\right)$ in saturation region

$$\text{Or } g_m = \sqrt{2I_{DS}\frac{\mu_n C_{ox} W}{L}} \qquad g_m = \sqrt{2I_{DS}k_n}$$

$$g_m = \sqrt{4I_{DS}k_n} \qquad g_m = \frac{2I_{DS}}{(V_{GS} - V_t)}$$

Where k_n and k_n are constants that depends on particular MOSFET.

All the formulae are same whether it is designed using FET or MOSFET

Common Source Configuration

- Common source configuration, including amplifiers with a source resistor.
- Common gate amplifiers.
- Common drain amplifiers.

$$A_V = \frac{V_o}{V_i} = -\frac{\mu R_D}{r_D + R_D + (1+\mu)R_S}$$

If $r_D \to \infty$

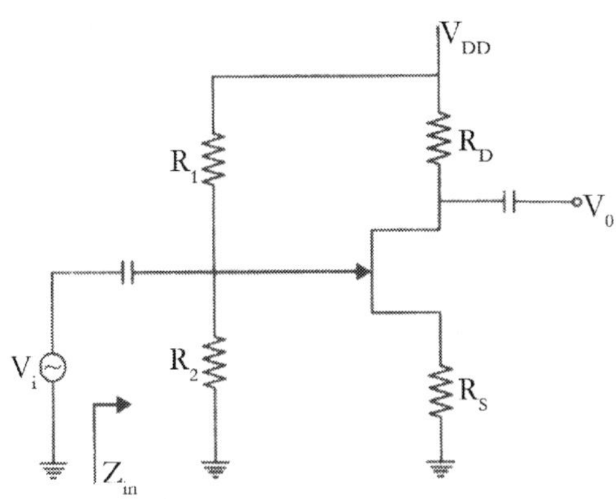

If $R_S=0$

Then $A_V = -gmR_D$ where, $A_V \rightarrow$ voltage gain

Input impedance $Z_{in} = R_1 \| R_2$

Output impedance $Z_0(\text{with load } R_D) = R_D \| [r_d + (\mu+1)R_S]$

Without load $R_D, Z_0 = (r_d + (\mu+1)R_S)$

If $r_d \rightarrow \infty$ then $Z_0=R_D$ where, $r_D \rightarrow$ drain resistance

Common Drain Amplifier

- Voltage gain

$$A_V = \frac{\mu R_S}{r_D + R_D + (\mu+1)R_S} = \frac{V_0}{V_i}$$

If $r_D \rightarrow \infty$

Then $A_V = \dfrac{g_m R_S}{1 + g_m R_S}$

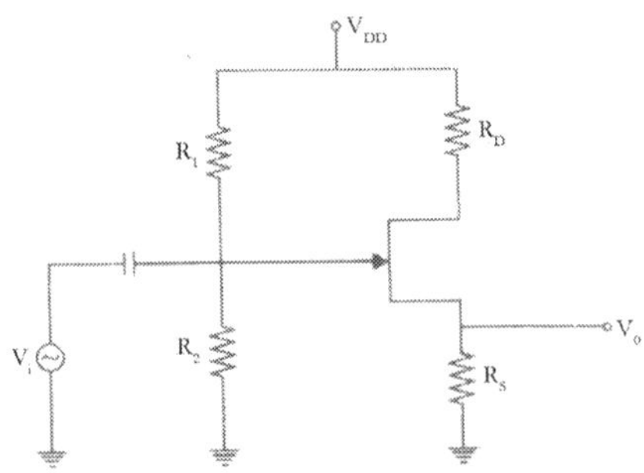

Output impedance without load $R_S \rightarrow Z = \left(\dfrac{r_D + R_D}{\mu+1}\right)$

If $r_D \to \infty$ then $Z = \frac{1}{g_m}$

- Output impedance with load $R_S \to$

$$Z_0 = \left(\frac{r_D + R_D}{\mu + 1}\right) || R_S \quad \text{if } r_D \to \infty \text{ then } Z_0 = \frac{1}{g_m} || R_S$$

Common Gate Amplifier

- Unloaded voltage gain
- Input signal is applied to the source, output is taken from the drain

$$A_V = g_m R_D$$

- Input impedance

$$Z_n = \frac{1}{g_m}$$

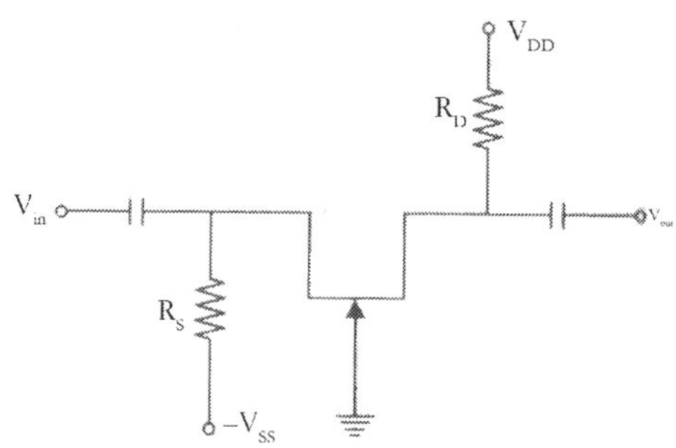

- DC on-state resistance

$$r_{DS(ON)} = \frac{V_{DS}}{I_D}$$

where, $r_{DS(ON)}$ →DC resistance in saturation region

V_{DS}→DC drain source voltage

I_D→DC drain current

Note:

- Common source amplifier is equivalent to CE amplifier of BJT.
- Common drain amplifier is equivalent to common collector amplifier of BJT.
- Common gate amplifier is equivalent to common base amplifier of BJT.
- Its circuit is as

$$V_{out} = A(V_1 - V_2)$$

Where V_{out} = Voltage between collectors

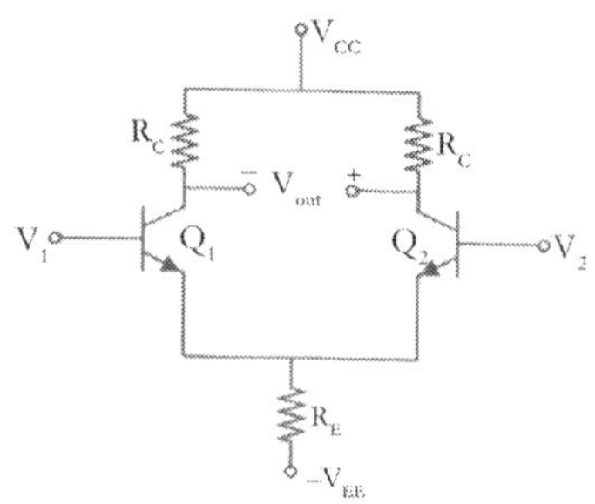

$$A = \frac{R_C}{r_e}$$

Where r_e → Emitter resistance

V_1 → Noninverting input voltage

V_2 → Inverting input voltage

DC Analysis of a Differential Amplifier

- Emitter bias current

$$I_E = \frac{V_{EE} - V_{BE}}{2R_E}$$

- Tail current

$$I_T = \frac{V_{EE} - V_{BE}}{R_E}$$

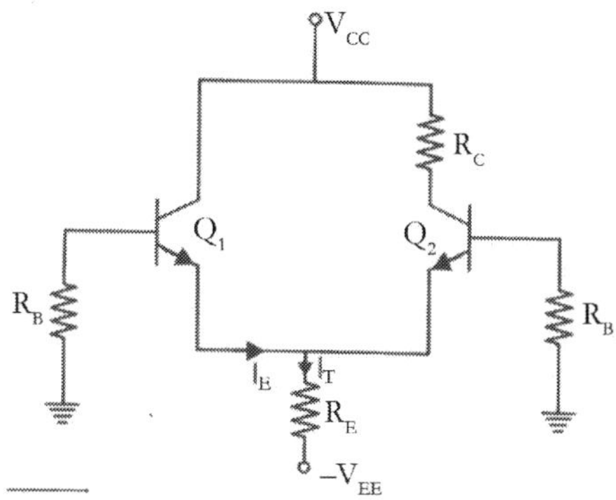

Input offset current $I_{in(off)} = I_{B1} - I_{B2}$

Input bias current $I_{in(bias)} = \frac{I_{B1} - I_{B2}}{2}$

Output offset voltage $V_C = V_{CC} - I_C R_C$

AC Analysis of a Different Amplifier

1 Noninverting input

a. AC emitter current

$$i_e = \frac{V_1}{2r_e} \quad \text{where, } r_e \rightarrow \text{emitter resistance}$$

AC output voltage is $V_{out} = i_c R_C = \frac{V_1}{2r_e} R_C$

Voltage gain for Noninverting input is $\frac{V_{out}}{V_1} = \frac{V_2}{2r_e}$

2 Inverting input

AC emitter current is $i_e = \frac{V_2}{2r_e}$

AC output voltage is $V_{out} = -i_c R_C = -\frac{V_2}{2r_e} R_C$

Voltage gain for inverting input is $\frac{V_{out}}{V_2} = \frac{R_c}{2r_e}$

Differential voltage gain $A = \frac{R_c}{2r_e}$

Input impedance $r_{in} = 2\beta r_e$

Common Mode Voltage Gain

$$A_{CM} = \frac{V_{out}}{V_{in(cm)}} = -\frac{R_C}{r_e + 2R_E} \cong -\frac{R_C}{2R_E}$$

$$R_E >> r_e$$

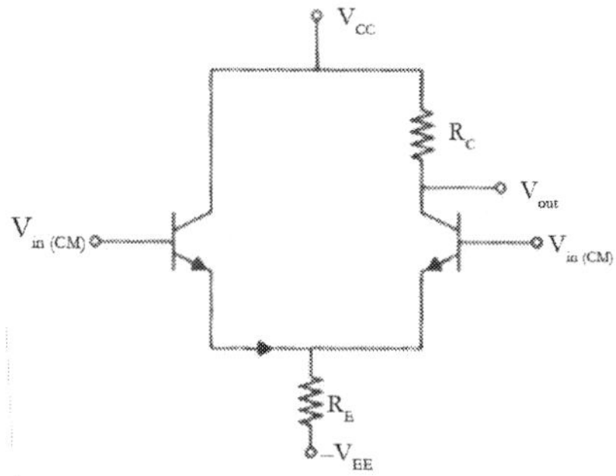

$$CMRR = \frac{A}{-A_{CM}}$$

Where, $A_{CM} \rightarrow$ common mode gain

AC phase Shift Oscillator Using Transistor

- Frequency of oscillation is

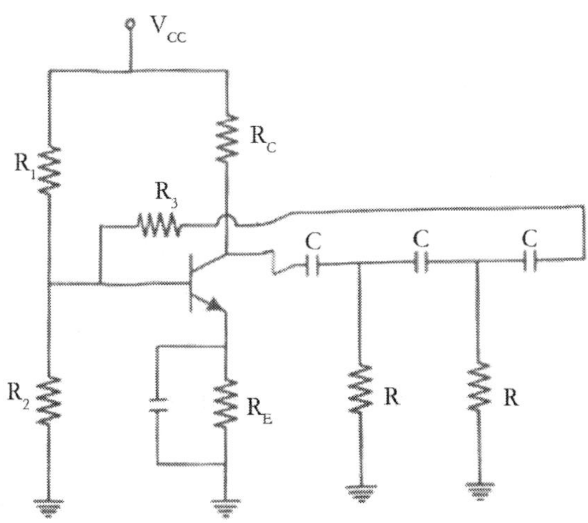

$$\omega = \frac{1}{R.C\sqrt{4k+6}}$$

$$f = \frac{1}{2\pi R.C\sqrt{4k+6}}$$

Where $k = \frac{R_C}{R}$

Note:

- For transistor to work as oscillator

$$h_{fe} \geq \left(4k + 23 + \frac{29}{k}\right)$$

$$h_{fe(min)} = 44.5$$

AC Phase Shift Oscillator using FET

- Frequency of oscillation

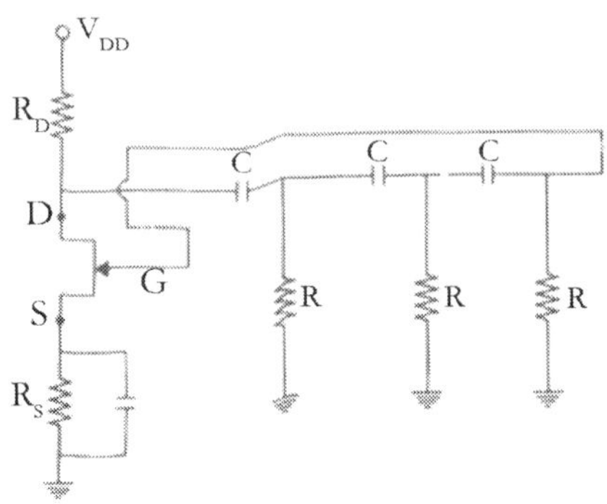

$$f = \frac{1}{2\pi RC\sqrt{6}} \quad R_f \geq 29R_1 \rightarrow \text{for oscillation}$$

$$\beta = -\frac{1}{29}(\text{feedback factor})$$

Note:

- Amplification factor for FET to work as an Oscillator is $\mu \geq 29$

RC Phase Shift Oscillator using OP-AMR

- Feedback factor

$$\beta = \frac{V_f}{V_0} - \frac{1}{29}(\text{for oscillation})$$

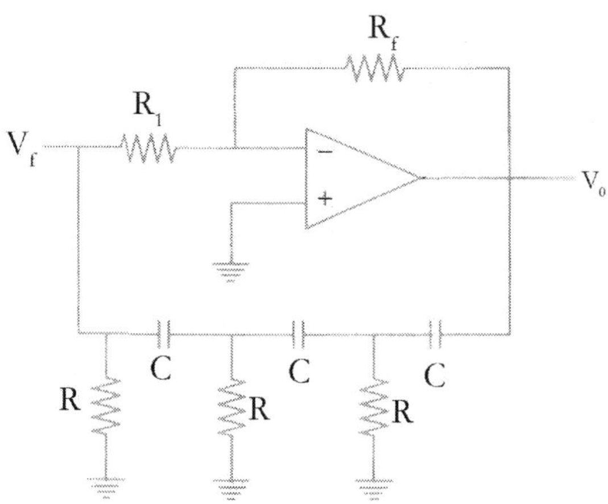

- Frequency of oscillation

$$f = \frac{1}{2\pi RC\sqrt{6}} \quad R_f \geq 29R_1 \rightarrow \text{for oscillation}$$

Wein Bridge Oscillator

A Wien bridge oscillator is a type of electronic oscillator that generates sine waves. It can generate a large range of frequencies.

- Frequency of oscillation

$$f = \frac{1}{2\pi\sqrt{R_3R_4C_1C_2}}$$

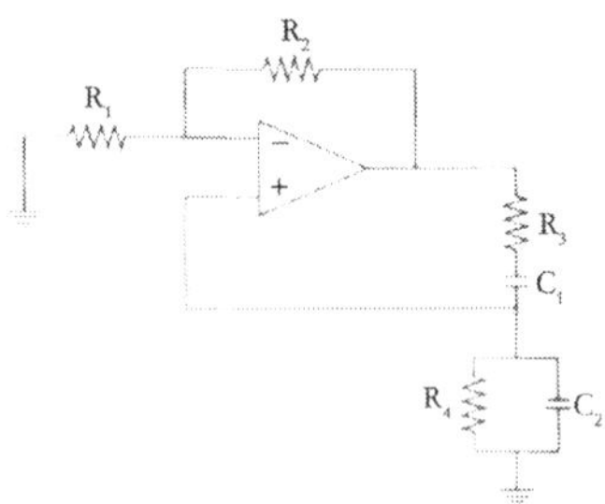

If $R_3=R_4=R$

And $C_1=C_2=C$

Then $f = \frac{1}{2\pi R.C}$

Note:

$\beta = \frac{1}{3}$ for oscillation

$R_2 \geq 2R_1$ for oscillation

Colpitt's Oscillator

The Colpitts Oscillator, named after its inventor Edwin Colpitts is another type of LC oscillator design.

- Frequency of oscillation

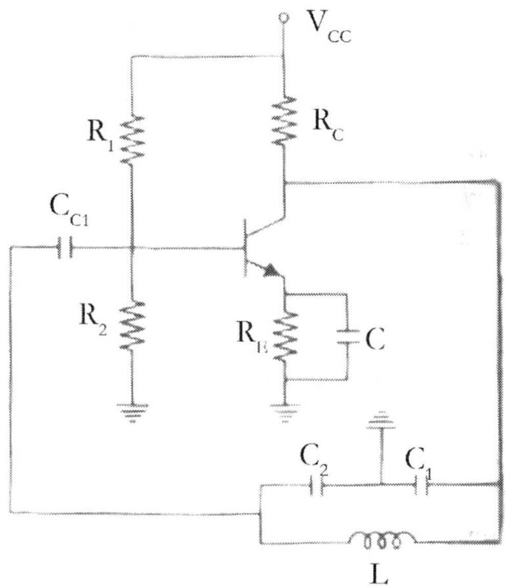

$$f = \frac{1}{2\pi\sqrt{L\left(\frac{C_1 C_2}{C_1 + C_2}\right)}}$$

and

$$g_m R_C \geq \frac{C_2}{C_1}$$

- If in place of C_1, C_2 there is L_1 and L_2 and in place of L there is C then it becomes Hartley Oscillator.
- Frequency of oscillator for Hartley Oscillator is

$$f = \frac{1}{2\pi\sqrt{C(L_1 + L_2)}} \text{ and } g_m RC \geq \frac{L_1}{L_2}$$

Clapp's Oscillator

The Clapp oscillator is one of several types of electronic oscillator constructed from a transistor and a positive feedback network, using the combination of an inductance (L) with a capacitor (C) for frequency determination, thus also called LC oscillator.

- If in Colpitts Oscillator circuit inductor is replaced by a variable capacitor (C_3) then frequency of oscillator becomes

$$f = \frac{1}{2\pi\sqrt{LC_3}} \text{ it happens if } \frac{C_1}{C} >> 1 \text{ and } \frac{C_1}{C} >> \frac{C_1}{C_2}$$

Crystal Oscillator

A crystal oscillator is an electronic oscillator circuit that uses the mechanical resonance of a vibrating crystal of piezoelectric material to create an electrical signal with a very precise frequency.

- It is electrical equivalent circuit is
- Series resonance frequency

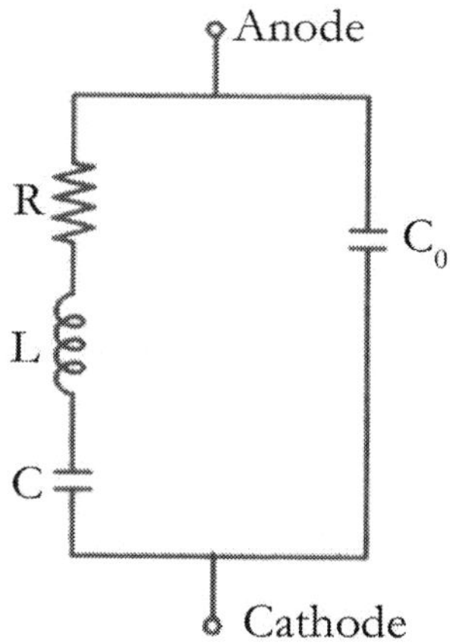

$$f_s = \frac{1}{2\pi\sqrt{LC}}$$

- Parallel resonance frequency

$$f_p = \frac{1}{2\pi\sqrt{LC_{eq}}} \text{ where } C_{eq} = \frac{CC_0}{(C + C_0)}$$

Note:

Frequency of oscillator is chosen between f_p and f_s

Multistage Amplifier

The performance obtainable from a single stage amplifier is often insufficient for many applications, hence several stages may be combined forming a multistage amplifier.

- Upper cutoff frequency of overall configuration is

$f_H^* = f_H\sqrt{2^{1/n} - 1}$ (For identical amplifiers in cascade)

Where n→number of amplifiers in cascade

f_H →upper cutoff frequency of one amplifier

- Lower cutoff frequency of overall configuration

$$f_l^* = \frac{f_l}{\sqrt{2^{1/n} - 1}}$$ where n→number of identical amplifiers in cascade

Note:

Approximate BW of amplifier in cascade is f_H^*

- When amplifiers are non-identical then

$$f_l^* = 1.1\sqrt{f_{l1}^2 + f_{l2}^2 + f_{l3}^2 + \dots\dots\dots + f_n^2} \text{ and}$$

$$\frac{1}{f_H^*} = 1.1\sqrt{\frac{1}{f_{H1}^2} + \frac{1}{f_{H2}^2} + \frac{1}{f_{H3}^2} + \dots\dots\dots + \frac{1}{f_{Hn}^2}}$$

Rise time $t_l^* = 1.1\sqrt{t_{r1}^2 + t_{r2}^2 + \dots\dots\dots + t_{rn}^2}$

Linear Wave Shaping

- Pulse response of high pass R.C circuit

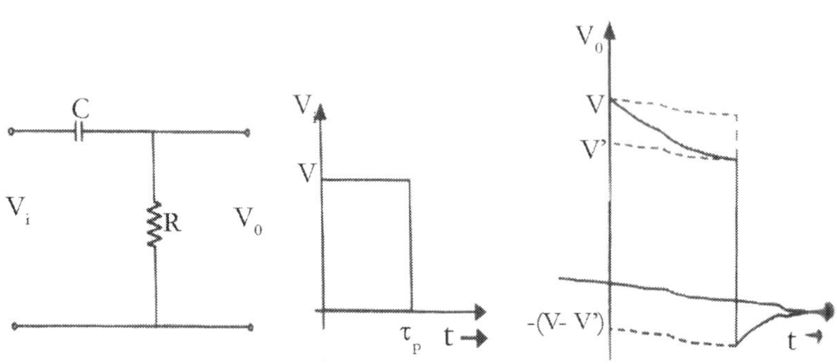

$(V - V')$ is called tilt or sag

$$\%\text{till} = \frac{V - V'}{V} x100\%$$

Sag or tilt will be obtained if $R.C >> \tau_p$

If input is a square wave of time period T then

$$\% \text{ till} = \frac{\pi f_L}{f} x100 \quad \text{where } f_L = \frac{1}{2\pi R.C} \quad \text{and } f \frac{1}{T}$$

Low Pass RC Circuit

A low-pass filter is a filter that passes signals with a frequency lower than a certain cut off frequency and attenuates signals with frequencies higher than the cut off frequency.

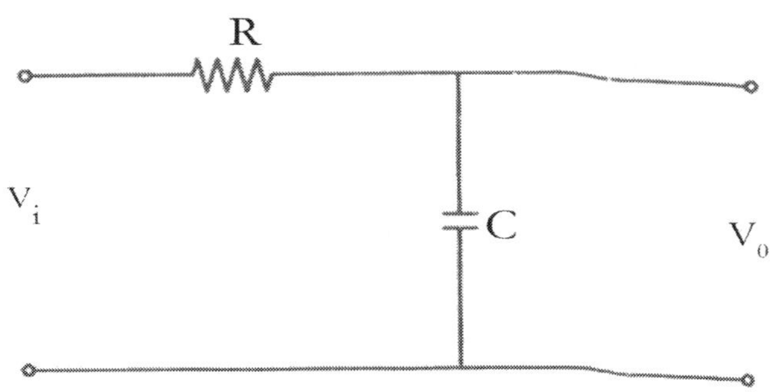

$$\text{Rise time } t_r = \frac{0.35}{f_H} (\text{rise from } 10\% \text{ to } 90\%)$$

$$\text{Where } f_H = \frac{1}{2\pi RC}$$

A Handbook on
Electronics and Communications Engineering

3

Signal & System

CONTENTS

1 Introduction

It is designed to cover analysis, types, convolution, sampling and operations performed on signals.

Standard Signals

In telecommunication, a standard test signal is a single-frequency signal with standardized level used for testing the peak power transmission capability and for measuring the total harmonic distortion of circuits or parts of an electric circuit.

Unit Step Function

In engineering applications, we frequently encounter functions whose values change abruptly at specified values of time t. One common example is when a voltage is switched on or off in an electrical circuit at a specified value of time t.

For continuous time

$$u(t) = \begin{cases} 1; t>0 \\ 0; t<0 \end{cases}$$

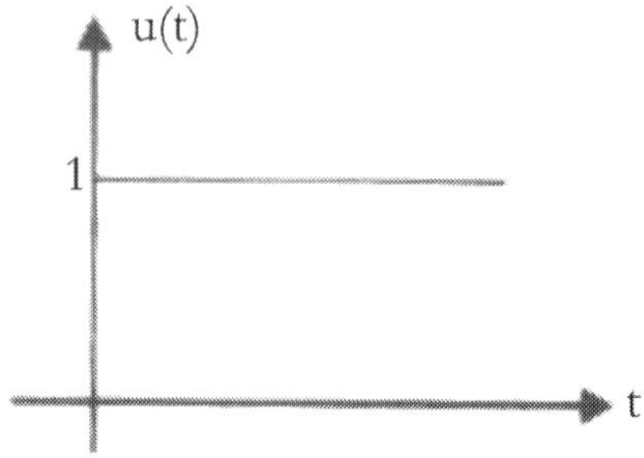

For discrete-time

$u[n] = \begin{cases} 1; n \geq 0 \\ 0; n < 0 \end{cases}$

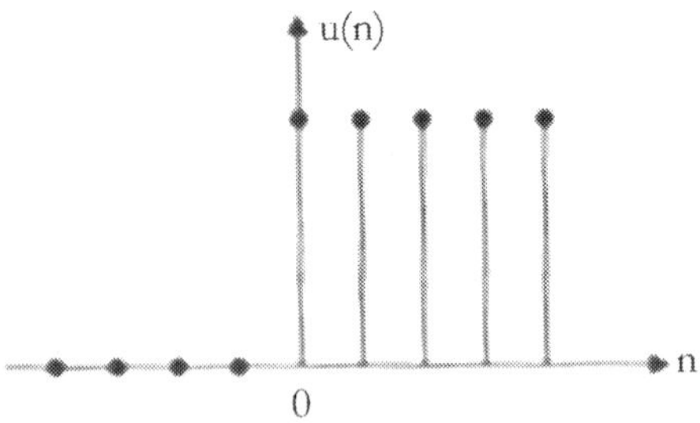

Unit step function in continuous-time domain is not continuous and differentiable at t = 0.

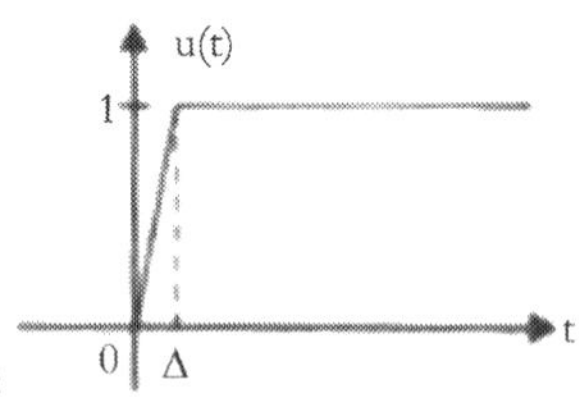

Mathematically $u(0) = \frac{1}{2}$; average value

Unit step function represent a change in step size at t = 0.

Unit Impulse Function

An impulse function is a special function that is often used by engineers to model certain events. An impulse function is not realizable, in that by definition the output of an impulse function is infinity at certain values

For continuous time

$$\delta(t) = \begin{cases} \infty; & t=0 \\ 0; & n \neq 0 \end{cases}$$

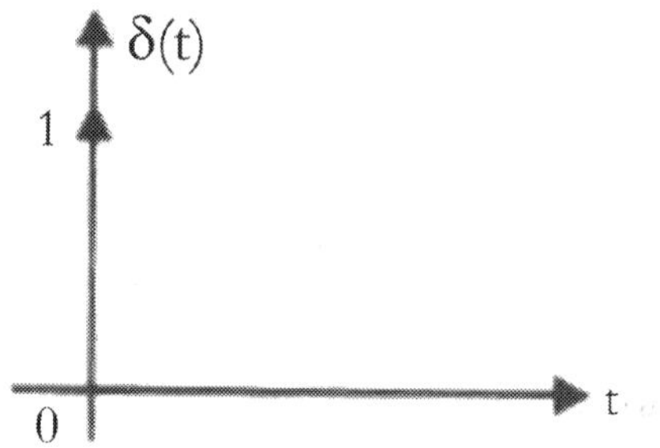

In figure of δ (t) 1 with arrow represents the area of the impulse.

$$\int_{-\infty}^{\infty} \delta(t)\,dt = \int_{0^-}^{0^+} \delta(t)\,dt = 1$$

Impulse function is a continuous function and the area under this function is equal to one.

Properties of the Impulse Function

$$\int_{-\infty}^{\infty} \delta(t-t_o)\phi(t)\,dt = \phi(t_o)$$

$$\int_{-\infty}^{\infty} \frac{d\delta(t)}{dt}\phi(t)\,dt = -\left.\frac{d\phi(t)}{dt}\right|_{t=0}$$

$$\delta(at-b) = \frac{1}{|a|}\delta\left(t-\frac{b}{a}\right); a \neq 0$$

$$\delta(j\omega) = \delta(\omega) \text{ and } \delta(t) = \delta(-t)$$

Relationship between Unit Impulse and Unit Step Function

$$u(t) = \int_{-\infty}^{t} \delta(t)\,d\tau \quad \text{and} \quad \delta(t) = \frac{du(t)}{dt}$$

As the unit step function is neither continuous nor differentiable at t = 0. The unit impulse function is defined as

$$\delta_\Delta(t) = \frac{du_\Delta(t)}{dt}$$

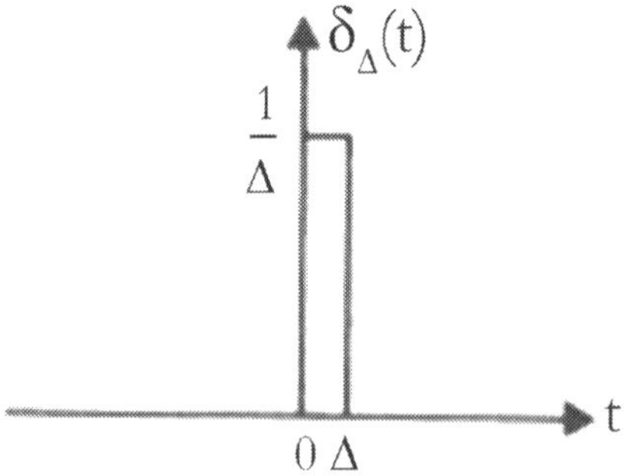

$\delta(t)$ is the limit as $\Delta \rightarrow 0$ of $\delta_\Delta(t)$

Sampling Property of Unit impulse Function

For continuous time

$$x(t)\delta(t-t_o) = x(t_o)\delta(t-t_o)$$

For discrete-time

$$\delta[n] = \begin{cases} 1; & n=0 \\ 0; & n\neq 0 \end{cases}$$

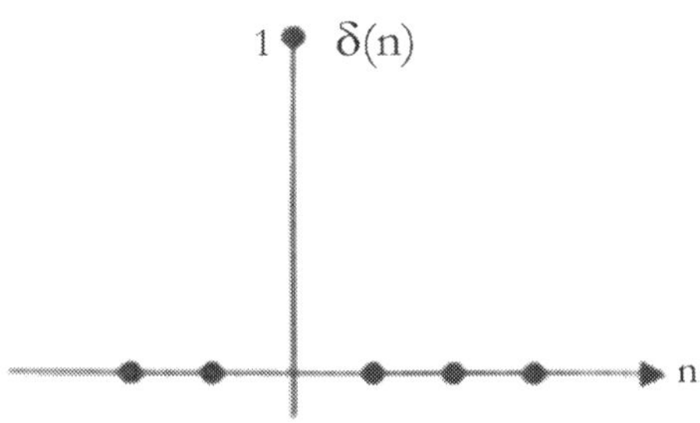

The discrete-time unit impulse is the first difference of the discrete-time unit step.

$$\delta[n] = u[n] - u[n-1]$$

Sampling property $x[n]\delta[n-n_o]=x[n_o]\delta[n-n_o]$

Relationship between Unit Impulse and Unit Step

$$u[n]=\sum_{m=-\infty}^{n}\delta[m]$$

Rectangular or Gate Function

$$x(t)=\begin{cases}A; & -\frac{T}{2}<t<\frac{T}{2}\\ 0; & \text{otherwise}\end{cases}$$

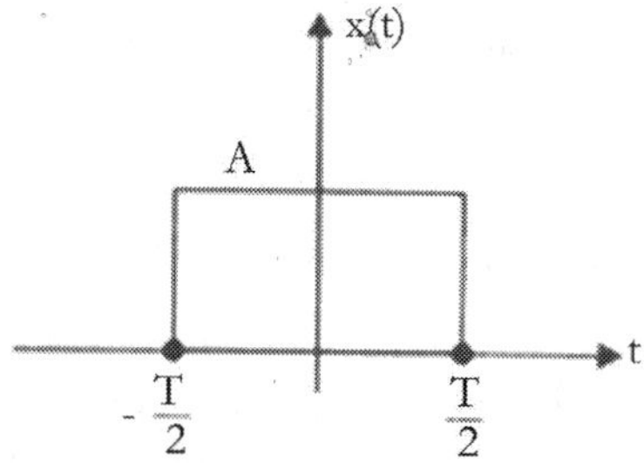

Unit Ramp Function

$$r(t)=\begin{cases}t; & t>0\\ 0; & t<0\end{cases}$$

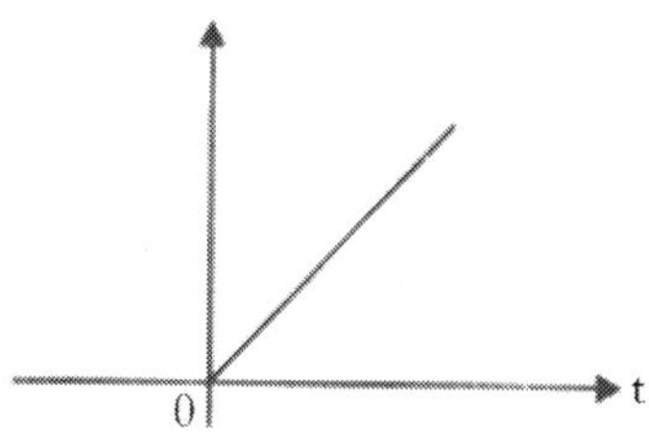

Ramp function represent a change in slope at t = 0.

Signum (or sgn) Function

The sign function or signum function is an odd mathematical function that extracts the sign of a real number. In mathematical expressions the sign function is often represented as sgn.

$$\text{sgn}(t) = u(t) - u(-t)$$
$$= 2u(t) - 1$$

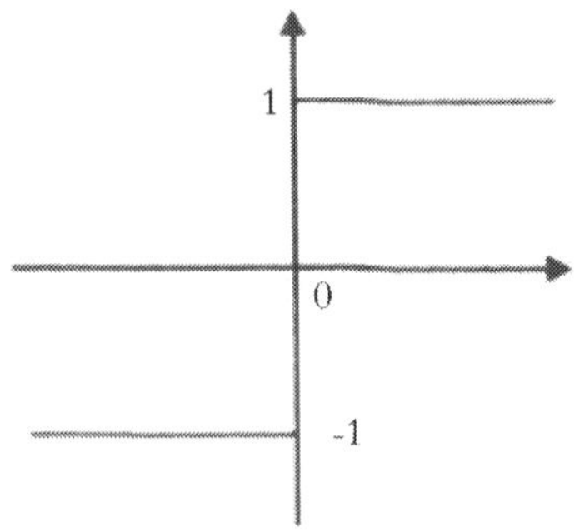

Sgn function is not defined at t = 0 and is chosen as 0 at t = 0.

Sinc and Sinc² Function

The sinc and sinc^2 functions are defined in terms of an independent variable λ.

$$\sin c(\lambda) = \frac{\sin(\pi\lambda)}{\pi\lambda}$$

Sinc (λ) is equal to zero for X = ±n (n 0), n an integer.

Sine Integral Function

Sine integral function is an odd function

$$\text{si}(y) = \int_0^y \frac{\sin(\alpha)}{\alpha} d\alpha; \text{si}(y) = \frac{y}{(1)!1} - \frac{y^3}{(3)3!} + \frac{y^5}{(5)5!} - \frac{y^7}{(7)7!} + \ldots\ldots$$

$$\text{Si}(a) = 0, \text{Si}(\pi) \cong 2.0123, \text{Si}(\infty) = \left(\frac{\pi}{2}\right)$$

Si function converges fast and only a few terms in above equation are needed for a good approximation.

Energy and Power Signals

A signal is said to be energy signal when it has finite energy.

Energy of the signal

$$E_x = \int_{-\infty}^{\infty} |x(t)|^2 \, dt$$

Average power of a signal

$$P_x = \lim_{T\to\infty} \frac{1}{T} \int_{-T/2}^{T/2} |x(t)|^2 \, dt$$

The signal x(t) is an energy signal if $0 < E_x < \infty$, that is Ex finite and $P_x = 0$. The signal x(t) is a power signal if $0 < P_x < \infty$, i.e., P_x is finite and therefore E_x, is infinite. If a signal does not satisfy one of these conditions, then it is neither an energy signal nor a power signal. The power and energy signals are mutually exclusive. Scaling does not change the power content of the signal. While scaling change the energy content of signal energy of sic f(at) will be

$E_a = \frac{E}{a}$ Where, F→ energy of f(t).

$e^{\cdots} u(t)$

where

E_0→ energy in odd part of signal.

E_e→energy in even part of signal.

$e^{-at}u(t)$ is an example of energy signal? All periodic signals are power signals but vice-versa is not true. Ramp signal is neither energy nor power signal.

Periodic Signal

A signal is said to be periodic if it satisfies the condition x(t) = x(t + T) or x(n) = x(n + N).

$$x(t+T)=x(t)$$

The smallest positive number T that satisfies above equation is call the fundamental period.

For periodic signal $x_T(t)$

Average value of the signal

$$x_{avg}=\frac{1}{T}\int_T x_T(t)\,dt$$

Average signal power

$$P_x=\frac{1}{T}\int_T |x_T|^2\,dt$$

Effective or rms value of the signal

$$x_{rms}=\sqrt{P_x}$$

If $x_{T_1}(t)$ and $x_{T_2}(t)$ are two periodic functions with periods T_1 and T_2, then $x(t)=x_{T_1}(t)+x_{T_2}(t)$ is periodic with period T if

$$T=nT_1=mT_2 \quad \text{or} \quad \left[\frac{T_1}{T_2}\right]=\left[\frac{m}{n}\right]$$

Where m and n are integers and $\left(\frac{T_1}{T_2}\right)$ is a rotational number.

The period of x(t) is equal the least common multiple (LCM) of T_1 and T_2. The LCM of two integers m and n; is the smallest integer divisible by both m and n.

Symmetries of Periodic Function

The unit circle has a circumference of $C = 2\pi r = 2\pi(1) = 2\pi$. Therefore, if a point P travels around the unit circle for a distance of 2π, it ends up

where it started. In other words, for any given value q, if 2π is added or subtracted, the coordinates of point P remain unchanged

$$x_T(t) = -x_T\left(t - \frac{T}{2}\right)$$

Where, T = Period of signal $x_T(t)$

Quarter-Wave Symmetry

A combination of half-wave symmetry with either even or odd symmetry results in a wave with quarter-wave symmetry. For example: the wave possesses half-wave symmetry as well as odd symmetry. The wave possesses both half-wave symmetry and even symmetry.

$$x_T(t) = \begin{cases} x_T(-t) = x_T(t) \\ x_T(t) = -x_T\left(t - \frac{T}{2}\right) \end{cases} ; \; x_T(t+T) = x_T(t)$$

For odd quarter-wave symmetry

$$x_T(t) = \begin{cases} x_T(-t) == -x_T(t) \\ x_T(t) = -x_T\left(t + \frac{T}{2}\right) \end{cases} ; \; x_T(t+T) = x_T(t)$$

Shifting does not change period or frequency of the periodic signal. Scaling changes the frequency. $e^{j\omega t}$ is always a periodic signal, but $e^{j\omega_0 n}$ is not always periodic signal. It is periodic if $\frac{2\pi}{\omega_0}$ is rational. Summation of periodic discrete time signals is always periodic.

Analog Signal

An analog or analogue signal is any continuous signal for which the time varying feature (variable) of the signal is a representation of some other time varying quantity, i.e., analogous to another time varying signal.

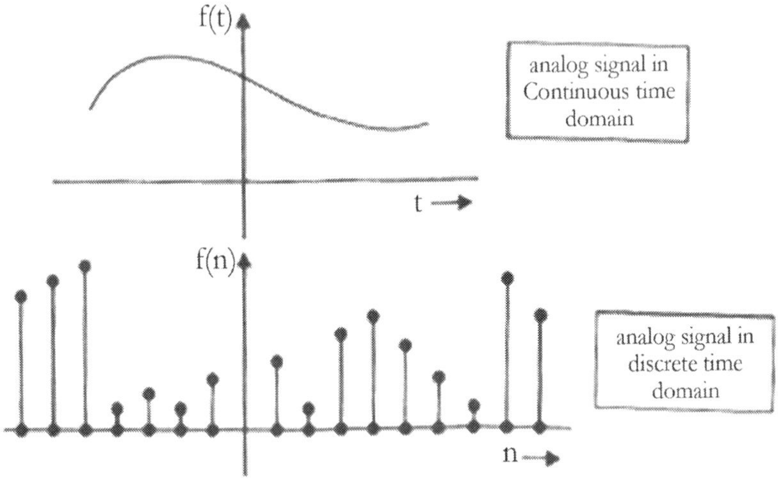

Digital Signal

A signal is said to be digital signal if it takes discrete values for its amplitude. This discrete set can be finite or infinite. Digital signal is defined in both time domains in continuous as well as discrete time domain.

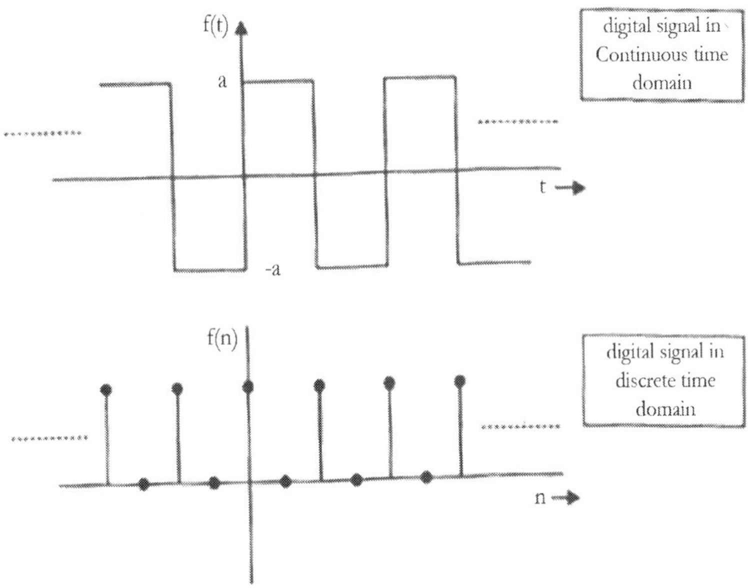

Some Useful Signal Operations

- f(t - T)→ signal is shifted towards right if T is positive. It means sign is delayed.
- f(t + T)→signal is shifted towards left if T is negative, it means signal is advanced.
- f(at)→scaling of signal. If lal > 1 then compression of signal an if lal < 1 then extension of signal in time domain.
- F(-t)→ represents inversion of signal in time domain that is mirror image about y-axis.

Even and Odd Function

If f(t)=f(-t) then function is said to be even.

If f(t) = - f (-t) then function is said to be odd.

Every function can be expressed as summation of even and odd components.

Even function x odd function = Odd function.

Odd function x odd function = Even function

Even function x even function = Even function

Systems

System is characterized by its input, its outputs and rules of operation.

Linear and Non-linear Systems

The concept of linear relationship suggests that two quantities are proportional to each other: doubling one causes the other to double as well. Linear systems satisfy homogeneity and superposition principle while non-linear system does not satisfy one of them or both. Linearity ensures the regeneration of input frequencies at output.

Time Invariant and Time Variant Systems

A system is said to be time invariant if its input output characteristics do not change with time. Otherwise it is said to be time variant system.

Causal and Non-causal Systems

A system is said to be causal system if its output depends on present and past inputs only and not on future inputs. A system whose present response depends on future values of the inputs is called as a non-causal system.

Dynamic and Static System

Dynamic means energetic, capable of action and/or change, or forceful, while static means stationary or fixed.

If f(t) = f*(- t) signal is conjugate symmetric signal or even conjugate signal. If f(t) = -1*(- t) signal is conjugate anti-symmetric signal or odd conjugate signal. For even conjugate signal real part is even in nature and imaginary part is odd in nature. For odd conjugate signal real part is odd in nature and imaginary part is even in nature.Excess carriers, essential for device operation, are created by optical excitation, electron bombardment, or injected across a forward-biased p-n junction. These excess carriers can dominate the conduction process in semiconductor materials.

2 Linear Time Invariant System

Linear time-invariant theory, commonly known as LTI system theory, comes from applied mathematics and has direct applications in NMR spectroscopy, seismology, circuits, signal processing, control theory, and other technical areas.

LTI systems are used to represent linear combinations of time invariant signals.

Basic System Properties

Basic System Properties are follows:

System with Memory and Without Memory

A system is said to be memory less if its output for each value independent variable at a given time is dependent only on the input' that same time.

Invertibility and Inverse Systems

A system is said to be invertible if distinct inputs lead to distinct output Causality

A system is causal if the output at any time depends only on values the input at the present time and in past.

Such system is referred as non-anticipative as the system output does not anticipate future values of input.

All memory fewer systems are causal, since output responds only to tt current value of the input.

Time Invariance

A system is time invariant if the behaviour and characteristics of t system are fixed over time.

A system is time invariant if a time shift in input signal results in identical time shift in the output signal.

Linearity

A system that possess the property of superposition.
The system will be linear if
Additive Property
The response to $x_1(t) + x_2(t)$ is $y_1(t) + y_2(t)$
Scaling or Homogeneity Property
The response to ax_1 (t) is a $y_1(t)$; $a \in$ any complex constant

$$\Rightarrow ax_1(t)+bx_2(t) \rightarrow ay_1(t)+by_2(t)$$

Convolution

Convolution is the most important and fundamental concept in signal processing and analysis. By using convolution, we can construct the output of system for any arbitrary input signal, if we know the impulse response of system.

Continuous-time LTI System

A linear continuous-time system obeys the following property:

Convolution of two functions, $x_1(t)$ and $x_2(t)$

$$y(t)=x_1(t)*x_2(t)=\int_{-\infty}^{\infty}x_1(\tau)x_2(t-\tau)\,d\tau$$

Some Important Result for Convolution

$$\int_{-\infty}^{\infty} x_1(\tau)x_2(t-\tau)\,d\tau = \int_{-\infty}^{\infty} x_2(\tau)x_1(t-\tau)\,d\tau$$

Special Cases

$$\phi(t)*\delta(t) = \int_{-\infty}^{\infty} \phi(\tau)\delta(t-\tau)\,d\tau = \delta(t)*\phi(t) = \phi(t)$$

$$\delta(t)*\phi(t) = \int_{-\infty}^{\infty} \delta(\tau)\delta(t-\tau)\,d\tau = \delta(\tau)$$

Properties of the Convolution Integral

Commutative Property

$$y(t) = x_1(t)*x_2(t) = x_2(t)*x_1(t)$$

Distribution Property

$$x_1(t)*\left[x_2(t)+x_3(t)\right] = x_1(t)*x_2(t)+x_1(t)*x_3(t)$$

Associative Property

$$x_1(t)*\left(x_2(t)*x_3(t)\right) = \left(x_1(t)*x_2(t)\right)*x_3(t)$$

Derivative of the convolution

$$\frac{dy(t)}{dt} = \frac{dx_1(t)}{dt}*x_2(t) = x_1(t)*\frac{dx_2}{dt}(t)$$

$$\frac{d^n y(t)}{dt^n} = \frac{d^p x_1(t)}{dt^p} * \frac{d^q x_2}{dt^q}(t)$$

Running integral

$$\int_{-\infty}^{t} y(\tau)d\tau = \int_{-\infty}^{t} \left[x_1(t) * x_2(t)\right] dt$$

Convolution of two delayed functions

If $y(t) = x_1(t) * x_2(t)$

$$x_1(t - t_1) * x_2(t - t_2) = y\left(\left(t - (t_1 + t_2)\right)\right)$$

Time scaling property

If $y(t) = x_1(t) * x_2(t)$

$$x_1(\alpha t) * x_2(\alpha t) = \frac{1}{|\alpha|} y(\alpha t); \ \alpha \neq 0$$

Discrete-time LTI system

$$y[n] = \sum_{k=-\infty}^{\infty} x_1(k) x_2[n-k] = x_1[n] * x_2[n]$$

Where-ever Two Discrete Time Signals are Convolve

Resultant of convolution will have a length which is equal to sum of individual lengths L1 and L2 of the signals being convolve minus one.

$$L = L_1 + L_2 - 1$$

Resultant of the convolution will have extends which is equal to sum of individual extends of the signal being convolved.

Sum of sample values of output is same as product of sum of sample values of individual signals being convolved.

$$\sum_{-\infty}^{\infty} y[n] = \sum_{-\infty}^{\infty} x[n] \times \sum_{-\infty}^{\infty} h[n]$$

Whenever Two Continuous Time Signals are convolved then

Width of resultant signal will be always sum of widths of i/p and impulse response that is

$$W = W_1 + W_2$$

- Lower extend of resultant will be equal to sum of lower extends of i/p and impulse response, that is 11=11+121
- Similarly upper extend of resultant signal (o/p signal) will be equal to sum of upper extends of i/p and impulse response that is 11,1 = 111 +U21
- Area of resultant function will be product of area under input and area under impulse response h(t) that is A = AI.A21

NOTE

- Convolution of two even signal results in even signal.
- Convolution of two odd signal results in a even signal.
- Convolution of an even and an odd signal results in odd signal.,
- For a LTI system to be stable

ChM' < *43 in continuous time domain.

E tt1nll < cm) in discrete time domain.

where h(t) and h[n] are impulse response of the system.

To Check Linearity and Time Invariance of a System Given in Differential Equation Form

- If in differential equation there are no product terms of input x(t) or o/p y(t) are their derivative that is no terms like

X20), X30 ‘), x(t) y(t) (—dY dt)2 ‘ —dx dt dt ‘ .—dy (dxd2y dx

dt J2

‘ l dt2 dt

then system is said o be linear

- If coefficient of the differentials is constant, not the function of time then this is time invariant system.

Eigen Function

If the response of a given LTI system is same as the i/p function except for a scalar multiple, then the function is called Eigen function.

$e^{j\omega_0 t}$ always forms an Eigen function to any LTI system.

3 Fourier series

A Fourier series is a way to represent a wave-like function as the sum of simple sine waves. More formally, it decomposes any periodic function or periodic signal into the sum of a (possibly infinite) set of simple oscillating functions, namely sines and cosines

Dirichlet Condition

The term "well-behaved" function, x(t) defined in the interval $(t_0, t_0 + T)$ is given in terms of following Dirichlet-condition. The function f(t) must be single value with given interval. The function x(t) can have at most a finite number of discontinuous and a finite number of maxima and minima. The function x(t) must be absolutely integrals on the interval

$$\int_{t_o}^{t_o+T} |x(t)|\,dt = \text{finite} < \infty$$

Fourier Series

Approximation of a given function by Fourier series gives a smooth function even when the function being approximated has discontinuity

Fourier Series Representation of Continuous-time

By using Fourier series, a non-sinusoidal periodic function can expressed as an infinite sum of sinusoidal function. Fourier series represents the information of the signal f(t) as amplitude and angles at different angular frequencies which are integer multiple of a fundamental angular frequency ω_0. It is a time domain to time domain transformation.

Trigonometric Fourier Series

Any practical periodic function of frequency ω_0. can be expressed as infinite sum of sine (or) cosine functions that are integral multiples of ω_0.

$$f(t) = a_o + \sum_{n=1}^{\infty} \left(a_n \cos n\omega_o t + b_n \sin n\omega_o t\right)$$

Where ω_0. = Fundamental frequency

a_0, a_n, b_n = Trigonometric Fourier series coefficient

$$a_n = \frac{2}{T}\int_0^T f(t) \cos n\omega_o t \, dt \; ;$$

$$b_n = \frac{2}{T}\int_0^T f(t) \sin n\omega_o t \, dt$$

Polar form of trigonometric Fourier series

$$f(t) = d_o + \sum_{n=1}^{\infty} d_n \cos\left(n\omega_o t - \phi_n\right)$$

Where, $d_0 = a_0, d_n = \sqrt{a_n^2 + b_n^2}$

Magnitude plot

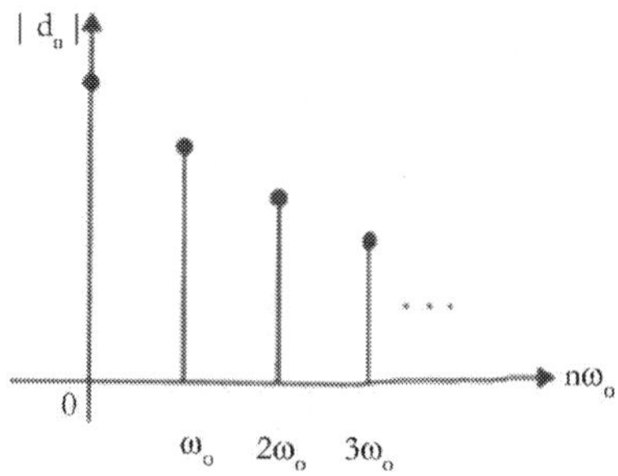

$$|d_n| = \sqrt{a_n^2 + b_n^2}$$

Phase plot

$$\phi_n = \text{ten}^{-1}\left(\frac{b_n}{a_n}\right)$$

Effect of Symmetry of Fourier co-efficient of Real Signals

Symmetry	Even	Odd	Half-Wave
Condition	F(t)=f(-t)	F(t)=-f(-t)	F(t)=-f(t$\pm$T/2)
a_o	?	0	0
a_n	?	0	$= \begin{cases} 0; n \text{ even} \\ ?; n \text{ odd} \end{cases}$
b_n	0	?	$= \begin{cases} 0; n \text{ even} \\ ?; n \text{ odd} \end{cases}$

Exponential Fourier Series

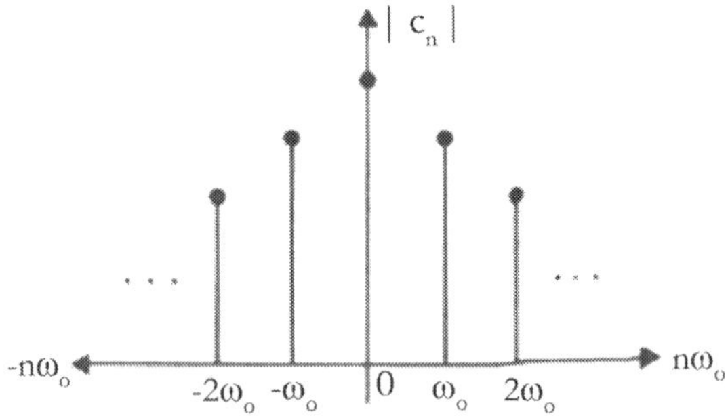

$$f(t) \sum_{n=-\infty}^{\infty} C_n e^{jn\omega_o t}$$

Where, $\boxed{C_o = a_o = \frac{1}{T}\int_0^T f(t)\,d(t)}$; $\boxed{C_n = \frac{1}{T}\int_0^T f(t)\,e^{-jn\omega_o t}dt}$

Relation between Exponential and Trigonometric Fourier Series

$$a_n = (C_n + C_{-n}) \text{ and } b_n = j(C_n - C_{-n})$$

Symmetry	C_n	C_{-n}
Even	Real	Real
Odd	Imaginary	Imaginary
Half wave	$= \begin{cases} 0; n \text{ even} \\ ?; n \text{ odd} \end{cases}$	$= \begin{cases} 0; n \text{ even} \\ ?; n \text{ odd} \end{cases}$

Magnitude spectrum of exponential Fourier series is even in nature.

Phase spectrum of exponential Fourier series is odd in nature.

If we add some constant a to a periodic signal then only dc term will change and other Fourier coefficients will remain same.

For rectangular pulse $C_n \propto \frac{1}{n}$.

For triangular pulse $C_n \propto \frac{1}{n^2}$.

If a function is varying as parabolic variation in a periodic means then

Fourier series coefficient $C_n \propto \frac{1}{n^3}$.

Parseval Power Relation for Periodic Signals

$$P = a_0^2 + \frac{1}{2}\sum_{n=1}^{\infty}\left(a_n^2 + b_n^2\right)$$

$$P = C_0^2 + \frac{1}{2}\sum_{\substack{n=-\infty \\ n\neq 0}}^{\infty} |C_n|^2$$

$$P = \sum_{n=-\infty}^{\infty} |C_n|^2$$

Fourier Series Representation of Discrete Time Periodic Signals

If x[n] is a periodic signal with period N then

$$x[n] = \sum_{k=\langle N\rangle} a_k\, e^{jk\omega_0 n}$$

Where $\omega_0 = \frac{2\pi}{N}$

$$\boxed{a_k = \frac{1}{N}\sum_{n=\langle N\rangle} x[n]\, e^{-jk\omega_0 n}} \quad \boxed{\omega_0 = \frac{2\pi}{N}}$$

Discrete time Fourier series coefficient; is also periodic with period that is

$$a_k = a_{k+N}$$

Properties of Discrete Time Fourier Series

Linearity $Ax[n]+By[n] \rightarrow A\,a_k + B\,b_k$

a_k — Fourier series coefficient of $x[n]$

b_k — Fourier series coefficient of $y[n]$

Time shifting $x[n-n_0] \rightarrow a_k\, e^{-jk\omega_0 n_0}$

Frequency shifting $e^{jM(2\pi/N).n}x[n] \rightarrow a_{k-M}$

Conjugation $x^*[n] \rightarrow a^*_{-k}$

Time Reversal $x[n] \rightarrow a_{-k}$

Periodic Convolution $\sum_{r=<N>} X[r]y[n-1] \rightarrow Na_k b_k$

Multiplication $x[n]y[n] \rightarrow \sum_{I=<N>} a_1 b_{k-1}$

For real and even signal a_k is real and even. For real and odd signal a_k is purely imaginary and odd. For even part of x[n], Fourier series coefficient is $R_e[a_k]$. For odd part of x[n], Fourier series coefficient is $I_m[a_k]$.

Parsevals Relation for Discrete Time Periodic Signals

$$\frac{1}{N}\sum_{n=<N>} |x[n]|^2 \sum_{k=<N>} |a_k|^2$$

4 Fourier Transform

Fourier transform provides a frequency domain description of time domain signals and is extension of Fourier series to non-periodic signals.

Dirchlets Conditions of Fourier Transform

For existence of Fourier transform

Fourier transform is defined for all stable signals $\int_{-\infty}^{\infty} |f(t)| dt < \infty$

Periodic signals, which are neither absolutely integral nor square integral over an infinite interval, can be considered to have Fourier transform if impulse functions are permitted in the transform. f(t) have a finite number of discontinuities and finite number of maxima and minima within any finite interval.

Fourier transform $F.T[f(t)] = F(\omega) = \int_{-\infty}^{\infty} f(t) e^{-j\omega t} dt$

Inverse fourier transform $I.F.T[F(\omega)] = F(t) = \frac{1}{2\pi} \int_{-\infty}^{\infty} f(\omega) e^{j\omega t} d\omega$

Fourier Transform Properties

Fourier transform properties given as following table:

Property	X(f)-form	$X(\omega)$ -form
Linerity $ax_1(t)+ bx_2(t)$	$ax_1(f)+ bx_2(f)$	$ax_1(\omega) + bx_2(\omega)$
Time-scaling X(at)	$\frac{1}{\|a\|}\times\left(\frac{f}{a}\right)$	$\frac{1}{\|a\|}\times\left(\frac{\omega}{a}\right)$
Time-reversal x(-t)	x(-f)	$x(-\omega)$
Time-shift $X(t-t_0)$	$e^{-j2\pi ft_0}x(f)$	$e^{-j\omega t_0}x(\omega)$
Frequency shift $x(t)e^{j\omega_0 t}$	$x(f-f_0)$	$x(\omega-\omega_0)$
Differentiation in time	$\frac{d}{dt}x(t)\leftrightarrow j2\pi fX(f)$	$\frac{d}{dt}x(t)\leftrightarrow j\omega X(\omega)$
Frequency Differentiation	$-j2\pi tx(t)\leftrightarrow\frac{d}{df}X(f)$	$-jtx(t)\leftrightarrow\frac{d}{d\omega}X(\omega)$
Convolution in time x(t)*h(t)	X(f)H(f)	$X(\omega)H(\omega)$
Frequency cinvolution $x_1(t)*x_2(t)$	$X_1(f)*X_2(f)$	$\frac{1}{2\pi}\left[X_1(\omega)^* X_2(\omega)\right]$
Integration $\int_{-\infty}^{t}x(t)d\tau$	$\frac{X(f)}{j2\pi f}+0.5X(0)\delta(f)$	$\frac{X(\omega)}{j\omega}+\pi X(0)\delta(\omega)$

Parseval's theorem $\int_{-\infty}^{+\infty} x^2(t)\,dt$	$\int_{-\infty}^{+\infty} \lvert x(f)\rvert^2\,df$	$\frac{1}{2\pi}\int_{-\infty}^{+\infty} \lvert X(\omega)\rvert^2\,d\omega$
$x^*(f)$	$X^*(-j2\pi f)$	$X^*(-j\omega)$

Duality Property

$f(t) \Leftrightarrow F(\omega)$

$F(t) \Leftrightarrow 2\pi f(-\omega)$

$F(t) \Leftrightarrow f(-f)$

Fourier Transform of Useful Signals

Fourier transform of useful signals given as following table:

Signal, x(t)	F(t) form	$F(\omega)$ form
$e^{-at}u(t), a > 0$	$\frac{1}{a+j2\pi f}$	$\frac{1}{a+j\omega}$
$e^{at}u(-t), a > 0$	$\frac{1}{a-j2\pi f}$	$\frac{1}{a-j\omega}$
$\delta(t)$	1	1
A, Constant	$A\delta(t)$	$2\pi A\delta(\omega)$
A rect.(t/T)	AT sinc(fT)	AT sinc $\left(\frac{\omega t}{2}\right)$
Sinc (t)	Rect.(t)	Rect. $(\omega / 2\pi)$

$e^{-a\|t\|}$	$\frac{2a}{a^2+4\pi^2f^2}$	$\frac{2a}{a^2+\omega^2}$
Sgn(t)	$\frac{1}{j\pi f}$	$\frac{2}{j\omega}$
u(t)	$\frac{1}{j2\pi f}+0.5\delta(f)$	$\frac{1}{j\omega}+\pi\delta(\omega)$
$\cos\omega_0 t$	$\frac{\delta(f-f_0)+\delta(f+f_0)}{2}$	$\pi\left[\delta(\omega-\omega_0)+\delta(\omega+\omega_0)\right]$
$\sin\omega_0 t$	$\frac{\pi}{j}\left[\delta(\omega-\omega_0)-\delta(\omega+\omega_0)\right]$	$\frac{\pi}{j}\left[\delta(\omega-\omega_0)-\delta(\omega+\omega_0)\right]$
e^{-at^2}	$\sqrt{\frac{\pi}{a}}e^{-\pi^2t^2/a}$	$\sqrt{\frac{\pi}{a}}e^{-\omega^2/4a}$
$t^n e^{-at}u(t)$	$\frac{\lfloor n}{(a+j2\pi f)^{n+1}}$	$\frac{\lfloor n}{(a+j\omega)^{n+1}}$

Fourier transform of a real signal is always even conjugate in nature. Fourier transform of real and even signal is purely real Ramesh. Fourier transfer of real and odd signal is purely imaginary and odd. Shifting in time domain only changes the phase spectrum of signal. Compression in time domain leads to extension in frequency and vice-versa. Time differentiation property is only valid if $f(t)\to 0$ as $|t|\to\infty$.

$\hat{f}(t)=f(t)*\frac{1}{\pi t}$ is energy spectral density of signal for energy signal

Hilbert Transform of a Signal

$$\hat{f}(t)=f(t)*\frac{1}{\pi t}$$

Where $\hat{f}(t)$ is Hilbert transform of f(t).

$* \rightarrow$ Represent convolution

Hilbert transform is a wideband phase shifter offering a shift of $\pi/2$.

$$H\left[\frac{1}{\pi t}\right] \rightarrow -\delta(t);\ H[\delta(t)] \rightarrow \frac{1}{\pi t} \rightarrow H[Hf(t)] \rightarrow -f(t)$$

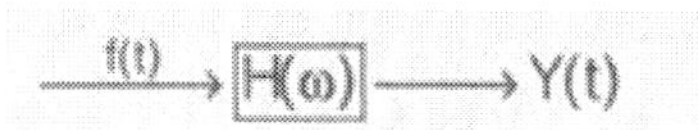

Energy spectral density of o/p

$$|Y(\omega)|^2 = |F(\omega)|^2 . |H(\omega)|^2$$

$$F[k] = \sum_{N=0}^{N-1} f[n] e^{-jk\left(\frac{2\pi}{N}\right)n}$$

k varies from 0 to N-1.

$$F[n] = \frac{1}{N} \sum_{k=0}^{(N-1)} F[k] e^{-jk\left(\frac{2\pi}{N}\right)n};\ 0 \le n \le N-1$$

$$F[k] = NC_k$$

where

C_k is Fourier coefficient of f[n] assuming it to be periodic with period N.

- For real f[n]

F[k] = F*[N-k]

F[0]= f[0] + f[1] + ... +f[N-1]

and if N is even then

F[N/2]= f[0] - f[1] +f[2] ....

Circular Convolution

The circular convolution, also known as cyclic convolution, of two aperiodic functions (i.e. Schwartz functions) occurs when one of them is convolved in the normal way with a periodic summation of the other

function. That situation arises in the context of the Circular convolution theorem.

$$F[n] \otimes_N h[n] = \sum_{k=0}^{N-1} f[k] h[(n-k)]_N$$

$\otimes_N$ represent circular convolution of order N.

$h[(n\text{-}k)]_N$ represent circular shift of order N to h[n].

Properties of DFT

- Time reversal of sequence

$$x[(-n)]_N \leftrightarrow X[(-k)]_N$$

$$x[[-n]]_N = x[N-n]$$

- Circular time shift

$$x[[n-l]]_N \leftrightarrow X(k)e^{-j2\pi kl/N}$$

- Circular frequency shift

$$x[n]e^{j2\pi ln/N} \leftrightarrow X[[k-l]]_N$$

- Complex conjugate property

$$x^*[n] \leftrightarrow X^*[[-k]]_N$$

- Circular Convolution

$$x_1[n] \leftrightarrow X_1[k]$$

$$x_2[n] \leftrightarrow X_2[k]$$

$$x_2[n] \otimes_N x_2[n] \leftrightarrow X_1[k].X_2[k]$$

- Multiplication of two sequences

$$x_1[n]x_2[n] \leftrightarrow \frac{1}{N} X_1[k] \otimes_N X_2[k]$$

Parseval's Theorem

Parseval's theorem usually refers to the result that the Fourier transform is unitary; loosely, that the sum (or integral) of the square of a function is equal to the sum (or integral) of the square of its transform.

$$\sum_{n=0}^{N-1} |x[n]|^2 = \frac{1}{N}\sum_{k=0}^{N-1} |X[k]|^2$$

where X(k) is discrete Fourier transform of x[n].

Circular convolution can be find using linear convolution method is divide the result of linear convolution into two parts such that first part has N samples and then add corresponding part harder and write as circular convolution. Here N is no. of samples in each signal.

Given two signals f[n] and h[n] each consisting of N-samples, if we zero pad each of the signal with N-zeros and carry out circular convolution of order 2 N this will be equal to linear convolution of two signals f[n] and h[n].

5 Laplace Transform

The Laplace transform is an integral transform perhaps second only to the Fourier transform in its utility in solving physical problems. The Laplace transform is particularly useful in solving linear ordinary differential equations such as those arising in the analysis of electronic circuits.

The main drawback of (continuous) Fourier transform is that F.T. can defined only for stable systems, whereas L.T. can be defined for both (Fourier Transform) stable and unstable systems.

Laplace transform of a general signal f(t)

Bilateral (or two sided) Laplace transform

$$F(s)=\int_{-\infty}^{\infty} f(t)e^{-st}dt$$

Unilateral (or unisided) Laplace transform

$$F(s)=\int_{0}^{\infty} x(t)e^{-st}dt$$

Inverse Laplace Transform

Two integrable functions have the same Laplace transform only if they differ on a set of Lebesgue measure zero. This means that, on the range of the transform, there is an inverse transform.

$$F(t) = \frac{-j}{2\pi} \int_{\sigma-j\infty}^{\sigma+j\infty} f(s) e^{st} ds$$

Relation between Laplace Transform and Fourier Transform

$$F(\sigma + j\omega) = \int_{-\infty}^{\infty} f(t) e^{-st} dt$$

where

$s = \sigma + j\omega$

for $\sigma = 0$, laplace transform turn out to be fourier transform.

Convergence of Laplace Transform

A necessary condition for convergence of Laplace transform is

$$\int_{-\infty}^{\infty} f(t) e^{-\sigma t} dt < \infty .$$

$e^{-\sigma t}$ may be decaying or growing depending on whether a is positive or negative.

Region of Convergence (ROC)

The range of value for which the integral $\int_{-\infty}^{\infty} f(t) e^{-\sigma t} dt$ converges is known as ROC.

Properties of ROC

- ROC does not contain any poles.
- ROC of F(s) contains of strips parallel to imaginary axis.
- If f(t) is of finite duration and is absolutely integrable than the ROC is entire s-plane.
- If f(t) is a right sided and if the line $\mathrm{Re}\{s\} = \sigma_o$ is in the ROC then all values of s for which $\mathrm{Re}(s) > \sigma_o$ will also be in ROC.
- If f(t) is left sided and if the line $\mathrm{Re}\{s\} = \sigma_o$ is in ROC then all values of s for which $\mathrm{Re}(s) < \sigma_o$ will also be in ROC.
- If F(s) is rational and signal f(t) is right sided than the ROC is the region in s-plane to the right of right most pole and if f(t) is left sided and ROC is the region in s-plane to the left of left most pole.

For unilateral laplace transform ROC is not necessary to mention because it is unique.

Properties of Laplace Transform

Linearity

$f_1(t) \xrightarrow{\text{L.T.}} F_1(s)$ with ROC = R1

$f_2(t) \xrightarrow{\text{L.T.}} F_2(s)$ with ROC = R2

$$af_1(t) + bf_2(t) \xrightarrow{\text{L.T.}} aF_1(s) + bF_2(s);\ \text{ROC} = R_1 \cap R_2$$

Time-shifting

$f(t) \xrightarrow{\text{L.T.}} F(s)$ with ROC = R

$$f(t - t_0) \xrightarrow{\text{L.T.}} e^{-st_0}F(s);\ \text{ROC} = R\begin{pmatrix} a_{11} & a_{12} & a_{13} \\ a_{21} & a_{22} & a_{23} \\ a_{31} & a_{32} & a_{33} \end{pmatrix}$$

Shift in S-domain

$$f(t) \xrightarrow{L.T.} F(s) \text{ with ROC} = R$$

Time-reversal

$$f(t) \xrightarrow{L.T.} F(s)$$

$$f(-t) \xrightarrow{L.T.} F(-s); \text{ROC} = -R$$

Differentiation in S-domain

$$f(t) \xrightarrow{L.T.} F(s) \text{ with ROC} = R$$

$$tf(t) \xrightarrow{L.T.} -\frac{d}{ds}F(s); \text{ROC} = R$$

Convolution in time

If $f(t) \xrightarrow{L.T.} F(s)$ with ROC = R1

and $h(t) \xrightarrow{L.T.} H(s)$ with ROC = R2

$$f(t) * h(t) \xrightarrow{L.T} F(s)H(s); \text{ROC} = R_1 \cap R_2$$

- LT impulse response is known as system or transfer function

Frequency integration

$$\frac{f(t)}{t} \xrightarrow{L.T} \int_s^{\infty} F(s)ds$$

Integration in time

$$\int_0^t f(\tau)d\tau \xrightarrow{L.T.} \frac{f(s)}{s}$$

Differentiation in time

$$\frac{d}{dx}f(t) \xrightarrow{L.T.} sF(s) - f(0)$$

$$\frac{d^2}{dt^2}f(t) \xrightarrow{L.T.} s^2F(s) - sf(0) - f'(0)$$

Initial value theorem

$$f(0)=\lim_{s\to\infty} sF(s)$$

Final value theorem

$$f(\infty)=\lim_{s\to\infty} sF(s)$$

Initial value theorem can be applied only if degree of numerator of L.T. is less than degree of denominator.

Final value theorem can be applied only if all the poles sF(s) lie in left half of the S-plane.

Laplace Transforms of Elementary Function

Transform pair	Signal	Transform	ROC
1	$\delta(t)$	1	All s
2	u(t)	$\frac{1}{s}$	$\text{Re}\{s\}>0$
3	u(-t)	$\frac{1}{s}$	$\text{Re}\{s\}<0$
4	$\frac{t^{n-1}}{(n-1)!}u(t)$	$\frac{1}{s^n}$	$\text{Re}\{s\}<0$
5	$-\frac{t^{n-1}}{(n-1)!}u(-t)$	$\frac{1}{s^n}$	$\text{Re}\{s\}<0$
6	$e^{-\alpha t}u(t)$	$\frac{1}{s+\alpha}$	$\text{Re}\{s\}>-\alpha$
7	$-e^{-\alpha t}u(-t)$	$\frac{1}{s+\alpha}$	$\text{Re}\{s\}<-\alpha$

8	$\frac{t^{n-1}}{(n-1)!}e^{-\alpha t}u(t)$	$\frac{1}{(s+\alpha)^n}$	$Re\{s\} > -\alpha$
9	$-\frac{t^{n-1}}{(n-1)!}e^{-\alpha t}u(-t)$	$\frac{1}{(s+\alpha)^n}$	$Re\{s\} < -\alpha$
10	$\delta(t-T)$	e^{-sT}	All s
11	$[\cos \omega_0 t]u(t)$	$\frac{s}{s^2+\omega_0^2}$	$Re\{s\} < 0$
12	$[\sin \omega_0 t]u(t)$	$\frac{\omega_0}{s^2+\omega_0^2}$	$Re\{s\} > 0$
13	$[e^{-\alpha t}\cos \omega_0 t]u(t)$	$\frac{s+\alpha}{(s+\alpha^2)+\omega_0^2}$	$Re\{s\} > -\alpha$
14	$[e^{-\alpha t}\sin \omega_0 t]u(t)$	$\frac{\omega_0}{(s+\alpha^2)+\omega_0^2}$	$Re\{s\} > -\alpha$
15	$u_n(t) = \frac{d^n\delta(t)}{dt^n}$	s^n	All s
16	$u_{-n}(t) = \frac{u(t)*\cdots*u(t)}{\text{n times}}$	$\frac{1}{s^n}$	$Re\{s\} > 0$

Characterization of LTI Systems Using Laplace Transform

Following are characterization of lti systems using Laplace transform:

Causality

- ROC associated Path the System function for a causal system is a right half plane.

For a system with a rational system function, causality of the system equivalent to the ROC berg the right-half plane to the right of rightmost pole and it includes ∞

A system is anticausal if its impulse response (t) = 0 for t > 0.

Stability

- An LTI system is stable A and only it the ROC of its system function includes the $j\omega$-axis, [i.e. $Re\{s\}=0$]
- A causal system with rational system functions H(s) is stable if and if all of the poles d H(s) lie in the left-half of the s-Plane-i.e. all of the poles have negative real parts.

For a green F(s) if there we n distinct poles then there will (n-1) two sided inverse laplace transform, one right sided and one sided inverse laplace transform exist. So total (n +1) in laplace transform exist.

6 Z-Transform

Analysis of continuous time LTI systems can be done using z-transforms. It is a powerful mathematical tool to convert differential equations into algebraic equations.

The bilateral (two sided) z-transform of a discrete time signal x(n) is given as

$$Z.T\left[x(n)\right] = X(Z) = \sum_{n=-\infty}^{\infty} x(n) z^{-n}$$

The unilateral (one sided) z-transform of a discrete time signal x(n) is given as

$$Z.T\left[x(n)\right] = X(Z) = \sum_{n=0}^{\infty} x(n) z^{-n}$$

Z-transform may exist for some signals for which Discrete Time Fourier Transform (DTFT) does not exist.

Z-transform converts a discrete-time signal, which is a sequence of real or complex numbers, into a complex frequency domain representation.

For a discrete-time LTI system with impulse response h[n], the response y[n] of the system to a complex exponential input of the form z^n is $y[n] = z^n H(z)$where H(z) o known as transfer function of the system.

Reason of Convergence (ROC)

ROC is the region of range of value for which the summation $\sum_{n=-\infty}^{\infty} x[n]z^{-n}$ converges.

Properties of ROC

- The ROC of X(z) consists of a ring in the Z-plane centered about the origin.
- ROC does not contain any pole.

If x[n] is of fate duration then ROC is entire z-plane except possibly z=0 and/or $z \to \infty$.

If x[n] is right sided sequence and if circle $|z| = a$ is in the ROC then all finite values of z for which $|z| > a$ will also be in ROC

If x[n] is a left sided sequence and if circle $|z|=a$ is the ROC then all finite values of z for which $|z| < a$ will also be in ROC.

If x[n] is two sided and if the circle $|z| = a$ is In the ROC, then the ROC will consist of a ring in the z-plane that includes the circle $|z| = a$.

Properties of z-transform

Linearity

$$x_1[n] \xrightarrow{Z.T.} X_1[z],\ ROC=R_1$$

$$x_2[n] \xrightarrow{Z.T.} X_2[z],\ ROC=R_2$$

$$ax_1[n]+bx_1[n] \xrightarrow{Z.T.} aX_1(z)+bX_2(z);\ ROC = R_1 \cap R_2$$

Time shift

$$x[n] \xrightarrow{Z.T.} x[z],\ ROC=R$$

$$x[n-n_0] \xrightarrow{Z.T.} z^{-n_0}X(z);\ ROC = R$$

Exponential multiplication or scaling in z-domain

$$x[n] \xrightarrow{\text{Z.T.}} X[z], \text{ ROC=R}$$

$$a^n x[n] \xrightarrow{\text{Z.T.}} X(z/a); \text{ROC} = |a|R$$

Time reversal

$$x[n] \xrightarrow{\text{Z.T.}} X[z]$$

Then

$$nx[n] \xrightarrow{\text{Z.T.}} -z\frac{d}{dz}X(z); \text{ROC} = R$$

Differential in z-domain

$$nx[n] \xrightarrow{\text{Z.T.}} -z\frac{d}{dz}X(z); \text{ROC} = R$$

Convolution in time

$$x[n] \xrightarrow{\text{Z.T.}} X[z], \text{ ROC=R1}$$

$$h[n] \xrightarrow{\text{Z.T.}} H[z], \text{ ROC=R2}$$

$$x[n]*h[n] \xrightarrow{\text{Z.T.}} X(z)H(z); \text{ROC} = R_1 \cap R_2$$

Accumulation

$$x[n] \xrightarrow{\text{Z.T.}} X[z], \text{ ROC=R}$$

$$\sum_{k=-\infty}^{n} x[k] \xrightarrow{\text{Z.T}} \frac{X(z)}{1-z^{-1}}; \text{ROC} = R \cap |z| > 1$$

Unilaterial Z.T.

$$x[z] \xrightarrow{\text{U.Z.T}} \sum_{n=0}^{\infty} x[n]z^{-n}$$

Left-shift

$$x[n+1] \xrightarrow{\text{Z.T}} zX(z) - zx(0)$$

Right-shift

$$x[n]-x[n-1]=\left(1-z^{-1}\right)X(z)$$

First difference

$$x[n]-x[n-1]=\left(1-z^{-1}\right)X(z)$$

Conjugation

$$x^*[n]=X^*(z^*)$$

Initial value theorem

If x[n] =0 for n<0

$$x(0)=\lim_{z\to\infty}X(z)$$

Final value theorem

$$\lim_{n\to\infty}x[n]=\lim_{Z\to1}\left(1-z^{-1}\right)X(Z)$$

Initial value theorem can be applied to any X(z). Final value theorem will be applicable if $(1-z^{-1})\,X(z)$ has all poles lie inside the unit circle.

Inverse z-transform

The inverse Z-transform is

$$x[n]=\frac{1}{2\pi j}\oint X(z)z^{n-1}dz$$

Where the symbol $\oint$ denotes integration around a counter clockwise circular contour centered at the origin with radius a.

Some common z-transform pairs

No.	Signal	Transform	ROC
1.	$\delta[n]$		All z
2.	u[n]	$\frac{1}{1-z^{-1}}$	$\lvert z\rvert > 1$
3.	-u[-n-1]	$\frac{1}{1-z^{-1}}$	$\lvert z\rvert < 1$
4.	$\delta[n-m]$	z^{-m}	All z, except 0(if m>0) or ∞ (if m<0)
5.	$a^n u[n]$	$\frac{1}{1-\alpha z^{-1}}$	$\lvert z\rvert > \lvert\alpha\rvert$
6.	$-\alpha^n u[-n-1]$	$\frac{1}{1-\alpha z^{-1}}$	$n\alpha^n u[n]$
7.	$n\alpha^n u[n]$	$\frac{\alpha z^{-1}}{\left(1-\alpha z^{-1}\right)^2}$	$\lvert z\rvert > \lvert\alpha\rvert$
8.	$-n\alpha^n u[-n-1]$	$\frac{\alpha z^{-1}}{\left(1-az^{-1}\right)^2}$	$\lvert z\rvert < \lvert\alpha\rvert$
9.	$[\cos\omega_0 n]u[n]$	$\frac{1-[\cos\omega_0]z^{-1}}{1-[2\cos\omega_0]z^{-1}+z^{-2}}$	$\lvert z\rvert > \lvert 1\rvert$
10.	$[\sin\omega_0 n]u[n]$	$\frac{[\sin\omega_0]z^{-1}}{1-[2\cos\omega_0]z^{-1}+z^{-2}}$	$\lvert z\rvert > \lvert 1\rvert$
11.	$[a^n\cos\omega_0 n]u[n]$	$\frac{1-[a\cos\omega_0]z^{-1}}{1-[2a\cos\omega_0]z^{-1}+a^2z^{-2}}$	$\lvert z\rvert > a$

12.	$[a^n \sin \omega_0 n]u[n]$	$\frac{[a\sin\omega_0]z^{-1}}{1-[2a\cos\omega_0]z^{-1}+a^2z^{-2}}$	$\lvert z\rvert > a$
13.	$n^2u[n]$	$\lvert z\rvert > a$	$\lvert z\rvert > a$

Characterization of LTI Systems Using Z-Transform

Characterization of LTI systems using Z-transform are following:

Causality

- A discrete-time LTI system is causal it and only if the ROC of its system function is the exterior of a circle, inducing infinity.
- A discrete-time LTI system with rational system function H (z) is causal if and only if.

  the ROC is the exterior of a circle outside the outermost pole; and with H(z) expressed as a ratio of polynomials in z, the order of the numerator cannot be greater than the order of the denominator.

Stability

- An LTI system is stable if and only it the ROC of its system function H (z) includes the unit circle, $|z|=1$
- A causal LTI system with rational system function H (z) is stable if and only if all of the poles of H (z) lie inside the unit circle i.e.. they must all have magnitude smaller than 1.

7 Miscellaneous

- IIR filters are filters having infinite samples in their impulse response.
- FIR filters are filters having finite samples in their impulse response.
- FIR filters have exact linear phase.
- FIR filters are always stable.
- FIR filters can be realized efficiently in hardware.
- While IIR filters involve fewer parameters, less memory requirement and lower computational complexity.
- Suppose we have a signal f[n] then f[an] where lal is an integer greater than 1 represent decimations.
- f[an] with lal <1 represents interpolation.

Derivative of Impulse function

$$\delta'(at) = \frac{1}{a|a|}\delta'(t)$$

- δ'(-t) is odd function of time while δ (t) is even function of time.
- All even derivative of δ(t) are even function of time and all odd derivative of δ(t) are odd function of time.

$$f(t)\delta'(t-a) = f(a)\delta'(t-a) - f'(t)\delta(t-a)$$

$$\int_{-\infty}^{\infty} f(t)\delta'(t-a)dt = -f'(a)$$

In order to find Fourier transform of a signal having N-samples it require 4 N^2 multiplication and 2N(N - 1) additions.

When we find Fourier transform using Cooley-Jucky algorithm multiplications are reduced to N/2 $\log_2 N$ and additions are reduced to N $\log_2 N$.

$$\text{Group Delay} = -\frac{d}{d\omega}\phi(\omega)$$

$$\text{Phase Delay} = -\frac{\phi(\omega)}{\omega}$$

where

$\phi(\omega) \rightarrow$ Phase of system

- For a distortion less system group delay and phase delay must be equal and equal to a constant value, that is

$$-\frac{d}{d\omega}\phi(\omega) = -\frac{\phi(\omega)}{\omega} = \text{Cons}\tan\text{t}$$

Group delay means delay offer by system to a group of frequencies and phase delay means delay offer by system to a single frequency.

A Handbook on Electronics and Communications Engineering

4

Communication System

CONTENTS

1 Basics

- The basic elements of communication system are transmitter, channel and receiver.
- Transmitter is a device that generates and amplifies a carrier wave, modulates it with a meaningful signal derived from speech or other sources, and radiates the resulting signal from an antenna.
- Channel refers either to a physical transmission medium such as a wire or to a logical connection over a multiplexed medium such as a radio channel.
- Receiver is a device that converts an electrical signal into usable information.

Modulation

Modulation is an important step of communication system. Modulation is defined as the process whereby some characteristic (line amplitude, frequency, phase of a high frequency signal wave (carrier wave) is varied in accordance with instantaneous value intensity of low frequency signal wave (modulating wave.)

Need of Modulation

(i) To separate signal from different transmitters: Audio frequencies are within the range of 20 Hz to 20 kHz. Without modulation all signals at same frequencies from different transmitters would be mixed up.

(ii) Size of the antenna

(iii) Multiplexing is possible only due to modulation.

(iv) Some modulation techniques reduce the effect of noise and signal.

(v) Some modulation techniques are power efficient.

Baseband transmissions typically use digital signaling over a single wire; the transmissions themselves take the form of either electrical pulses or light.

Broadband uses analog signals in the form of optical or electromagnetic waves over multiple transmission frequencies. For signals to be both sent and received, the transmission media must be split into two channels. Alternatively, two cables can be used: one to send and one to receive transmissions.

Spectrum is the frequency domain representation of the signal.

Noise mainly added to signal in the channel.

Bandwidth is a measure of the width of a range of frequencies, measured in hertz.

2 Amplitude Modulation

In this modulation, the amplitude of the carrier signal varies in accordance with the message signal, and other factors like phase and frequency remain constant. Here are the three signals in mathematical form:

Information: $_{Vm}$ (t)

Carrier: $V_C(t) = V_{co} \sin(2pf_c t + f)$

AM: $V_{AM}(t) = \{V_{co} + V_m(t)\} \sin(2pf_c t + f)$

Amplitude Modulation Technique

Amplitude modulation (AM) is a modulation technique used in electronic communication, most commonly for transmitting information via a radio carrier wave. In amplitude modulation, the amplitude (signal strength) of the carrier wave is varied in proportion to the waveform being transmitted.

Let

$v_c = V_c \cos \omega_c t$

$v_m = V_m \cos \omega_m t$

v_c – Instantaneous value of the carrier

V_c – Peak value of the carrier

ω_c – Angular velocity of the carrier

v_m – Instantaneous value of the modulating signal

V_m – Maximum value of the modulating signal

ω_m – Angular velocity of the modulating signal

f_m – Modulating signal frequency

Then amplitude modulated signal is

$$s(t) = V_c\left(1+\frac{V_m}{V_c}\cos\omega_m t\right).\cos\omega_c t$$

$\frac{V_m}{V_c}$ is called modulation index (m)

$$m = \frac{V_m}{V_c}$$

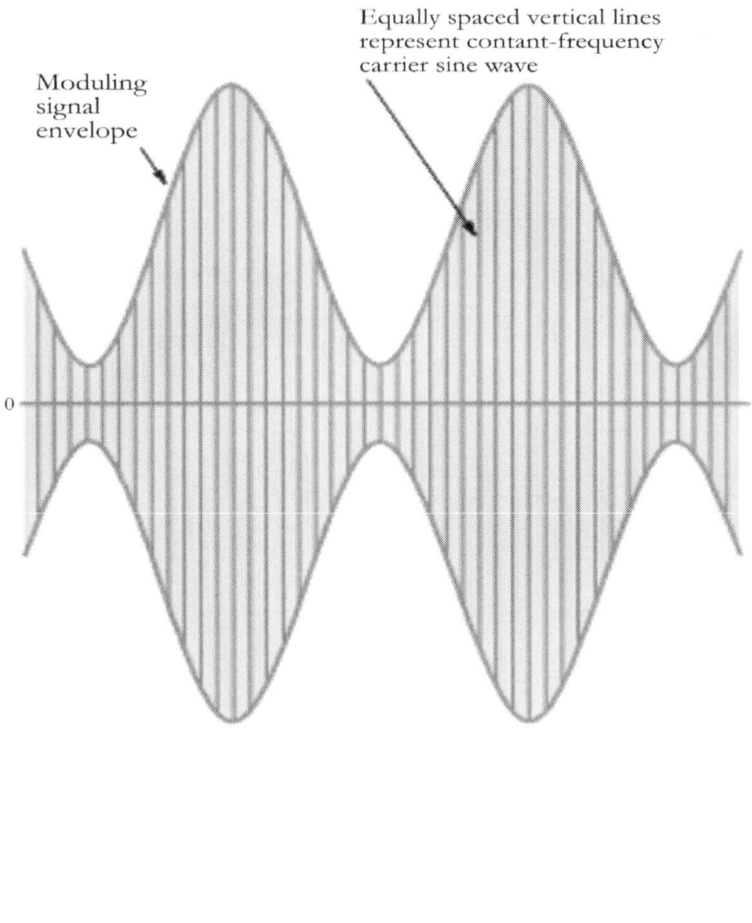

Frequency spectrum of Amplitude Modulation wave

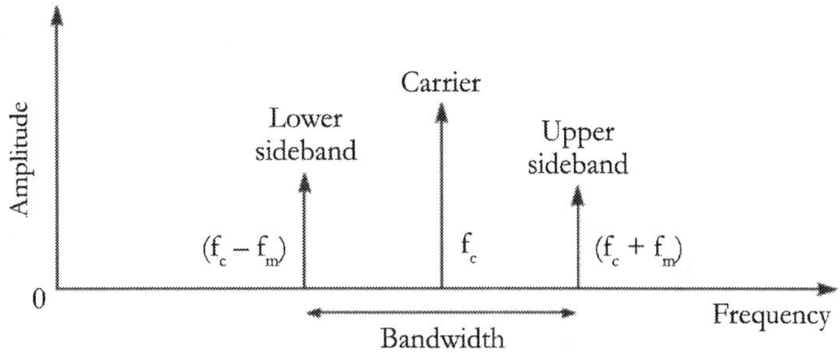

$$\text{B.W.} = (f_c + f_m) - (f_c - f_m) = 2f_m$$

Note:

- Significant frequencies from f_c to $(f_c + f_m)$ is called upper sideband.
- Significant frequencies from $(f_c - f_m)$ to f_c is called lower sideband.

Minimum value of amplitude of amplitude modulated wave is

$V_{min} = (V_c - V_m)$

Maximum value of amplitude of amplitude modulated wave is

Now modulation index in terms of V_{max} and V_{min} will be

$$m = \frac{V_m}{V_c} = \left(\frac{V_{max} - V_{min}}{V_{max} + V_{min}}\right)$$

Note:

For avoiding phase reversal $|m| < 1$

Power Relation of AM Wave

A modulated wave has more power than had by the carrier wave before modulating. The total power components in amplitude modulation can be written as:

$$P_{total} = P_{carrier} + P_{LSB} + P_{USB}$$

$$P_{carrier} = \left[\left(V_c / \sqrt{2}\right) / R\right]^2 = V^2{}_C / 2R$$

$$P_{LSB} = P_{USB} = \left(mV_c / 2\sqrt{2}\right)^2 / R = m^2 / 4 * V^2C / 2R = m_2 / 4\, P_{carrier}$$

$$P_{total} = V^2{}_C / 2R + \left[m^2 / 4 * V^2C / 2R\right] + \left[m^2 / 4 * V^2C / 2R\right]$$

$$= V^2{}_C / 2R \left(1 + m^2 / 2\right) = P_{carrier} \left(1 + m^2 / 2\right)$$

$$Mt = \sqrt{\left(m1^2 + m2^2 + m3^2 + m4^2 + \ldots..\right)}$$

If I_c and It are the r.m.s values of unmodulated current and total modulated current and R is the resistance through which these current flow, then

$$M_t = \sqrt{\left(m_1{}^2 + m_2{}^2 + m_3{}^2 + m_4{}^2 + \ldots..\right)}$$

$$P_{total} / P_{carrier} = \left(I_t.R / I_c.R\right)^2 = \left(I_t / I_c\right)^2 P_{total} / P_{carrier}$$

$$= \left(1 + m^2 / 2\right) I_t / I_c = 1 + m^2 / 2$$

Maximum power dissipated in the AM wave is P_{total}=1.5P_c for m=1 and this is maximum power that amplifier can handle without distortion.

Current Relation of Amplitude Modulation wave

$$P_{total} / P_{carrier} = \left(I_tR / I_cR\right)^2 = \left(1 + m^2 / 2\right)$$

$$\left(It / Ic\right) = \sqrt{\left(1 + m^2 / 2\right)}$$

Where Ic = Unmodulated current

It = Total current

Double Sideband Suppressed Carrier (DSBSC) Technique

The modulation and demodulation technique discussed last lecture require the existence of high quality multipliers (usually called mixers in communication applications). The use of multipliers is generally undesirable for reasons that are beyond the scope of this course. So, we need to find DSBSC modulation techniques that do not depend on multipliers.

Let S(t) is DSBSC modulated signal then

$s(t) = V_m \cos \omega_c t . \cos \omega_m t;$

$s(t) = mV_c \cos \omega_c t . \cos \omega_m t$

Frequency Spectrum of DSBSC Wave

The frequency spectrum of DSBSC wave is follows:

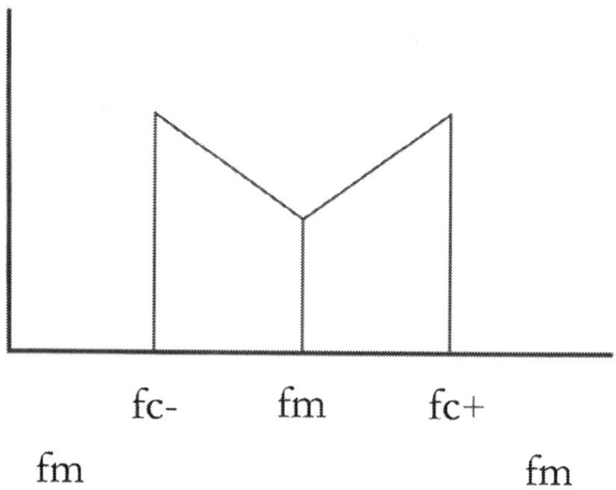

$$\text{B.W.} = 2f_m$$

$$P_{DSBSC} = P_{USB} + P_{LSB}$$

$$P_{USB} = P_{LSB} = m^2 V_c^2 / 8R$$

$$P_{DSBSC} = m^2 V_c^2 / 4R$$

$$P_{DSBSC} = m^2 P_c / 2$$

Single Sideband Technique

A single sideband (SSB) AM signal is represented mathematically as

$$S(t) = V_C m(t)\cos 2\pi f_c t \mp A_c \hat{m}(t)\sin 2\pi f_c t$$

Where $\hat{m}(t)$ is Hilbert transform of message m(t)

Where (-) sign represent USB

(+) sign represent LSB

Frequency Spectrum of SSB Wave

$$S(t) = \frac{mV_C}{2}\cos[\omega_c - \omega_m)] + F\left(\frac{mV_c}{2}\cos(\omega_c - \omega_m)t\right)$$

Where F = representation of fraction

Frequency Spectrum of VSB Wave

The frequency spectrum of VSB wave diagram is follows:

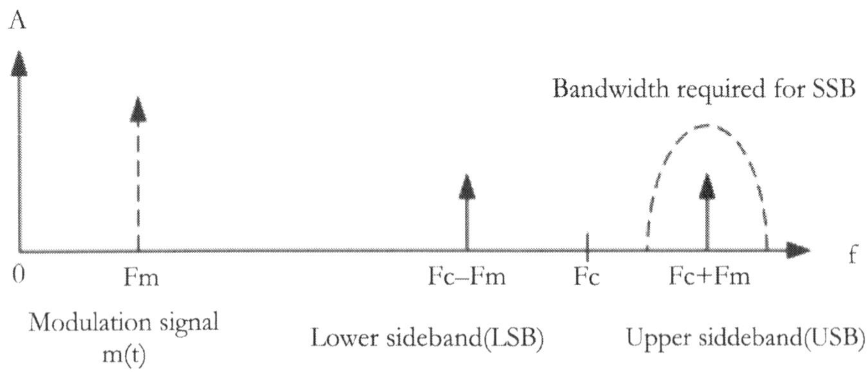

Power Relation in SSB Wave

$$P_{SSB} = \frac{V^2_{LSB}}{R} = \frac{V^2_{USB}}{R}$$

$$P_{LSB} = P_{USB} = \frac{V^2_{SB}}{R}$$

$$P_{SSB} = \frac{m^2}{4} P_C$$

Vestigial Sideband (VSB) Modulation Technique

Vestigial sideband (VSB) is a type of amplitude modulation (AM) technique (sometimes called VSB-AM) that encodes data by varying the amplitude of a single carrier frequency. Portions of one of the redundant sidebands are removed to form a vestigial sideband signal so-called because a vestige of the sideband remains.

Mathematical representation of VSB wave is

$$S(t) = \frac{mV_C}{2}\cos[\omega_c - \omega_m)] + F\left(\frac{mV_c}{2}\cos(\omega_c - \omega_m)t\right)$$

Power Relation in VSB Wave

$$P_{VSB} = \frac{m^2}{4} P_c + F(\frac{m^2}{4} P_c)$$

Envelope Detector

It is used for detection of AM wave

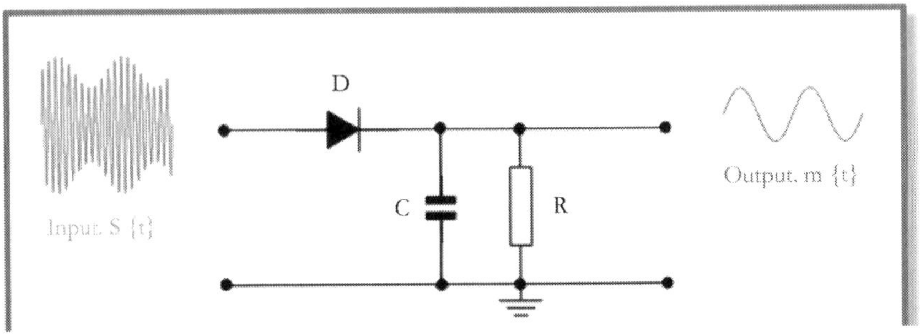

$1/f_c << RC << 1/W$

Where W is bandwidth of message signal

Envelop detector can also be used to recover message by passing received VSB signal through it.

Comparison of Different AM Techniques

In standard AM systems the sidebands are transmitted in full, accompanied by the carrier.

Suppressed-carrier modulation system have an advantages over standard AM system in that they require much less power to transmit the same amount of information, which makes the transmitter for such a systems less expensive than those required for standard AM.

Single-sideband modulation requires the minimum transmitter power and minimum transmission bandwidth possible for conveying a message signal from one point to another. Where vestigial sideband required a transmission bandwidth that is intermediate between that required for SSB and DSBSC modulation.

3 Angle Modulation

Angle modulation is a method of analog modulation in which either the phase or frequency of the carrier wave is varied according to the message signal. In this method of modulation the amplitude of the carrier wave is maintained constant.

Relationship between FM and PM

An angle modulated signal in general can be written as

$$u(t) = A_c \cos[\theta(t)]$$

Where $\theta(t)$ the phase of the signal and its instantaneous frequency is given by

$$f_i(t) = \frac{1}{2\pi} \cdot \frac{d}{dt}[\theta(t)]; u(t) = A_c \cos[2\pi f_c t + \phi(t)];$$

$$f_i(t) = f_c + \frac{1}{2\pi}\frac{d}{dt}\phi(t)$$

If m(t) is the message signal, then in PM system we have

$$\phi(t) = k_p m(t)$$

And in FM system we have

$$f_i(t) - f_c = k_f m(t) = \frac{1}{2\pi}\frac{d}{dt}\phi(t)$$

Where kp and kf are phase and frequency deviation constant

$$\phi(t) = \begin{cases} k_p m(t) & \text{PM} \\ 2\pi k_t \int_{-\infty}^{t} m(\tau)d(\tau) & \text{FM} \end{cases} ; \frac{d\phi}{dt} = \begin{cases} k_p \dfrac{dm(t)}{dt} & \text{PM} \\ 2\pi k_f m(t) & \text{FM} \end{cases}$$

PM and FM modulated signal for sinusoidal Message signal

$$m(t) = a\cos(2\pi f_m t)$$

$$u(t) = A_c \cos[2\pi f_c t + k_p a\cos(2\pi f_m t)] \rightarrow \text{PM}$$

$$u(t) = A_c \cos[2\pi f_c t + \frac{k_f a}{f_m}\sin(2\pi f_m t)] \rightarrow \text{FM}$$

$\beta_p = k_p a$ and $\beta_f = \dfrac{k_f a}{f_m}$ are known as modulation indices for PM and FM system respectively.

We can extend the modulation index definition for a general signal m(t) as

$$\beta_p = k_p \max[m(t)]$$

$$\beta_f = \frac{k_f \max[m(t)]}{W}$$

Where, W is bandwidth of message

$\beta_f = \dfrac{\Delta f_{max}}{W}$ where Δf_{max} represents maximum frequency deviation

Note

For low value of β angle modulation scheme is called' narrow band angle modulation scheme. Narrow band FM is similar to AM. It is used in mobile communication.

For large value of β angle modulation scheme is called wideband angle modulation scheme.

Another Representation of PM and FM Signals for Sinusoidal Message

$$u(t) = \sum_{n=-\infty}^{\infty} A_c J_n(\beta)\cos\left[2\pi(f_c + nf_m)t\right]$$

Where $J_n(\beta)$ is known as Bessel function of first kind of order n.

Note: Total bandwidth requirement for angle modulated signal is ∞.

Bandwidth of Angle Modulated Signal

In general the effective bandwidth of an angle modulated signal, which contains at least 98% of the signal power is given by

$$B_c = 2(\beta+1)f_m \quad \text{Carson's formulae}$$

where, β = modulation index.

For $m(t) = a\cos 2\pi f_m t$

$$B = 2(\beta+1)f_m = \begin{cases} 2(k_p a+1) & \text{PM} \\ 2\left(\frac{k_f a}{f_m}+1\right) & \text{FM} \end{cases}$$

Number of harmonics in the bandwidth (including the carrier) is

$$M_c = 2[\beta] + 3$$

$$M_c = 2k_p a + 3 \quad \text{for PM}$$

$$M_c = \frac{2k_f a}{f_m} + 3 \quad \text{for FM}$$

Note: By seeing the graph of angle modulation signal we cannot tell whether it is PM or FM.

Angle Modulators and Demodulators

- FM wave can be generated using VCO called direct method.
- FM wave can be generated using varacter diode called indirect method
- FM and PM both can be generated using reactance tubes.
- Another method is first generating narrow band FM then pass through frequency multiplier.

Demodulation is normally done using PLL.

4 Effect of Noise on Analog Communication System

SNR is defined as ratio of signal power to noise power.

Effect of Noise on a Baseband System

Its SNR is given by

$$\left(\mathrm{N} \right)_b$$

where, PR --> is received signal phase.

No/2 - two sided noise spectral density.

W —> message signal bandwidth.

$\left(\frac{S}{N}\right)_b$ —>SNR of baseband system.

Effect of Noise on DSB-SC AM

For coherent receiver, SNR at the o/p is

$$\left(\frac{S}{N}\right)_o = \frac{A_c^2 P_m}{2WN_o}$$

$$P_R = \frac{A_c^2 P_m}{2}$$ —> received power for DSB-SC receiver

Where P_m —> message signal power.

A_c —> Carrier signal amplitude

Note: In DSB-SC the o/p SNR is the same as the SNR for a baseband system. Therefore DSB-SC does not provide any SNR improvement over a baseband communication system.

Effect of Noise on SSB AM

For coherent receiver

$$\left(\frac{S}{N}\right)_{OSSB} = \frac{A_c^2 P_m}{N_o W} \rightarrow \text{SNR at the o/p}$$

$$P_R = A_c^2 P_m; \quad \left(\frac{S}{N}\right)_{OSSB} = \frac{P_R}{N_o W} = \left(\frac{S}{N}\right)_b$$

Note: SNR in case of SSB is same as that of DSB-SC and baseband system.

Effect of Noise on Conventional AM

SNR for coherent modulator is

$$\left(\frac{S}{N}\right)_{OAM} = \frac{\frac{1}{4} A_c^2 a^2 P_{M_n}}{\frac{1}{4} P_{n_c}}$$

$$= \frac{A_c^2 a^2 P_{M_n}}{2N_0 W}$$

$$= \frac{a^2 P_{M_n}}{1 + a^2 P_{M_n}} \frac{\frac{A_c^2}{2}\left[1 + a^2 P_{M_n}\right]}{N_0 W}$$

$$= \frac{a^2 P_{M_n}}{1 + a^2 P_{M_n}} \frac{P_R}{N_0 W}$$

$$= \frac{a^2 P_{M_n}}{1 + a^2 P_{M_n}} \left(\frac{S}{N}\right)_b$$

$$= \eta \left(\frac{S}{N}\right)_b,$$

Where A_c is the amplitude of career wave

k = modulation index

$$P_R = \frac{A_c^2}{2}[1 + k^2 Pm_n]$$

$$\left(\frac{S}{N}\right)_{OAM} = \frac{k^2 Pm_n}{1 + k^2 Pm_n} \times \frac{P_R}{N_0 W}$$

$$\left(\frac{S}{N}\right)_{OAM} = \eta\left(\frac{S}{N}\right)_b$$

Where $$\eta = \frac{k^2 Pm_n}{1 + k^2 Pm_n}$$

SNR of conventional AM is always less than the SNR of the baseband system

Effect of Noise on Angle Modulation

The performance of angle-modulated signals when contaminated by additive white Gaussian noise; we will also compare this performance with the performance of amplitude-modulated signals. Recall that in amplitude modulation, the message information is contained in the amplitude of the modulated signal; since noise is additive, the noise is directly added to the signal. However, in a frequency-modulated signal, the noise is added to the amplitude and the message information is contained in the frequency of the modulated signal. Therefore, the message is contaminated by the noise to the extent that the added noise changes the frequency of the modulated signal.

The frequency of a signal can be described by

$$S_{Y_n}(f) = \begin{cases} \frac{N_0}{A_c^2} & |f| \leq \frac{B_c}{2} \\ 0 & \text{otherwise} \end{cases}.$$

SNR at o/p for angle modulation systems is

$$\left(\frac{S}{N}\right)_o = \frac{k_p^2 A_c^2}{2} \frac{P_M}{N_0 W} \rightarrow \text{for PM}$$

$$\left(\frac{S}{N}\right)_o = \frac{3k_f^2 A_c^2}{2W^2} \frac{P_M}{N_0 W} \rightarrow \text{for FM}$$

Ac2/2 = received signal power PR

$$\left(\frac{S}{N}\right)_o = P_M \frac{\beta_p^2}{\left(\max\left[m(t)\right]\right)^2} \left(\frac{S}{N}\right)_b \rightarrow \text{for FM system}$$

For FM system $\left(\frac{S}{N}\right)_o = 3P_M \frac{\beta_f^2}{\left(\max\left[m(t)\right]\right)^2} \left(\frac{S}{N}\right)_b$

In both PM and FM systems output SNR is proportional to square of modulation index.

With increase in β without increasing the transmitter power we can increase SNR at output. Increasing R will increase the bandwidth requirement for transmission so we are increasing SNR by increasing bandwidth.

Increasing β signal power at transmitter will increase the SNR at o/p but in case of AM it is due to increase in signal power but in case of angle modulation it is due to decrease in noise power (noise power at the o/p of receiver).

In case of FM high frequency component suffers from more noise so pre-emphasis and de-emphasis is used to increase the SNR for FM system.

In pre-emphasis we amplify high frequency component but low frequency component's amplitude are unchanged.

In de-emphasis we de-amplify higher frequencies amplitude keeping low frequency amplitude same.

Pre-emphasis and de-emphasis Filter Characteristics

Transfer function of pre-emphasis filter is

$$H_p(f) = 1 + j\frac{f}{f_0}$$

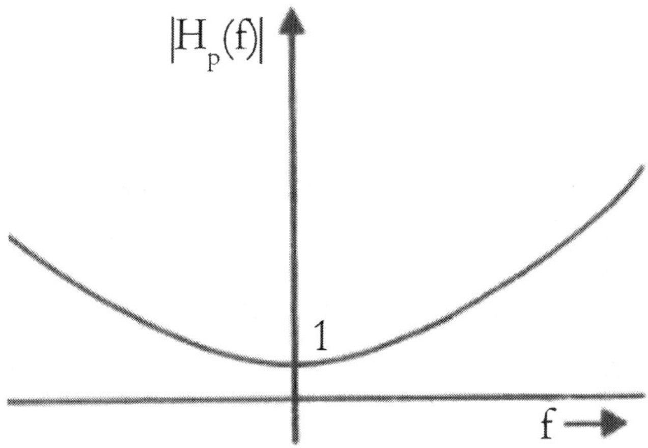

Transfer function of de-emphasis filter is

$$H_d(f) = \frac{1}{\left(1 + j\frac{f}{f_0}\right)}$$

5 Random Variables

A random variable X is in simplest terms a variable which takes on values at random; it may be thought of as a function of the outcomes of some random experiment.

Cumulative Distribution Function (CDF)

Cumulative distribution function (CDF) of a random variable X is defined as

$$F_x(x) = P(X \leq x)$$

Properties of CDF

$$0 \leq F_x(x) \leq 1$$

FX(x) is non-decreasing function.

$$\lim_{x \to -\infty} F_x(x) = 0 \text{ and } \lim_{x \to +\infty} F_x(x) = 1$$

$$P(a < X \leq b) = F_x(b) - F_x(a)$$

$$P(X = a) = F_x(a) - F_x(a^-)$$

Note: For discrete random variable $F_X(x)$ is a stair case function,

For continuous random variable CDF is continuous.

Random Signals and Noise

A random process is a process (i.e., variation in time or one dimensional space) whose behavior is not completely predictable and can be characterized by statistical laws.

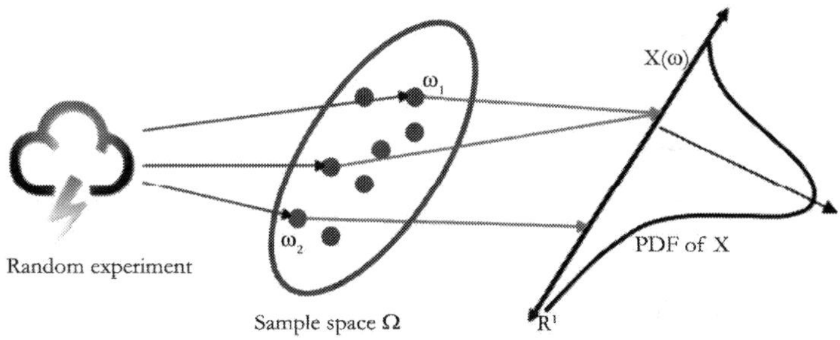

Probability Density Function

The derivative of the distribution function is the probability density function (PDF)

$$P_X(x) \equiv \frac{d}{dx}\left(F_x(X)\right)$$

Probability density can also be defined by

$$P_X(x)dx = P[x < X \leq x + dx]$$

Properties of PDF

$f_x(x) \geq 0$

$\int_{-\infty}^{\infty} f_x(x)dx = 1$ total probability

$$\int_{a^+}^{b^-} f_x(x)\,dx = P(a < X \le b)$$

$$f_x(x) = \int_{-\infty}^{x^t} f_x(u)\,du$$

- For discrete random variables it is more common to define the probability mass function (PMF) which is defined as PMF = $\{P_i\}$ where $P_i = P(X = x_i)$

Important Random Variable

The all important random variable are follows:

Bernoulli Random Variable

This is a discrete random variable taking two values one and zero with proabilities p and (1 - p).

Binomial Random Variable

This is a discrete random variable giving the number of in a sequence of II independent Bernoulli trials.

The PMF is given by nk p)° k, 051(P(X = k) = 0 . otherwise

$$P(X = k) = \begin{cases} {}^nC_k p^k (1-p)^{n-k}, & 0 \le k \le n \\ 0 & \text{,otherwisw} \end{cases}$$

Note: In this random variable models the total number of bits received in error when a sequence of n bits is transmitted over a channel with bit error probability p.

Uniform Random Variable

This is a continuous random variable taking values between a and b with equal probability over intervals of equal length. The density function is given by

$$f_x(x) = \begin{cases} \frac{1}{(b-1)} & a<x<b \\ 0 & \text{otherwisw} \end{cases}$$

Note: When the phase of a sinusoid is random it is usually modeled as a uniform random variable between 0 and 2n.

Gaussian or Normal Random Variable

This is a continuous random variable described by the density function

$$f_x(x) = \frac{1}{\sqrt{2\pi\sigma^2}} e^{-\frac{(x-m)^2}{2\sigma^2}}$$

where, m = 0 mean of random variable

σ = standard deviation

Note: Gaussian random variable is the most important and frequently' encountered random variable in communications.

The CDF for the Gaussian random variable with m = 0 and $\sigma = 1$ is denoted by $\varphi(x)$ and given by

$$\phi(x) = P(X \le x) = \int_{-\infty}^{x} \frac{1}{\sqrt{2\pi}} e^{-t^2/2} dt$$

A closely related function is Q(X) = 1-φ(x). This function is well tabulated and frequently used in communication Q(x) which satisfies the following relations.

$$Q(-x) = 1 - Q(x); Q(0) = \frac{1}{2}; Q(\infty) = 1$$

For a Gaussian random variable having mean m and standard deviation

$$P(X > x) = Q\left(\frac{x-m}{\sigma}\right)$$

Function of a Random Variable

A function of a random variable Y =g(X) is itself a random variable.

$$f_Y(y) = \sum_i \frac{f_X(X_i)}{|g'(x_i)|}$$

Where, $\{X_i\}$ are countable number of solutions of equation g(X) = Y

Statistical Averages

Expected value of random variable X is defined as

$$E(X) = \int_{-\infty}^{\infty} x f_X(x)\,dx$$

nth order moment of a random variable X is defined as

$$m_X^n = \int_{-\infty}^{\infty} X^n f_X(X)\,dx$$

Expected value of random variable Y is defined as

$$E(Y) = \int_{-\infty}^{\infty} g(X) f_X(X)\,dx$$

For discrete random variable X is defined as

$$E(X) = \sum_i x_i P(X = x_i);\ E[g(X)] = \sum_i g(x_i) P(X = x_i)$$

Properties of Expected Value

- E(CX) = CE(X), where C is a constant.
- E(C) = C
- E(X + C) = E(X) + C

Variance (σ^2)

It is defined as

$$E[\{X - E(X)\}^2] = \sigma^2;\ \sigma^2 = E[X^2] - \{E(X)\}^2$$

Note: Variance represents A.C. power content of random signal.

Properties of Variance

$$V_{ar}(CX) = C^2 V_{ar}(X)$$
$$V_{ar}(C) = 0$$
$$V_{ar}(X + C) = V_{ar}(X)$$

Characteristic Functions

Characteristic Functions of random variable X is denoted by $\psi_X(V)$ and defined as

$$\psi_X(V) = \int_{-\infty}^{\infty} f_X(X) e^{jvx} dx$$

To obtain nth moment of the random variable X we use

$$m_X^n = \frac{1}{j^n} \frac{d^n}{dv^n} \psi_X(V)\Big|_{v=0}$$

Multiple Random Variables

Joint CDF for two random variables X and Y is given by

$$F_{X,Y}(X, Y) = P(X \le x, Y \le y)$$

Joint PDF is defined as

$$F_{X,Y}(x, y) = \frac{\partial^2}{\partial x \partial y} F_{X,Y}{}^{(x,y)}$$

Basic Properties of Joint PDF and CDF

$$F_X(x) = F_{X,Y}(x,\infty)$$

$$F_Y(y) = F_{X,Y}(\infty,y)$$

$$F_X(x) = \int_{-\infty}^{\infty} F_{X,Y}(x,y)dy$$

$$F_Y(y) = \int_{-\infty}^{\infty} F_{X,Y}(x,y)dx$$

$$\int_{-\infty}^{\infty}\int_{-\infty}^{\infty} f_{X,Y}(x,y)\,dxdy = 1$$

$$F_{X,Y}(x,y) = \int_{-\infty}^{x}\int_{-\infty}^{y} f_{X,Y}(u,v)\,dudv$$

Conditional PDF of random variable γ, given that the value of the random variable X is equal to is denoted by $F_{Y/X}(y/x)$ and defined as

$$f_{Y/X}(y/x) = \begin{cases} \frac{f_{X,Y}(x,y)}{f_X(x)}, & f_x(x) \neq 0 \\ 0 & , \text{otherwise} \end{cases}$$

The expected value of g (X,Y) where g(X,Y) is any function of X and Y, is obtained from

$$E[g(X,Y)] = \int_{-\infty}^{\infty}\int_{-\infty}^{\infty} g(x,y) f_{X,Y}(x,y)\,dxdy$$

Note: If X and Y are independent random variable then the COV (X,Y)= $\rho_{x,y}$ but if $\rho_{x,y} = 0$ it does not imply necessarily that random variables are independent.

$$V_{ar}\left(\sum_i C_i X_i\right) = \sum_i C_1^2 V_{ar}(X_i) + \sum_i \sum_{j \neq i} C_i C_i COV(X_i, X_j)$$

Multiple Functions of Multiple Random Variable

If we define two function of the random variable X and Y by

$$\begin{cases} Z=g(X,Y)............(1) \\ W=h(X,Y)...........(2) \end{cases}$$

then joint PDF of Z and W is

$$f_{Z,W}(Z,W)=\sum_i \frac{f(x_i,y_i)}{|\det J(x_i,y_i)|}$$

Where

$$\rho_{x,y}=\frac{COV(X,Y)}{\sigma_X\sigma_Y}$$

In the special case where g(X,Y) = (X-mY) we obtain E(g(X,Y)) = E(XY)-mxmy, which is called covariance of X and Y.

The normalized version of covariance

$$\rho_{x,y}=\frac{COV(X,Y)}{\sigma_X\sigma_Y}$$

Note: If X and Y are independent random variable then the COV(X,Y)= ϱx,y=0 it does not imply necessarily that random variables are independent.

$$V_{ar}\left(\sum_i C_iX_i\right)=\sum_i C_i^2V_{ar}(X_i)+\sum_i\sum_{j\neq i}C_iC_jCOV(X_i,X_j)$$

Multiple Functions of Multiple Random Variables

If we define two functions of the random variable X and Y by

$$\begin{cases} Z=g(X,Y)............(1) \\ W=h(X,Y)...........(2) \end{cases}$$

then joint PDF of Z and W is

$$f_{Z,W}(Z,W)=\sum_i \frac{f(x_i,y_i)}{|\det J(x_i,y_i)|}$$

Where

$$J(x,y)=\begin{bmatrix}\frac{\partial z}{\partial x} & \frac{\partial z}{\partial y}\\ \frac{\partial w}{\partial x} & \frac{\partial w}{\partial y}\end{bmatrix}$$

Central limit Theorem

This theorem says that some of random variables (which generally occurs practical systems) take the form of Gaussian random variable.

Random Processes

Random process is a collection of time functions or signals correspond to various outcomes of a random experiment.

Autocorrelation Function

This represents the relation of a function with its shifted version.

The auto correlation function of the random process X(t), denoted by

Rxx(t_1, t_2) is defined by

$$R_{XX}(t_1,t_2)=E\left[X(t_1)X(t_2)\right]$$

Stationary Process

A process whose statistical properties are independent of choice of time origin is called stationary process.

A process X(t) is called wide - sense stationary if the following conditions are satisfied m_x,(t) = E[X(t)] is independent of t.

Rxx(t_1 and t_2) depends only on the time difference $\tau = t_1 - t_2$) and not on t_1, and t_2 individually.

Note: If mx(t) and Rxx t+ τ, t) are periodic with period T then process is called cyclostationary process.

Properties of auto correction function Rx(τ)

Rx(τ). the auto correction function of stationary process X(t) has following properties :

Rx(τ) is an even function i.e. Rx(τ) = Rx(- t).

The maximum absolute value of Rx(x) is achieved at T = 0 i.e. Rx(KT_0) Rx(0).

If for some To we have Rx(To) = Rx(0), then for all integers K, Rx(KT_0) Rx(0)

Power and Energy of a Random Signal

Power content of a random signal X is given by

$$P_X = E\left[\lim_{T\to\infty}\frac{1}{2T}\int_{-T}^{T} X^2(t)\,dt\right]; P_X = \lim_{T\to\infty}\frac{1}{2T}\int_{-T}^{T} E\left[X^2(t)\right]dt$$

$$P_X = \lim_{T\to\infty}\frac{1}{2T}\int_{-T}^{T} R_{XX}(t,t)\,dt$$

Energy content of Random Signal is given by

$$E_x = E\left[\int_{-\infty}^{\infty} X^2(t)\,dt\right]; E_x = \int_{-\infty}^{\infty} R_{XX}(t,t)\,dt$$

For the case of stationary processes, only power type process of theoretical and practical interest.

Cross Correlation

Cross correlation function between two random processes X(t) and Y(t)

$$R_{XY}(t_1,t_2) = E\left[X(t_1)Y(t_2)\right]; R_{XY}(t_1,t_2) = R_{YX}(t_2,t_1)$$

Random Processes and Linear Systems

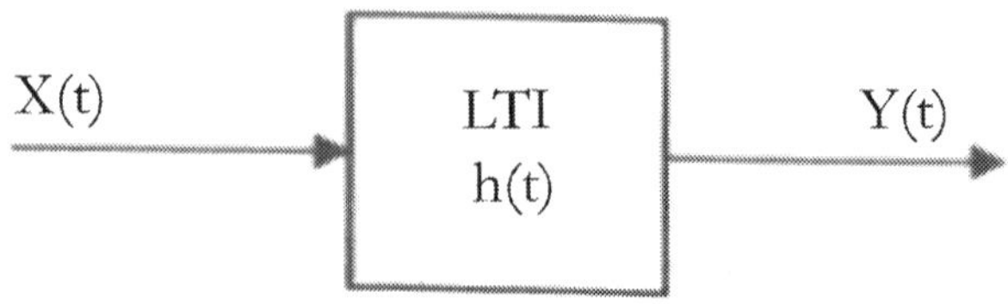

Let random process X(t) is input to an LTI system having imp response h(t) and Y(t) is o/p of LTI system then If X(t) is stationary process then X(t) and Y(t) will be jointly station with

$$E[Y]=E[X(t)]\int_{-\infty}^{\infty}h(t)dt; \quad R_{XY}(\tau)=R_X(\tau)^* h(-\tau)$$

$$R_Y(\tau)=R_X(\tau)^* h(-\tau)^* h(\tau); \quad S_Y(f)=S_X(f)|H(f)|^2$$

Where * represent convolution

Power spectral Density

Power spectral density (PSD) of the signal, when multiplied by the appropriate factor, describes the power contributed to the wave, by a frequency, per unit frequency.

$$S_X(f)\,\text{F.T.}\lim_{T\to\infty}\frac{1}{2T}\int_{-T}^{T}R_X(t+\tau,t)dt$$

Note: If the two processes X(t) and Y(t) are uncorrected then $R_{xy}(\tau)$ = E[X(t)] E[Y(t)]

Where X(t) and Y(t) are jointly stationary processes.

6 Pulse Modulation Techniques

There are two types of pulse modulation techniques, pulse analog modulation techniques and pulse digital modulation techniques.

Pulse Analog Modulation Technique

Pulse analog modulation techniques are of three types namely PAM, PWM and PPM.

Pulse Amplitude Modulation

In this modulation scheme amplitude of a high frequency carrier is yarned in accordance with the sampled values of message signal.

Time domain representation of PAM is Where m(t) is message signal, C(t) is carrier signal and SO) is PAM modulated signal.

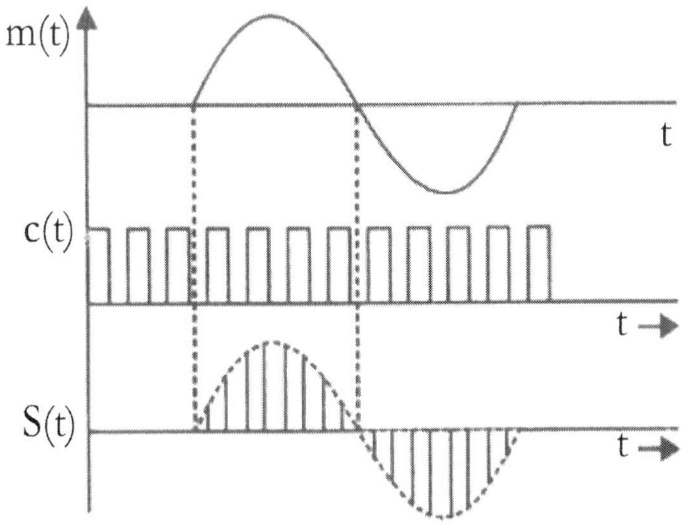

Pulse Width Modulation

In this modulation scheme width of carrier pulses are yarned in accordance with the sampled values of message signal. Time domain representation of SO PWM is Speed control of DC motor is done using PWM.

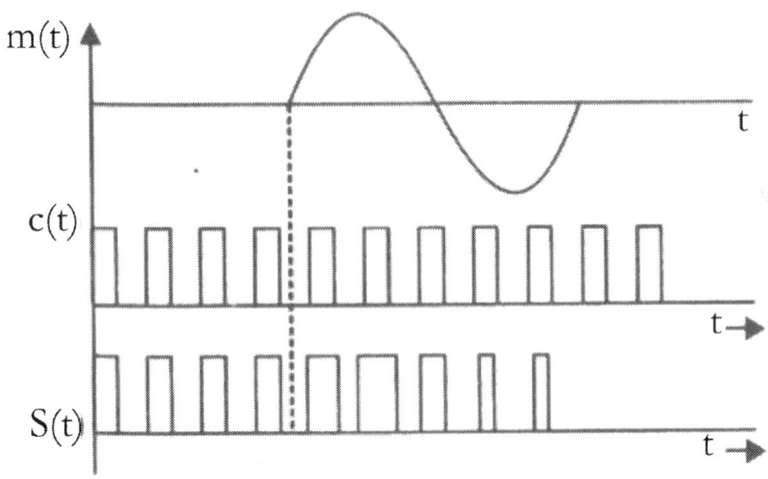

Pulse Position Modulation

In this modulation scheme position of high frequency carrier pulses is changed in accordance with the sampled value of message signal. Time domain representation of PPM is

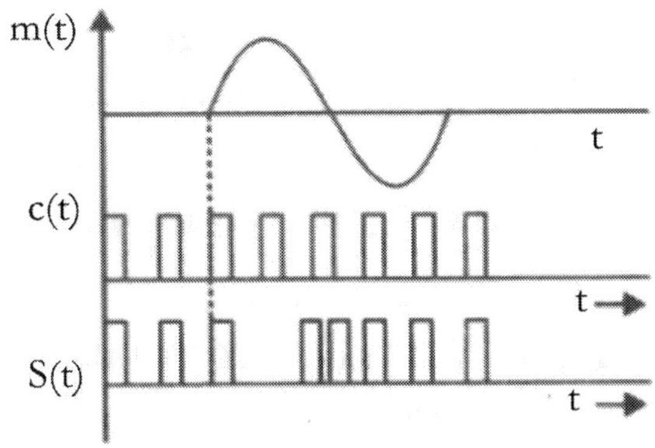

PWM is generated using 555 timer in mono-stable signal in mono-stable multi-vibrator mode.

PPM is generated using 555 timer by using PWM mode.

Comparison of PAM, PWM, PPM

	B.W	Power efficiency (SNR)	Complexity
PAM	Less	Lowest	Lowest
PWM	High	Moderate	Moderate
PPM	High	High-test	High-test

PPM is most effective pulse analog modulation scheme

Pulse Digital Modulation Techniques

Important pulse digital modulation techniques are pulse code m PCMJ, delta modulation, DPCM.

Pulse Code Modulation (PCM)

In this modulation scheme we first sample the analog signal quantize it to convert it into levels and encoded and then send it form of digital codes.

If there is n bit quantize and sampling rate is fs, then bit rate will

$R_b = nf_s$ bits / sec.

Sampling must be done at a rate greater than or equal quist rate in order to avoid aliasing. Higher will be sampling rate easy will be reconstructs signal at receiver.

B.W. requirement is minimum Rb/2 and maximum Ro.

If we are using uniform quartier with step size Δ.

$$\Delta = \frac{V_{max} - V_{min}}{L} = \frac{V_{max} - V_{min}}{2^n}$$

Then maximum quantizers error will be $\Delta/2$

Normally used uniform quantizers are mid tread type and midrise type.

Signal to quantization noise ratio for PCM for Sinusoidal input is (for uniform quantizer)

$(SQNR = 6n + 1.7)$ dB

Where, n is bit of quantizer.

As number of bits of quantizer increase SQNR increases but at the same time bandwidth requirement for transmission also increase.

Companding

Companding is compression of signal at transmitter end and expansion of signal at receiver end. (here are two type of companding techniques μ -law companding and μ-law companding. Formulae for it-law companding is

$$|V_{out}| = \frac{In[1+\mu|V_{in}|]}{In[1+\mu]}$$

$\mu_{max} = 255$

Formula for A-law companding

$$|V_{out}| = \frac{A|V_{in}|}{1+\text{In}A} \text{ for } 0 \le |V_{in}| \le \frac{1}{A}$$

$$= \frac{1+\text{In } A|V_{in}|}{1+\text{In } A} \text{ for } \frac{1}{A} \le |V_{in}| \le 1$$

Where

$A_{max} = 87.6$

Note: Companding is done to avoid nonlinear distortion of channel.

Companding is widely used in telephone systems to reduce nonlinear distortion and also to compensate for signal level difference between soft and loud talkers.

Inferential Pulse Code Modulation

When the samples of a signal are highly correlated then we go for DPCM in order to save bandwidth or using the same bandwidth at higher data rate.

Note: In DPCM principle used is prediction.

Delta Modulation

Delta modulation is a special case of differential pulse code modulation. It is called one bit DPCM as it transmits only one bit per sample. In delta modulation problem of slope overload occurs if input is changing very fast that is

$$\left|\frac{\Delta}{T_S}\right| < \left|\frac{dm(t)}{dt}\right|$$ slope overload occurs.

To overcome slope overload error we choose optimum size of Δ such that

$$\frac{\Delta_{opt}}{T_S} = \left|\frac{dm(t)}{dt}\right|_{max}$$

For m(t) = Amcos 2πfmt

$$\frac{\Delta_{opt}}{T_S} = 2\pi f_m A_m$$

Second problem occurs is hunting, which occurs when message is a constant.

Adaptive Delta Modulation

In adaptive delta modulation step size is chosen in accordance message signal sampled value to overcome slope overload error hunting.

If message is varying at a high rate then step size is high and if message is varying slowly then step size is small.

Multiplexing

There are many types of multiplexing out of which two are FDM TDM.

In FDM we divide whole bandwidth of channel in small segments a lot it to different users so that they can access channel at the s time by using their allotted bandwidth.

In TDM, whole time slot is divided among different users so at a time only one user is accessing the channel.

Bandwidth requirement in TDM and FDM is almost same for number of users.

If we are using TDM + PCM and there are N-messages having s bandwidth and are sampled at same rate fs then bit rate will be

$R_b = N.nf_s$ where, n — bits in quantizer.

If B.W. of messages and sampling rates are different then overall rate will

$$R_b = R_{b1} + R_{b2} + R_{b3} + \dots\dots\dots\dots R_{bN}$$

where Rb1,, Rb2, Rb3 RbN are bit rate of respective messages

B.W. requirement if raised cosine pulses having fall rate a is used

$$B.W. = \frac{R_b}{2}(1+\alpha)$$

Where, a -4 lies between 0 and 1.

Note: In case of ADM and DM bandwidth required is almost same

7 Digital Modulation Schemes

Commonly used digital modulation schemes are amplitude shift keying. frequency shift keying. and phase shift keying.

Amplitude Shift Keying

In Iris scheme amplitude of a high frequency carrier is varied accordance with digital data (0 or 1).

For digital input 1 amplitude level is high and for digital input 0 amplitude level is low.

Signaling used is on-off signaling.

B.W. requirement for ASK is

$$\text{B.W.} = \frac{2}{T_b} = 2R_b$$

Note: In ASK probability of error (Pe) is high.

SNR is less.

Frequency Shift Keying

In frequency shift keying frequency of carrier is varied in accordance with digital data (1 or 0).

For digital data 1 we use frequency f_1 and for digital data 0 vie use frequency f_2. normally $f > f_2$

NRZ signaling is used.

Bandwidth requirement in case of FSK is

$$B.W. = 2R_b + (f_1 - f_2)$$

Note: In case of FSK Pe. is less and SNR is high.

Multiplexing is difficult.

Used in modem

PSK

In phase shift keying phase of high frequency carrier is varied in accordance with digital data 1 or O. NRZ signaling is used.

Bandwidth requirement for PSK is $B.W. = 2R_b$ = 2x bit rate

- In case of PSK probability of error is less. SNR is high.
- Mainly used technique in wireless transmission.

'Scheme	S,(t) & S2(1)	B.W.	P8	SNR	
Coherent ASK Non-coherent ASK	Sp = A cos reel S2(t) = 0 S(t) = A cos wet S,(t) = 0	2 Rh 2 Rb	high high	low low	high low
Coherent FSK Non-coherent FSK	S,(t) = A cos 2R tit S2(1)= A cos 2p f2t ill > t2) S,(t) = A cos 2rt tit S2(t) = A cos 2x 12t	2 FI,+(f , 42) >2 nt,	moderate moderate	high high	high low

Coherent PSK Non-coherent PSK	S.(1) = A cos coct S2(t) = - A cos okt S,(t) = A ooS oct S2(t) = - A cos wct	2R, 2 Rb	low low	high high	high low

- Complexity of coherent schemes is always high compare to noncoherrent schemes.
- Coherent schemes have less probability of error com noncoherrent schemes under same conditions. ASK scheme has lowest noise immunity against noise.
- PSK is best among all shift keying schemes.
- Coherent schemes require lesser power compare to noncoherrent schemes for same probability of error.

M-array Signaling

In M-array signaling we use M number of phases (in case of P number of frequencies (in case of FSK). M number of amplitude (incase of ASK) for sending digital data.

By using M-array scheme we can send data at higher rate at the exd of increased probability of error. Bandwidth requirement for M-array scheme is

$$\text{B.W.} = \frac{\text{B.W. for binary scheme}}{\log_2 M}$$

For M-array PSK

$$\text{B.W.} = \frac{2R_b}{\log_2 M}$$

Baud rate is data rate at the o/p of the modulator.

Baud rate is given by

$$\text{Baud rate} = \frac{\text{Bitrate}}{\log_2 M}$$

Quadrature Amplitude Modulation (QAM)

In QAM digital information is contained in both amplitude and phase of the signal. • It is used in both digital modulation scheme and analog modulation scheme.

In digital cable television and in cable modem applications QAM is used.

Noise in Digital Communication

In digital communication for better SNR matched filter is used whose Impulse response h(t) is Ih(t)

$$h(t) = S^*(T_b - t)$$

* (Tb — where, * -, represent complex conjugate. Tb —> bit duration. S(t) —> input signal to filter.

Probability of error Pe is

$$P_e = Q\left[\sqrt{\frac{(\text{signal difference})^2}{4\text{ noise spectral density}}}\right]; \quad P_e = Q\left[\sqrt{\frac{E_d}{2N_0}}\right]$$

Where

$$E_d = \int_{T_b} [S_1(t) - S_2(t)]^2 dt$$

N0/2 is two sided noise power spectral density.

Probability of Error for ASK (Coherent)

Let

$$S_1(t) = A_c \cos 2\pi f_c t \qquad S_2(t) = 0$$

$$E_d = \int_{T_b} [S_1(t) - S_2(t)]^2 dt \qquad E_d = T_b \frac{A_c^2}{2}$$

$$\boxed{P_e = Q\left[\sqrt{\frac{A_c^2 T_b}{4N_0}}\right]}$$

$$P_e = Q\left[\sqrt{\frac{A_c^2 T_b}{4N_0}}\right]$$

Probability of Error for PSK (Coherent)

In PSK

$$S_1(t) = A_c \cos 2\pi f_c t$$

$$S_2(t) = -A_c \cos 2\pi f_c t$$

$$P_e = Q\left[\sqrt{\frac{A_c^2 T_b}{N_0}}\right]$$

For FSK probability error is

$$P_e = Q\left[\sqrt{\frac{A_c^2 T_b}{2N_0}}\right]$$

Pe for Non-coherent ASK and Non-coherent FSK

Probability of error for non-coherent ASK is

$$P_e = \frac{1}{2}\exp\left(-\frac{A_c^2}{8N_0}\right) \quad \text{if } A_c^2 >> N_0$$

Probability of error for non-coherent FSK is

$$P_e = \frac{1}{2}\exp\left(-\frac{A_c^2}{4N_0}\right)$$

$$Q(X) = \frac{1}{2}\text{erf}_c\left(\frac{x}{\sqrt{2}}\right); \; Q(X) \cong \frac{1}{\sqrt{2\pi x^2}} e^{-x^2/2}$$

Probability of error for differential PSK is

$$\frac{1}{2}\exp\left(-\frac{A_c^2 T_b}{2N_0}\right)$$

In case of FSK f1 and f2 are chosen such that f1 = mfs and f2 = kfs where m, k are integers

Bandwidth efficiency for PSK is

$$\text{B.E.} = \frac{\text{Transmission rate}}{\text{B.W. of signal at o / p}} = \frac{R_b}{2R_b} = 0.5$$

8 Information Theory

Information content in a signal will be high if its probability of occurrence will be less If probability of occurrence of message is p then its information content I will be

$I = \log_2\left(\frac{1}{p}\right)$	$I = \log_{10}\left(\frac{1}{p}\right)$	$I = \ln\frac{1}{p}$
bits	Hartley or decit	nat

Entropy

Average information content per symbol in a group of symbols is known as entropy and represented by H. If we have M symbols whose probability of occurrence is p,. p2.

then entropy H will be

$$H = \sum_{i=1}^{M} p_i \log_2 \frac{1}{p_i} \text{ bits / symbols}$$

Entropy will be maximum it probability of occurrence of all M Symbols will be equal and I-1,8,, = logy M bits/symbols.

Source Information Rate

If source is emitting symbols at a rate r, symbols/sec then source information rate R is given by IR $R = r_s H \rightarrow \text{bits / sec}$

Entropy of Mark-off Sources

Entropy of Mark-off source is

$$H = \sum_{i=1}^{n} P_i \sum_{j=1}^{n} P_{ij} \log_2 \left(\frac{1}{P_{ij}} \right)$$

Where Pi probability that source is in Pij=probability that it going from state i to j.

Note: Mark-off source is sourcing which emits the symbols dependently.

- Source Encoding
- Fixed length coding.
- Variable length coding
- Shannon fano coding
- Huffman coding

Efficiency of Huffman coding is high as compared to Shannon coding.

Source coding is done for data compression in order to tr 1/07 it more efficiently.

Channel coding is done to reduce probability of error.

Efficiency of Encoding Technique

Efficiency is given by $e = \frac{H}{\overline{L}}$

where. H -* is entropy. [e- [is average number of bits per system according to encoding technique.

Capacity of a Channel

Capacity of a channel is defined as maximum data rate that can supported by channel with minimum probability of error.

Capacity of a channel having bandwidth B and additive Gaussian b limited white noise is

$$C = B\log_2\left(1+\frac{S}{N}\right) \rightarrow \text{bits / sec}$$

this is Shannon Hartley theorem.

When B is tending to then capacity does not become co as tends to zero due to high noise power, so where, S signal power.

$$\lim_{B\rightarrow\infty} C = 1.44\left(\frac{S}{N_0}\right)$$

N0\2 is two sided noise spectral density.

9 Optical Communication

General model of an optical communication system is as

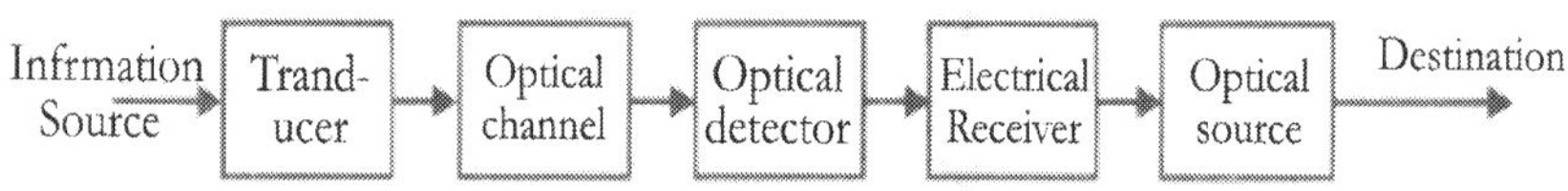

Transducer basically converts information from source into electrical signal. Optical source (laser or LED) converts electrical signal into optical (light) signal. Optical channel is optical fiber. Optical detector converts optical signal into electrical signal. Mainly avalanche photo diode is used. When we have to compare the behavior of light in two medium then we go for Ray theory. When we deal with transmission of optical signal or reception of optical signal then we go for Wave theory. If there is interaction between a material and light then we go for Photon Theory.

Refractive Index (n).

- Refractive index is defined as ratio of velocity of light in free space to velocity of light in medium for which we are calculating refractive index, that is r=_

Note: Value of refractive index is always greater than or equal to 1.

Total Internal Reflection

Total internal reflection is a phenomenon which occurs when a propagating wave strikes a medium boundary at an angle larger than a particular critical angle with respect to the normal to the surface. If the refractive index is lower on the other side of the boundary and the incident angle is greater than the critical angle, the wave cannot pass through and is entirely reflected. The critical angle is the angle of incidence above which the total internal reflection occurs. This is particularly common as an optical phenomenon, where light waves are involved, but it occurs with many types of waves, such as electromagnetic waves in general or sound waves.

By snell's law

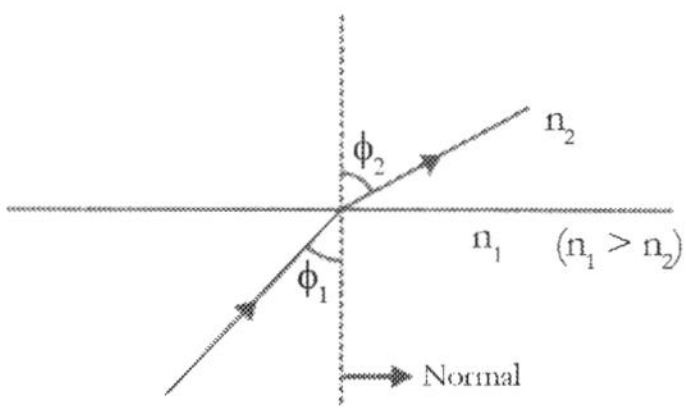

$$n1 \sin \phi_1 = n_2 \sin \phi_2$$

$$\text{if } \phi_2 = 90; \text{ int ernal reflection takes place}$$

$$\phi_1 = \sin^{-1} \frac{n_2}{n_1}$$

Acceptance Angel (θ_a)

The acceptance angle of an optical fiber is defined based on a purely geometrical consideration (ray optics): it is the maximum angle of a ray (against the fiber axis) hitting the fiber core which allows the incident light to be guided by the core. The sine of that acceptable angle is called the numerical aperture, and it is essentially determined by the refractive index contrast between core and cladding of the fiber, assuming that the incident beam comes from air or vacuum.

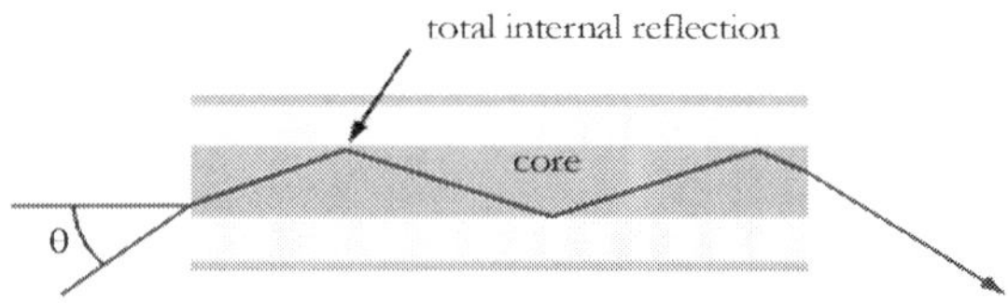

Numerical Aperture

Numerical Aperture (also termed Object-Side Aperture) is a value (often symbolized by the abbreviation NA) originally defined by Abbe for microscope objectives and condensers. It is given by the simple expression:

$$\text{Numerical aperture}\,(\text{NA}) = n \times \sin(\mu) \text{ or } n \times \sin(\alpha)$$

Many authors use the variable μ to designate the one-half angular aperture while others employ the more common term α, and in some instances, θ. In the numerical aperture equation, n represents the refractive index of the medium between the objective front lens and the specimen, and μ or α is the one-half angular aperture of the objective.

Relative Refractive Index Difference (Δ)

It is a measure of the relative difference in refractive index of the core and cladding. Refractive index contrast, Δ, is given by $\Delta = (n_1^2 - n_2^2)/(2n_1^2)$, where n_1 is the maximum refractive index in the core and n_2 is the refractive index of the homogeneous cladding. Normal optical fibers have very low refractive index contrast ($\Delta << 1$) hence are weakly guided medium. The weak guiding will cause more of the Electrical field to "leak" and travel through the cladding (as evanescent waves) as compared to the strongly guided waveguides.

$$\Delta = \frac{n_1^2 - n_2^2}{2n_1^2}; \Delta \cong 1 - \frac{n_2}{n_1}; \text{N.A} = n_1 (2\Delta)^{1/2}$$

Meridonal and Skew Rays

A meridional ray is a ray that passes through the axis of an optical fiber. A skew ray is a ray that travels in a non-planar zigzag path and never crosses the axis of an optical fiber.

Numerical Aperture for Meridonal and Skew Rays

For meridonal rays N.A. is $\text{N.A.} = n_0 \sin\theta_a = (n_1^2 - n_2^2)^{1/2}$.

For skew rays N.A is $\text{N.A.} = n_0 \sin\theta_{as} \cos\gamma = (n_1^2 - n_2^2)^{1/2} \cos\gamma$.

Where

θ_{as} →Acceptance angle for skew rays.

γ →Is the angle between angle of incidence and normal at the point of incidence.

θ_{as} →is always greater than θ_a.

Skew rays will be in large number inside fiber.

Normalized Frequency

It is a dimensionless quality and is given by

$$v = \frac{2\pi}{\lambda} \times a \times (\text{N.A.}) = \frac{2\pi}{\lambda} \times a \times n_1 \times (2\Delta)^{1/2}$$

Where, a= is core radius

λ= wave length of operation.

Step index Fiber

Step-index fibers have a uniform core with one index of refraction, and a uniform cladding with a smaller index of refraction. (Air serves as the cladding in the simple glass tube example.) When plotted on a graph as a function of distance from the center of the fiber, the index of refraction resembles a step-function. The figure to the left illustrates how the index of refraction varies with location in a cross-section of a step-index fiber.

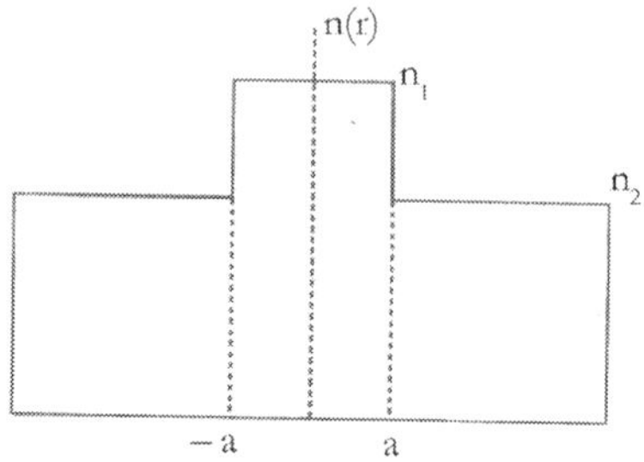

$$n(r) = \begin{cases} n_1 & r < a \\ n_2 & r \geq a \end{cases}$$

The refractive index profile for this type of fiber makes a step change 111 the core-cladding interface.

Modes

Modes are

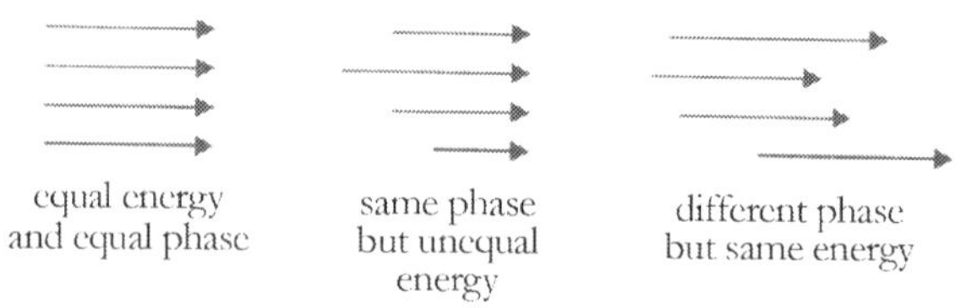

Number of modes or mode volume M is

$$M = \frac{V^2}{2}$$

If we reduce the radius such that single mode is only transmitted through core so no dispersion and so no different delays.

Note:

- Due to multimode transmission different waves reach destination at different time.
- This multimode transmission is known dispersion.

Graded Index Fiber

In fiber optics, a graded-index or gradient-index fiber is an optical fiber whose core has a refractive index that decreases with increasing radial distance from the optical axis of the fiber.

- Graded index fiber has variable refractive index inside core.
- Refractive index is maximum at center of core and decreases as radius distance increases from the core.
- This type of profile for refractive index is used for propagation of multimode with equal delays.
- Refractive index profile is

$$n(r) = n_1 \left[1 - 2\Delta\left(\frac{r}{a}\right)^{\alpha}\right]^{1/2} \qquad r \leq a$$

$$n_2 \;= n_1 [1 - 2\Delta]^{1/2} \qquad r \geq a$$

Where α is called profile of the fiber.

- Number of modes or mode volume Mg is given by

$$M_g = \left(\frac{\alpha}{\alpha + 2}\right) \cdot \frac{v^2}{2}$$

Note:

- Value of α Profile

Value of α	Profile
∞	step index
1	triangular
2	parabolic

- As α increases graded index fiber will tend towards step index fiber behavior.
- If we want transmit only single mode then normalized frequency must be in the range $0 \leq v \geq v_C$, where $v_C = 2.405$

- For signal mode transmission minimum value of wave length λ is given by

$$\lambda_{min} = \frac{2\pi a(N.A.)}{2.405}$$

Losses in Optical Fiber

Loss in optical fibers basically refers to the loss of power. During transit, light pulse loses some of their photons, thus reducing their amplitude. Attenuation for a fiber is usually specified in decibels per kilometer. The degree of attenuation depends on the wavelength of light transmitted.

- Absorption losses

  Intrinsic (due to interaction of one or major component of glass)

  Extrinsic (due to OH- ion or due to transition element impurity)
- Linear scattering losses

  Rayleigh

  Mie
- Nonlinear scattering losses
- Fiber bend losses

Critical Radius of Curvature

For multimode critical radius of curvature R_c is given by

$$R_c = \frac{3n_1^2\lambda}{4\pi\left(n_1^2 - n_2^2\right)^{3/2}}$$

For single mode, critical radius of curvature is given by

$$R_c = \frac{20\lambda}{\left(n_1^2 - n_2^2\right)^{3/2}}\left(2.748 - 0.996\frac{\lambda}{\lambda_{min}}\right)^{-3}$$

Optical Sources

- Normally used optical sources are Laser and LED.
- Laser is source of monochromatic and coherent light.
- LED is source of monochromatic and non-coherent light.
- In case of LED total recombination rate R_t is directly proportional to forward biased current and is given by

$R_t = R_{nr} + R_r$

Where, $R_{nr} \rightarrow$ Non radiative recombination

$R_r \rightarrow$ radiative recombination

Note:

- In case of He-Ne laser for some O/P optical power, less input power is required as compared to Ruby laser.

Internal Quantum efficiency of LED

The internal quantum efficiency of LED is just the ratio of the photons emitted from a device active region to the carriers in the active region; it is typically estimated by taking the ratio of photoluminescence intensities at room and low temperatures, and assuming 100% efficiency at the lowest temperature (~8K).

$$\left(\eta_{int} = \frac{R_r}{R_t}\right) R_t \alpha\ i; \left(R_t = \frac{i}{e}\right); \left(\eta_{int} = \frac{R_r \times e}{i}\right)$$

Where, e is charge on electron

- Internal power Pint is given by Pint= Photon rate × hf

$$P_{int} = R_r \times hf; P_{int} = \eta_{int} \times \frac{i}{e} \times \frac{hc}{\lambda}$$

- External power is somewhat lesser than internal power and is given by

$$P_{ext} = \frac{P_{int} F}{4} \cdot \frac{n^2}{n_x^2}$$

Where, $F \rightarrow$ is transmission coefficient and lies between (0, 1). 0 for opaque and 1 for transparent medium

$n \rightarrow$ refractive index of the medium

$n_x \rightarrow$ refractive index of the crystal

- External quantum efficiency is equal to P_{ext} / P where P is input electrical power.

Relation between overall mean life time (τ_t), mean life time (τ_r) corresponding to radiative recombination and mean life time (τ_{nr}) corresponding to non-radiative recombination is

$$\frac{1}{\tau_t} = \frac{1}{\tau_r} + \frac{1}{\tau_{nr}} \text{ and } n_{int} = \frac{R_r}{R_t} = \frac{\tau_t}{\tau_r}$$

Coupling Efficiency

Coupling efficiency is a parameter which defines the amount of optical power coupled to the optical fiber from LED. It is given by

$$\eta_c = \frac{\int_0^{\theta} I(\theta) \sin\theta d\theta}{\int_0^{\pi/2} I(\theta) \sin\theta d\theta}$$

$$\eta_c = \sin^2\theta_a = \frac{(NA)^2}{(n_1^2 - n_2^2)} = \frac{(NA)^2}{n_0^2}$$

For I (θ) =I_0 cosθ

Optical Detector

- Optical detector used is normally avalanche photo diode or PIN diode.
- Optical detector must have high sensitivity, high fidelity, short response time and stability within the range of operation.

Quantum efficiency of Optical detector

Quantum efficiency of Optical detector is given by

$$\eta_D = \frac{\text{electron generation rate}}{\text{incident photon rate}}$$

$$\eta_D = \frac{r_e}{r_p}$$

Where, r_e and r_p are electron and photon generation rate respectively.

Responsivity

- It gives the transfer characteristics of detector and is equal to

$$R = \frac{I_P}{P_O} \ A/w$$

Where, Ip→is photo current

P0→is incident photon power

$$P_O = r_p \times hf; \ r_e = \frac{\eta P_O}{hf}$$

Po= photon rate × hf

Now

$$R = \frac{I_P}{P_O} = \frac{er_e}{P_O} = \frac{\eta_D e}{hf}$$

Where,

$$I_p = er_e$$

Note:

There is always a tradeoff between speed of operation and sensitivity of detector.

10 Propagation of Wave

Wave propagation is any of the ways in which waves travel. With respect to the direction of the oscillation relative to the propagation direction, we can distinguish between longitudinal wave and transverse waves. For electromagnetic waves, propagation may occur in a vacuum as well as in a material medium.

- In an earth environment, electromagnetic waves propagate in ways that depend not only on their own properties but also on those of environment itself.
- Frequency ranges of wave propagations are ground wave propagation $\rightarrow < 2$MHz

  Sky wave / Ionospheric wave propagation $\rightarrow$ 2 MHz $< f <$ 30 MHz
  Space wave propagation / LOS $\rightarrow >$ 30 MHz.

Ground Wave Propagation

Ground Wave propagation is a method of radio frequency propagation that uses the area between the surface of the earth and the ionosphere for transmission. The ground wave can propagate a considerable distance over the earth's surface particularly in the low frequency and medium frequency portion of the radio spectrum.

Field Strength at a Distance for Ground Wave

- The field strength in volts/m is given as by taking into account the gain of transmitting antenna

$$E = \frac{120\pi h_t I}{\lambda d}$$

- If a receiving antenna is now placed at this point, the received signal will be, in volts.

$$V = \frac{120\pi h_t h_r I}{\lambda d}$$

Where, $120\pi \rightarrow$ characteristics impedance of free space

ht$\rightarrow$ effective height of transmitting antenna

hr$\rightarrow$ effective height of receiving antenna

I$\rightarrow$ antenna current

d$\rightarrow$ distance from the transmitting antenna

$\lambda \rightarrow$ wavelength

Note:

- If the distance between the two antennas is fairly long, the reduction of field strength due to ground and atmospheric absorption reduces the value of the voltage received.
- At low frequencies absorption is less due to ground so wave propagation can take place for a large distance.

Sky wave

Sky-wave or skip refers to the propagation of radio waves reflected or refracted back toward Earth from the ionosphere, an electrically charged layer of the upper atmosphere.

- Ionization of the upper parts of earth's atmosphere played a part, propagation of radio waves particularly at high frequencies.

- There are several layers of ionization which under certain condition reflects back to earth the high frequency waves that would otherwise escaped into space.
- Ionosphere is the upper portion of the atmosphere which absorbs large quantities of radiant energy from the sun becoming heated and ionized.

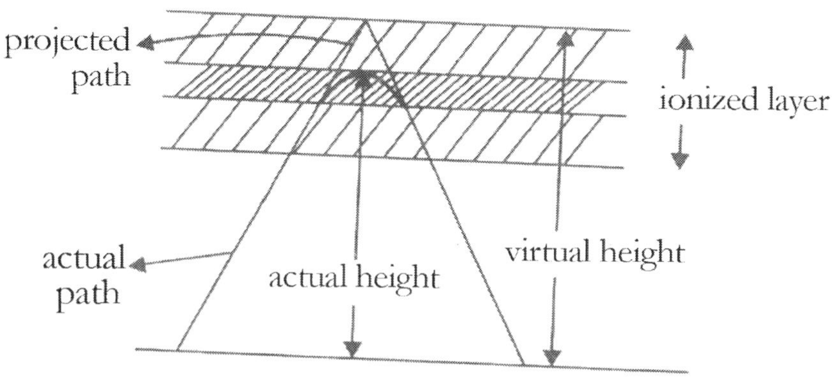

This figure shows that as wave is refracted, it is bent down gradually rather than sharply. However below ionized layer, the incident and refracted rays follow path that are exactly the same as they would be if reflection had been taken place from a surface located at greater height called Virtual Height.

Critical Frequency (f_c)

Critical frequency f_c for a given layer is the highest frequency that will be returned down to earth by that layer after being been beamed straight up at it.

Note: Critical frequency range for F_2 layer is from 5 MHz to 12 MHz.

Maximum Usable Frequency (MUF)

Maximum usable frequency is a limiting frequency, but for some specific angle of incidence other than the normal and it is given by

$$MUF = \frac{\text{Critical frequency}}{\cos\theta}; \quad MUF = f_c \sec\theta$$

Above question is also called secant law.

Note:

Normal values of MUF may range from 8 to 35 MHz

Skip Distance

A skip distance is the distance a radio wave travels, usually including a hop in the ionosphere. A skip distance is a distance on the Earth's surface between the two points where radio waves from a transmitter, refracted downwards by different layers of the ionosphere, fall.

- Fading is the fluctuation in signal strength at a receiver and may be rapid or slow, general or frequency selective.
- Fading can occur because of interference between the lower and upper rays of a sky wave.
- AM signals suffer from frequency selective fading.

Note:

- A frequency which makes a given distance correspond to the skip distance is the MUF for that pair of points.
- Transmission path is limited by the skip distance at one end and the curvature of the earth at other end.

Space Wave

- Space waves travels (more or less) in straight lines.
- Space waves are limited in their propagation by the curvature of the earth.

Radio Horizon

- The radio horizon for space waves is about four thirds as far as optical horizon.
- The radio horizon of an antenna is given by

$$d_t = 4.12\sqrt{h_t}\,\text{km}$$

Where, $d_t \rightarrow$ distance from transmitting antenna

$h_t \rightarrow$ height of transmitting antenna above ground in meters.

- The total distance 'd' between transmitting and receiving antenna is given by

$$d = d_1 + d_2$$

$$d = 4.12\left[\sqrt{h_t} + \sqrt{h_r}\right]\text{km}$$

Where, hr is height of receiving antenna above earth in meters.

Line of Sight (LOS) Propagation

This is also called Friis transmission.

- P_t is power transmitted by antenna and P_r is power received by receiving antenna.
- G_t, G_r are gains of transmitter and receiver respectively.
- Received power is given by

$$P_r = \frac{P_t G_t \times A_e}{4\pi d^2}$$

Where, $d \rightarrow$ is direct distance between two antennas

Ae→ effective area of receiver

Now

$$G_r = \frac{4\pi A_e}{\lambda^2}$$

So received power Pr will be

$$P_r = \frac{P_t G_t G_r}{\left(\frac{4\pi d}{\lambda}\right)^2}$$

- Received power in dB is

$$10\log_{10} P_r = 10\log_{10} P_t + 10\log_{10} G_t + 10\log_{10} G_r - 20\log\left(\frac{4\pi d}{\lambda}\right)$$

Note:

Term $20\log\frac{4\pi d}{\lambda}$ corresponds to path loss.

11 Radar and Satellite Communication

Radar consists of a transmitter and receiver each connected to a directional antenna. Transmitter is capable of sending out a large UHF or microwave power through the antenna. Receiver collects as much energy as possible from the echoes reflected in its direction by the target. Basic radar systems diagram and its timing diagram are as

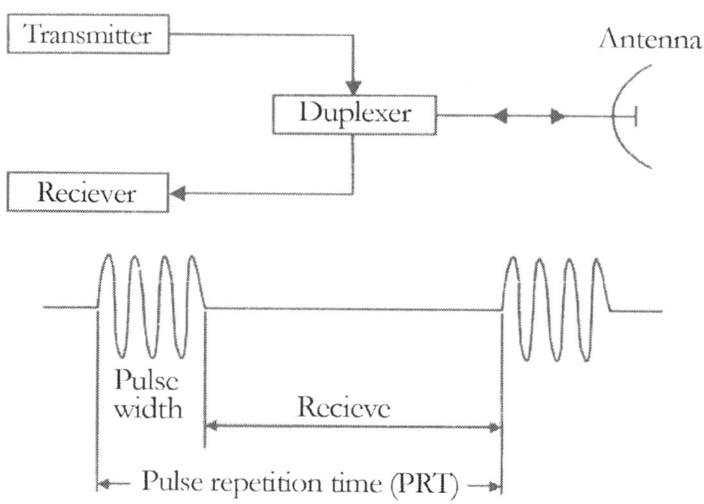

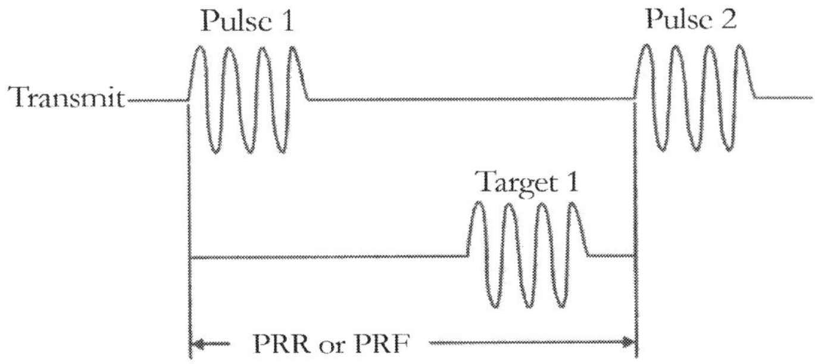

Pulse reception frequency (PRF) or pulse repetition rate (PRR) is number of pulses per unit time. These pulses are transmitted by a highly directional parabolic antenna at the target, which can reflect (echo) some of the energy back to the same antenna. The reflected energy is received and time measurements are made, to determine the distance of the target. After the radar pulse has been transmitted, a sufficient rest time must be allowed for echo to return so as not to interfere with next transmit pulse. Time difference between two successive pulses is called pulse repetition time (PRT).

Pulse repetition time determines the maximum distance to the target to be measured, that is range. Range is given by the formulae $\text{Range} = \frac{\Delta t}{12.36} \text{ miles}$. Where, $\Delta t \rightarrow$ time from transmitter to receiver in microseconds.

For higher accuracy and shorter ranges

$$\text{Range} = \frac{328\Delta t}{2} = 164\Delta t \text{ yards}$$

The range beyond which objects appear as second return echoes is called the maximum unambiguous range (MUR) and is given by

$$\text{mur} = \frac{\text{PRT}}{12.2} \text{ miles}$$

where, PRT is in μsec.

In timing diagram distance between pulse 1 and pulse 2 is called maximum range.

Minimum range is given by

$$\text{Minimum range} = 164\text{PW yards}$$

Where, PW –pulse width in μsec.

Duty cycle is given by

$$\text{duty cycle} = \frac{\text{PW}}{\text{PRT}}$$

- Average power is given by

$$\text{Average power} = \text{peak power xduty cycle}$$

Note: PRT= 1/PRF

Maximum Range r_{max}

Let P_t is peak transmitted power from transmitter, A_p is maximum power gain of transmitter, A_o is capture area of receiving antenna. Power intercepted by the target depends upon effective area, let us assume that it is S, then maximum range r is given by

$$r_{max} = \left(\frac{P_t A_0^2 S}{4\pi\lambda^2 P_{min}}\right)^{1/4} \quad \text{or} \quad r_{max} = \left(\frac{P_t A_0^2 \lambda^2 S}{(4\pi)^3 P_{min}}\right)^{1/4}$$

Where, λ→ wave length of EM radiation.

Pmin→ minimum power at receiver which can be recognizes.

Note:

- By increasing P_t by 16 times r_{max} becomes just double.

Effect of Noise on Maximum Range

- Noise affects the maximum range insofar as it determines the minimum power that the receiver can handle.
- Taking noise into account maximum range r_{max} is given by

Where, k→ Boltzmann's constant

$T_0 \rightarrow$ Standard ambient temp

$=290K$

$f\Delta \rightarrow$ Bandwidth of receiver

$kT_0\ f\Delta \rightarrow$ noise input power of receiver

$F \rightarrow$ Noise figure of receiver

- Maximum Range in kilometers is given by

$$r_{max} = 48\left[\frac{P_t D^4 S}{\Delta f \pi^2 (F-1)}\right]^{1/4} \text{ km}$$

Where, $D \rightarrow$ antenna diameter in meters

$S \rightarrow$ effective cross- section area in m^2

$\Delta f \rightarrow$ in Hz

$\lambda \rightarrow$ in m

Note:

- Bandwidth required for receiver is n/T, where T is the pulse duration and n is number whose value ranges from 1 to 10 depending on circumstances.

Blind Speed

The blind speed is dependent on the transmitted frequency and on the pulse repetition frequency of the radar unit.

- Blind speed is a radial speed of the airplane at which the phase shifting of echo-signal has the value $\pm n \times 2\pi$ between two pulse periods.
- It is given by

$$v_{blind} = \frac{\lambda}{2T_s}$$

Where, vblind $\rightarrow$ blind speed

$\lambda \rightarrow$ wave length of transmitted EM wave

$T_s \rightarrow$ pulse repetition time (PRT)

Doppler Effect

The Doppler Effect causes the received frequency of a source (how it is perceived when it gets to its destination) to differ from the sent frequency if there is motion that is increasing or decreasing the distance between the source and the receiver.

The Doppler Effect observer is moving with respect to source he observers change in frequency of the wave emitted by sources.

- Observed frequency f is given by

$$f = \frac{f_0}{\left(1 - \frac{v_{s,r}}{c}\right)} \text{ and } |\Delta f| = \frac{v_{s,r}}{\lambda} \text{ doppler shift}$$

Where, $v_{s,r} \rightarrow v_s - v_r$ is the relative velocity sources with respect to receiver.

$f_0 \rightarrow$ is frequency of wave radiated by sources.

- In radar technology Doppler Effect is using the following tasks

  Speed measuring

  MIT- moving target indication

  In air or space based radar system for precise determination of lateral distances.

- Doppler Frequency (F_D) is given by

$$f_D = \frac{2v}{\lambda}$$

Where, $F_D \rightarrow$ Doppler frequency in Hz

$v \rightarrow$ speed of the wave source (m/sec)

$\lambda \rightarrow$ wave length (m)

This equation is valid if the speed of the source of wave is like the radial speed.

Another formulae for Doppler frequency is given by

$$f_D = \frac{2v_r}{\lambda} = \frac{2v_r.f_{tx}}{C_0}$$

Where, $v_r \rightarrow$ radial speed of aim

$f_{tx} \rightarrow$ frequency of transmitter

$C_0 \rightarrow$ velocity of light in free space

$\lambda \rightarrow$ wavelength of wave

Note:

- Doppler Frequency occurs twice at radar once on the way from radar to aim and then for the reflected (already affected by a Doppler shift) energy on the way back.
- Radar beacons are normally used on aircraft to identify it.
- Another use of radio beacons is rather similar to that of light houses but it can operate over much large distances.

Satellite Communication

In satellite communication, signal transferring between the sender and receiver is done with the help of satellite. In this process, the signal which is basically a beam of modulated microwaves is sent towards the satellite.

- Satellite communication system exists because earth is a sphere.
- Repeater is needed to convey signals over long distances.
- A repeater is simply a receiver linked to a transmitter.
- Satellite systems operate in the microwave and millimeter wave frequency bands using frequencies between 1 and 50 GHz.
- There are four satellite system whose orbital velocity height and period is given as

Satellite system	Orbit height (km)	Orbit velocity (km/sec)	Orbital period		
			hr	Minn	sec
Intelsat (GEO)	35786.03	3.0747	23	56	4.1
New-ICO (MEO)	10255	4.8954	05	55	48.4
Skybridge (LEO)	1469	7.1272	01	55	17.8
Iridium	780	7.4624	01	40	27.0

- Time period of satellite orbit is given by

$$T = \frac{2\pi r^{3/2}}{\mu^{1/2}}$$

Where, r→ is distance of satellite's orbit from center of earth.

μ→ is constant which product of gravitational constant G and mass is of

earth ME.

- Kepler's third law gives the relation between orbital period and semi-major axis of elliptical path and is

$$T^2 = \frac{4\pi^2 a^3}{\mu}$$

Where, a→ semi major axis ellipse and replaced by r if path is circular.

μ→Kepler's constant.

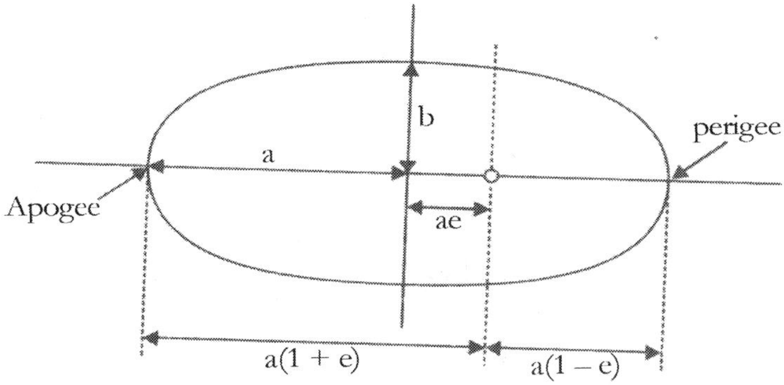

Semi major and minor axis length are given by

$$a = \frac{p}{\left(1-e^2\right)}$$

$$b = \frac{a}{\sqrt{1-e^2}}$$

Where, p is semilatus rectum of ellipse; e eccentricity of ellipse.

Geostationary and Geosynchronous Satellite

A geosynchronous satellite is a satellite whose orbital track on the Earth repeats regularly over points on the Earth over time. If such a satellite's orbit lies over the equator, it is called a geostationary satellite. The orbits of the satellites are known as the geosynchronous orbit and geostationary orbit.

- Geostationary satellite is in an orbit whose period is same as that of earth's rotation period.
- It remains stationary with respect to any point on earth at any time.
- Geosynchronous Satellite will be at fixed position with respect to a point on earth at the same time every day but not all the time.
- Average angular velocity of satellite is given by

$$\eta = \frac{\mu^{1/2}}{a^{3/2}} = \frac{2\pi}{T}$$ where, $\eta \rightarrow$ average angular velocity.

- Mean anomaly M is the arc length (in radians) that the satellite would have transversed since the perigee passage if it were moving on the circumscribed circle at the mean angular velocity η and is given by

$M = E\text{-}e\sin E$ where, $E \rightarrow$ is eccentric anomaly.

M is also given by

$M = \eta\,(t\text{-}t_p)$ where, $t_p \rightarrow$ is eccentric anomaly.

- Major bands used for satellite communication are 6/4 GHz, 14/11 GHz, 30/20 GHz. The uplink frequency is written first by convention.
- As frequency increases attenuation in atmosphere increases.
- All communication links are designed to meet certain performances usually a bit error rate (BER) in a digital link or a signal to noise rate (S/N) in an analog link.
- The baseband channel BER or S/N ratio is determined by the carrier to noise ratio (C/N) at the input to the demodulator in receiver.
- In most satellite communications applications, the (C/N) ratio at demodulator input must be greater than 6 dB.
- Overall (C/N) at earth station receiver depends on both uplink downlink.
- If a transmitting source, in free space, radiating a total power P_t, watts uniformly in all directions and G_t, is gain of antenna then Pt, Gt, is call effective is called effective isotropically radiated power (EIRP).
- Now received power by an antenna at a distance R from transmitt4 having gain G_r is

$$P_r = \frac{P_t G_t G_r}{\left(\frac{4\pi R}{\lambda}\right)^2}; \quad P_r = \frac{\text{EIRP} \times \text{Re ceving antenna gain}}{\text{patt loss}} \text{ watt}$$

- In decibel received power will be

$$(p_r)_{dB} = (\text{EIRP})_{dB} + (G_r)_{dB} - (L_p)_{dB}$$

Where, $(LP)dB \rightarrow$ path loss in dB

- In general received power is gain as

Where, $(L_a)_{dB} \rightarrow$ loss in dB in atmosphere

$(L_{ta})_{dB} \rightarrow$ loss associated with transmitter in dB

$(L_{ra})_{dB} \rightarrow$ loss associated with receiver in dB

System Noise Temperature

At microwave frequencies a black body with a physical temperature T_P degrees kelvin, generates electrical noise power in a wide bandwidth.

The noise power is given by

$$P_n = kT_P B_n$$

Where, $k \rightarrow$ Boltzmann's constant

$T_P \rightarrow$ physical temperature of source in kelvin degree

$B_n \rightarrow$ noise bandwidth in which noise power is measured in Hertz

- To determine the performance of a receiving system we need to be able to find the total thermal noise power against which the signal must be demodulated, this is done by determining system noise temperature.
- System noise temperature T_S is the noise temperature of a noise source located at input of noiseless receiver measured at the output of receiver and usually includes noise from the antenna.

Note:

- kTP is noise power spectral density in Watt/Hz.
- The power spectral density is constant for all radio frequencies up to 300 GHz.
- If the overall end to end gain of receiver is Grx and its narrowest bandwidth is B_n Hz the noise power at demodulator i/P is given by

$$P_{no} = kT_s B_n G_{rx}$$

Where, $T_S \rightarrow$ system noise temperature

$P_{no} \rightarrow$ noise power at demodulator i/P

G_{rx} → is the gain of receiver from RF input demodulator i/P

- The noise power referred to the input of receiver is P_n where

$$P_n = kT_sB_n$$

If antenna delivers a signal power Pr to the receiver RF input, the carrier to noise ratio at the demodulator is given by

$$\frac{C}{N} = \frac{P_r}{kT_sB_n}$$

System noise temperature TS is given by

$$T_s = T_{in} + T_{RF} + \frac{T_{in}}{G_{RF}} + \frac{T_{IF}}{G_mG_{IF}}$$

Where, Tin→ is the noise temperature of the antenna measured at its o/p port.

TRF→ equivalent noise temperature of RF amplifier

TIF→ equivalent noise temperature of IF amplifier

Tm→ equivalent noise temperature of mixer

GRF→ gain of RF amplifier

Gm→ gain of mixer

GIF→ gain of IF amplifier

- Succeeding stages of receiver contribute less and less noise to the total noise temperature and in this case system noise temperature is simply the sum of antenna noise temperature and LNA (Low Noise Amplifier) noise temperature that is

$$T_S = T_{antenna} + T_{LNA}$$

Where, Tantenna→ noise temperature of antenna.

TLNA→ noise temperature of LNA

Noise model for an equivalent output noise source produces a noise temperature Tno given by

$$T_{no} = T_P(1 - G_1)$$

Where, $G_1 \rightarrow$ is the linear gain less than unity of attenuating device or medium.

$T_P \rightarrow$ Physical temperature of device or medium in Kelvin degrees

Noise Figure and Noise Temperature

- Noise figure is frequently used to specify the noise generates within a device. The operational Noise Figure (NF) is defined as ratio of SNR at i/p to the SNR at o/p of a device and given by

$$NF = \frac{(S/N)_{in}}{(S/N)_{out}}$$

- Noise figure in terms of equivalent noise temperature is given by

Where, Tno→ equivalent noise temperature.

T0→ reference temperature and 290k

- If several devices are cascaded then overall noise figure can be found using Friis formulae as

$$F = F_1 + \frac{(F_2 - 1)}{G_1} + \frac{(F_3 - 1)}{G_1 G_2} + \frac{(F_4 - 1)}{G_1 G_2 G_3} + \ldots\ldots\ldots \frac{(F_n - 1)}{G_1 G_2 G_3 \ldots\ldots G_{n-1}}$$

Where, →F1, F2, F3 .... Fn are respective noise figure of devices in cascade.

→G1, G2, G3 .... Gn are respective power gains of devices in cascade.

G/T Ratio for Earth Station

- It is the ratio receiver gain Gr to system noise temperature TS.
- The link equation can be written in terms of (C/N) at earth station.

$$\left(\frac{C}{N}\right) = \left[\frac{P_t G_t G_r}{kT_s B_n}\right] \times \left(\frac{\lambda}{4\pi R}\right)^2 = \left(\frac{P_t G_t}{kB_n}\right)\left(\frac{\lambda}{4\pi R}\right)^2 \frac{G_r}{T_S}$$

Note:

So higher the (G/T) ratio higher will be (C/N) radio and better will be reception of signal at receiver so (G/T) ratio is figure of merit for receiver. Sometimes it is given that satellite terminal has -ve G/T which means it is below 0 dB/k means numerical value of Gr is smaller than numerical value of T_s.

Link Budget

A link budget is a tabular method for evaluating the received power and noise power in a radio link.

Link budgets invariably use decibel units.

Relation between uplink attenuation and downlink attenuation is given by

$$A_{up} = A_{down} \times \left(\frac{t_{up}}{f_{down}} \right)^2$$

Where, A_{up} → uplink attenuation in atmosphere (mainly due to rain)

A_{up} → downlink attenuation in atmosphere (mainly due to rain)

f_{up}, f_{down} → are uplink and downlink frequencies respectively.

- Relation between overall (C/N) ration and (C/N) for uplink and (C/N) for downlink is as

$$\frac{1}{(C/N)_{over\ all}} = \frac{1}{(C/N)_{up}} + \frac{1}{(C/N)_{down}}$$

Where, (C/N)overall→ overall (C/N) ratio

(C/N)up→ (C/N) ratio for uplink

(C/N)down→ (C/N) ratio for downlink

12 Miscellaneous

Frequency range of voice signal is 300 Hz to 3.4 kHz. Frequency range of audio signal is 20 Hz to 20 kHz. Frequency range of TV signal is 0 to 4.5 MHz. Frequency range of data is based on operation. In case of guided channel

Copper cable	-	kHz
Coaxial cable	-	MHz
Wave guide	-	GHz
Optical fiber	-	Tera Hz

Amplitude distortion

This type of distortion results from production of new frequency in the output which are not there in the input signal these new frequencies are caused due to non-linearity of active devices.

Frequency distortion

This type of distortion exist when signal component of different frequencies are amplified differently. It is caused by stray capacitances and circuit capacitances.

Phase distortion

It results from unequal delays of the signals of different frequencies

Super Heterodyne Receiver

- Super heterodyne receiver is used in AM receiver circuit
- AM radio broadcasting utilizes the frequency band 535 kHz to 1605 kHz for transmission of voice and music.
- Carrier frequency allocations range from 540 kHz 1600 kHz with 10 kHz spacing.
- The baseband message signal is band limited to 5kHz
- Intermediate frequency used for AM is 455 kHz.
- Frequency of local oscillator is given by

  $f_{Lo} = f_C + f_{IF}$

  Where f_c is the carrier frequency of the desired AM radio signal.
- Polarization is vertical and wave propagation is ground wave.
- The tuning range of local oscillator is 995 kHz to 2055 kHz.

An AM signal transmitted at a frequency image frequency is $(f_{Lo} + f_{IF})$ is called im frequency for an AM signal transmitted at frequency $f'c = f_{Lo} + f_{IF}$.

In order to reject image frequency bandwidth of RF amplifier should

$B_C < B_{RF} < 2f_{IF}$

Where,

B_C is bandwidth of AM radio signal.

B_{RF} is bandwidth of RF amplifier.

f_{IF} is intermediate frequency

Note: Coaxial cable is also used for connection making as connectors BNC (Bay Net Navy Connector) and TNC (Terminated Navy Connectors). Capacitors and inductors cause frequency and phase distortion 1 but not the amplitude distortion as they are linear elements.

FM Radio Broadcasting

- FM radio broadcasting utilizes the frequency band 88 - 108 MHz.
- The carrier frequencies are separated by 200 kHz and peak frequency deviation is fixed at 75 kHz. Intermediate frequency f_{IF} in case of FM is 10.7 MHz.
- Polarization is horizontal.
- Wave propagation is space wave propagation.
- In case of TV signal picture carrier is VSB-FC modulated, color sub carrier is QAM modulated and sound is FM modulated.

EM Spectrum

Radio Frequency	3 Hz to 300 GHz
Infrared	3×10^{12} to 4.3×10^{14} Hz
Visible	4.3×10^{14} to 7.5×10^{14} Hz
Ultraviolet	7.5×10^{14} to 3×10^{17} Hz
X-ray	3×10^{17} to 3×10^{19} Hz
y-rays	$> 3 \times 10^{19}$ Hz

Radio Frequency Spectrum

Band	Range		Application
EL	3	30 Hz	Ground wave propagation
SLF	30	300 Hz	→ Submarines
ULF	300 Hz	3 KHz	
VLF	3 KHz	30 KHz	
LF	30 KHz	300 KHz	→ Radio beacons
MF	300 KHz	3 MHz	→ AM
HF	3 MHz	30 MHz	→ Sky wave propagation
VHF	30 MHz	300 MHz	→ FM and TV; Space wave propagation
UHF	300 MHz	3 GHz	
SHF	3 GHz	30 GHz	

- Short wave has range of frequencies from 6 MHz - 25 MHz and it is an application of HF band.
- Digital communication has advantage that it can use repeaters over analog communication.
- To overcome inter symbol interference we use raised cosine pulses or duo binary signaling.

A Handbook on Electronics and Communications Engineering

5

Network Theory

CONTENTS

1 Circuits Element and Signal Waveform

Electric circuit has many elements like resistor, capacitor, inductor, battery etc. Circuit analysis is the process by which voltage or current is measured across the element. A waveform is a representation of how alternating current (AC) varies with time. The most familiar AC waveform is the sine wave, which derives its name from the fact that the current or voltage varies with the sine of the elapsed time. Other common AC waveforms are the square wave, the ramp, the saw tooth wave, and the triangular wave.

CIRCUIT ELEMENT

Circuit elements are inclusive passive elements such as Resistors, capacitors, inductors, transformers, diodes etc. Transistors, IC's and the like are active elements which require additional power for their operation.

Resistor

Resistor is an electrical component that reduces the electric current. The resistor's ability to reduce the current is called resistance and is measured in units of ohms.

Resistance

Resistance is the opposition that a substance offers to the flow of electric current. It is represented by the uppercase letter R.

$$R = \rho \frac{l}{A}, \text{ohm} \, (\Omega)$$

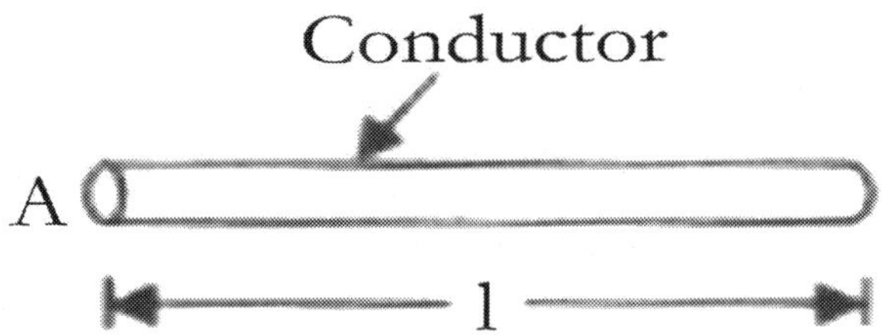

Where l= Length of conductor, m

A=Area of cross-section, m^2

ρ =Resistivity of the material, $\Omega - m$

Extension of wire results, increase in resistance and compression of wire result, decrease in resistance.

Capacitor

- A capacitor can store energy, so capacitors are often found in power supplies.
- A capacitor has a voltage that is proportional to the charge that is stored in the capacitor, so a capacitor can be used to perform interesting computations in op-amp circuits.
- Circuits with capacitors exhibit frequency-dependent behavior so that circuits that amplify certain frequencies selectively can be built.

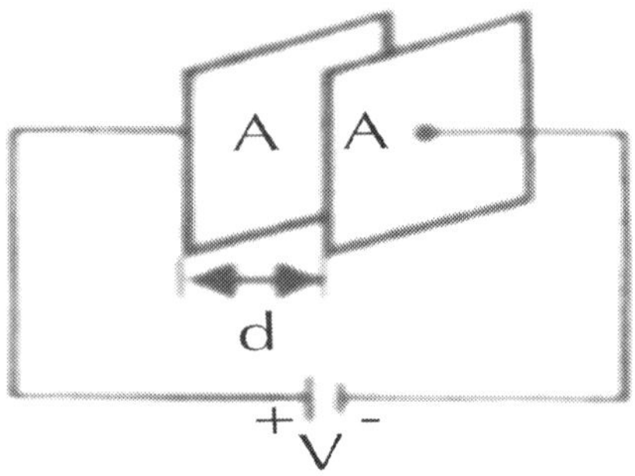

Capacitance

Capacitance is the ability of a body to store an electrical charge. Any object that can be electrically charged exhibits capacitance.

$$C = \frac{\epsilon_0 \epsilon_r A}{d}, \text{Farad}$$

Where, ϵ_0 =Permittivity of free space, F/m

ϵ_r =Relative permittivity of the dielectric.

A = Cross-sectional area of parallel plate, m^2

d = Separation of plates, m

- Under steady state condition for D.C. supply capacitor ac as open circuit.
- Capacitor does not allow sudden change of voltage, until unless impulse of current is applied.
- If unit impulse of current is applied then at $t = 0^+$ it stores $\frac{1}{2C}$ joule energy.
- It stores energy in the form of electric field.
- Power dissipation in ideal capacitor is zero.

INDUCTOR

An inductor, also called a coil or reactor, is a two-terminal electrical component which resists changes in electric current passing through it. It consists of a conductor such as a wire, usually wound into a coil.

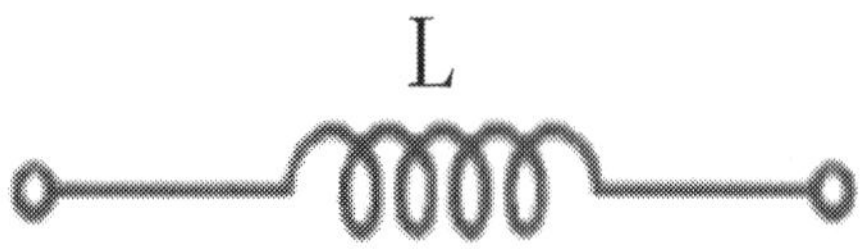

Inductance

In electromagnetism and electronics, inductance is the property of a conductor by which a change in current flowing through it induces a voltage.

$$L = \frac{\mu_0 N^2 A}{l}, \text{Henry}$$

Where, μ_0 = Permeability of free space; H/m

N=Total No. of turns in coil

A=Area of cross-section of coil, m^2

l=Length of the coil, m

- Under steady state condition for D.C. supply inductor acts as short circuit.
- Inductor does not allow sudden change of voltage, until unless impulse of voltage is applied.

If unit impulse of voltage is applied across it then at $t = 0^+$ it stores the energy $\frac{1}{2L}$ joule.

It stores energy in the form 01 magnetic field.

Power dissipation in ideal inductor is zero.

If an element is capable of delivering energy independently, then it is called active element, when it is not then called passive element.

Voltage source, current source, transistor. OPAMP are active elements and Resistor, inductor, capacitors are passive elements.

If voltage and current across an element are related to each other through constant coefficient then the element is called linear element otherwise nonlinear. Active elements (voltage source, current source) has nonlinear characteristic.

Resistance R, inductor L and capacitance are linear element under normal conditions.

When element's characteristics are independent of direction of current then element is called bidirectional element otherwise unidirectional.

Every linear element should obey the bidirectional property but vice versa is not necessary.

Diode is example of unidirectional element.

R L, C are bidirectional elements and voltage and current sources are unidirectional elements.

Internal resistance of ideal voltage source is zero.

Internal resistance of ideal current source is infinite.

Relationship of Parameters

Relationship of parameters discuss in following table:

Parameter	Basic Relationship	V-I relaionship	Energy
$R, G = \frac{1}{R}$	V=RI	v=Ri, i=GV	$\int_{-\infty}^{t} vidt$

L	$\phi = Li$	v=L(di/dt) $i = \left(\frac{1}{L}\right)\int_{-\infty}^{t} vdt$	$\frac{1}{2}Li^2$
C	q=Cv	$v = \left(\frac{1}{C}1\right)\int_{-\infty}^{t} vdt$ I=C(dv/dt)	$\frac{1}{2}Cv^2$

- A linear network is one in which the principle of super position holds well.
- A network containing circuit elements without any en source is passive network.
- A network containing energy source together with other a elements is called active network.

Ideal voltage source

- Voltage always remains constant for any value of current p through it

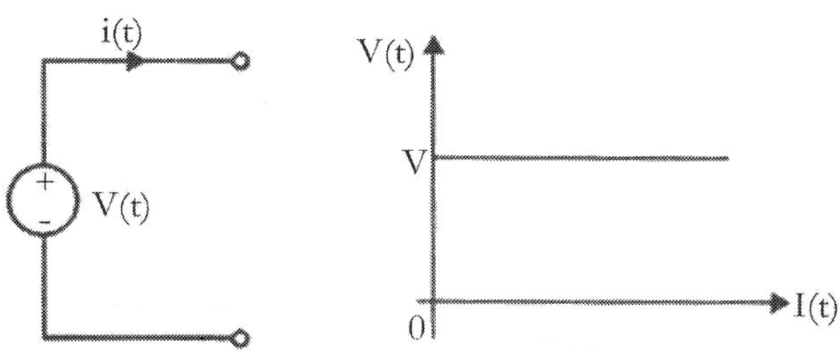

Practical voltage source

It has small internal resistance. Voltage across the element varies with respect to current .

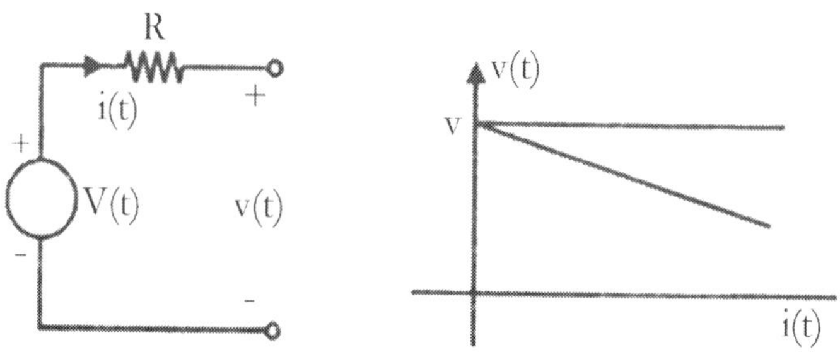

Idea current source

Current always remains constant for any value of voltage across it.

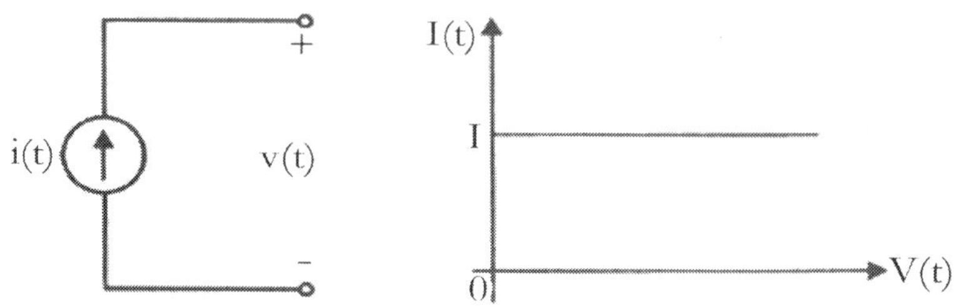

Practical current source

It has large value of internal resistance. Current varies with respect to the voltage across the element.

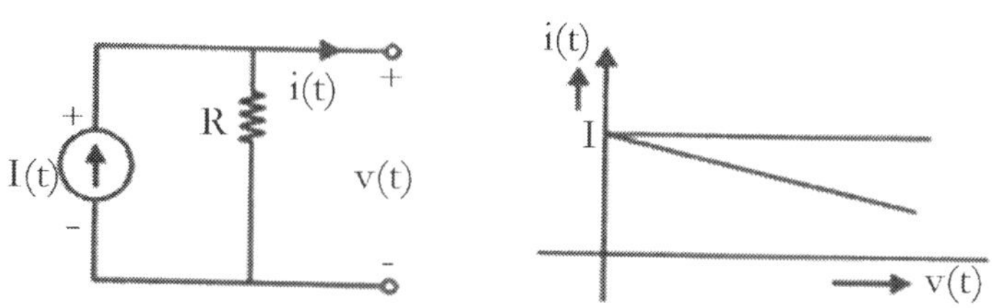

A network in which R, L, C can be separated electrically is called lumped network and network in which elements R, L, C cannot be separated is called distributed network.

- Transmission line is example of distributed network.
- Network theory concept can be applied only for low frequency application.

Standard Input Signals

Common input signals in system identification

- step function
- Get Function
- Ramp Function
- Impulse Function

Step Function

A step function is given by

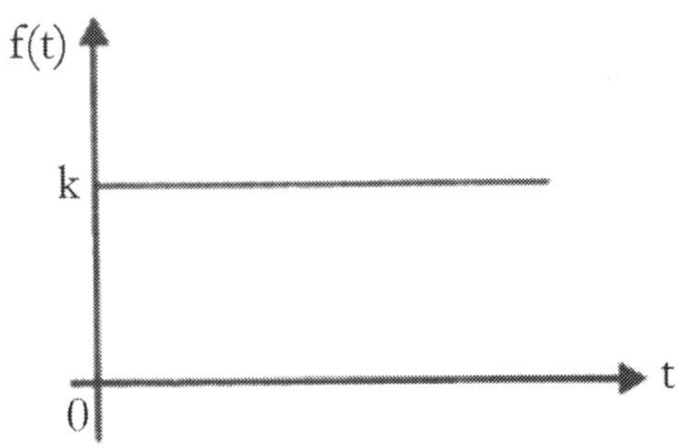

$$f(t)=\begin{cases}0 \ ; \ t \leq 0 \\ k \ ; \ t > 0\end{cases}$$

Get Function

A get function is given by

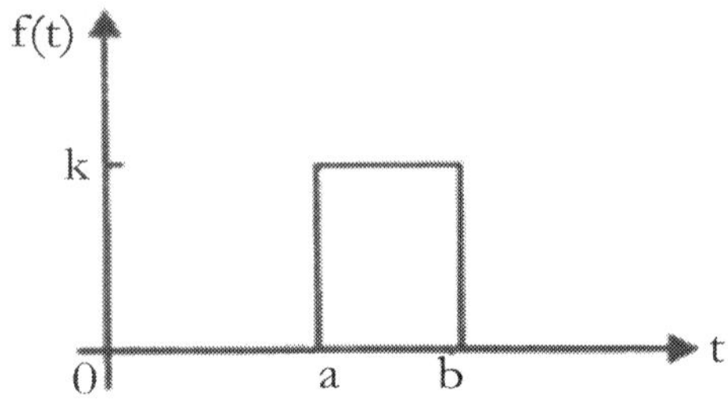

$$f(t)=k\left[u(t-a)-u(t-b)\right]$$

Ramp Function

A ramp function is given by

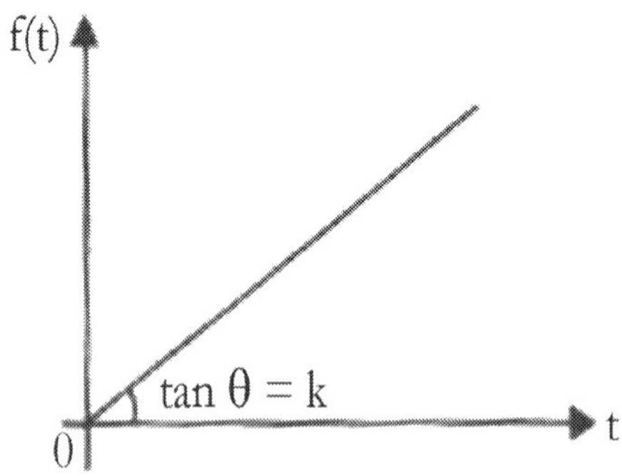

$$f(t)=\begin{cases}0\ ;\ t\leq 0\\ kt\ ;\ t>0\end{cases}$$

Impulse Function

An impulse function is given by

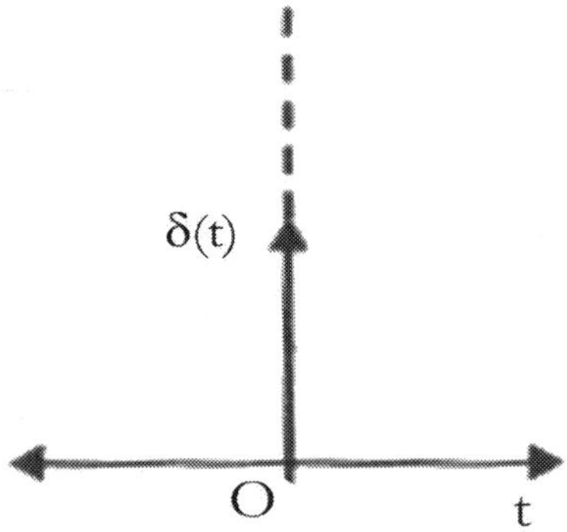

$$\delta(t)=\begin{cases} 0 \ ; \ t \neq 0 \\ \infty \ ; \ t = 0 \end{cases}$$

- Area under the impulse function is unity

$$\int_{-\infty}^{\infty} \delta(t)\,dt = 1$$

- Sampling property of impulse function

$$\int_{-\infty}^{\infty} f(t).\delta(t - t_0)\,dt = f(t_0)$$

Miscellaneous

Miscellaneous electric loads (MELs) in buildings are electric loads resulting from electronic devices not responsible for space heating, cooling, water heating, or lighting

Average Value

A average value is given by

$$Y_{av} = \frac{1}{T}\int_0^T y(t)dt$$

Where T is the time period of periodic function y(l)

- RMS or effective value

$$Y_{rms} = \sqrt{\frac{1}{T}\int_0^T [y(t)]^2\, dt}$$

RMS value of sin ωt and cos ωt is $\frac{1}{\sqrt{2}}$

If

$$y(t) = a_0 + (a_1 \cos\omega t + a_2 \cos 2\omega t + ...) + (b_1 \sin\omega t + b_2 \sin 2\omega t + ...)$$

$$Y_{rms} = \sqrt{a^2{}_0 + \frac{1}{2}(a^2{}_1 + a^2{}_2 + ...) + \frac{1}{2}(b^2{}_1 + b^2{}_2 + ...)}$$

Form Factor

A form factor is given by

$$FF = \frac{Y_{rms}}{Y_{av}} = \frac{\sqrt{\frac{1}{T}\int_0^T [y(t)]^2\, dt}}{\frac{1}{T}\int_0^T y(t)dt}$$

$$\sin \omega t = \frac{e^{jwt} - e^{-jwt}}{2j} \text{ and } \cos \omega t = \frac{e^{jwt} + e^{-jwt}}{2}$$

$$\sinh \omega t = \frac{e^{wt} - e^{-wt}}{2} \text{ and } \cosh \omega t = \frac{e^{wt} + e^{-wt}}{2}$$

- Average value is used for D.C. power calculation.
- RMS value is used for A.C. power calculation.
- Average value and RMS value do not depend on frequency but it depends on shape of the waveform.

2 Network Laws and Theorem

In electric network analysis, the fundamental rules are Ohm's Law and Kirchhoff's Laws. While these humble laws may be applied to analyze just about any circuit configuration, there are some "shortcut" methods of analysis to make the math easier for the average human.

Kirchhoff's Low

Although useful to be able to reduce series and parallel resistors in a circuit when they occur, circuits in general are not composed exclusively of such combinations. For such cases there are a powerful set of relations called Kirchhoff's laws which enable one to analyze arbitrary circuits

Kirchhoff's Voltage Law (KVL)

KVL is a fundamental law, as fundamental as Conservation of Energy in mechanics, for example, because KVL is really conservation of electrical energy.

- KVL and KCL are the starting point for analysis of any circuit.
- KCL and KVL always hold and are usually the most useful piece of information you will have about a circuit after the circuit itself.

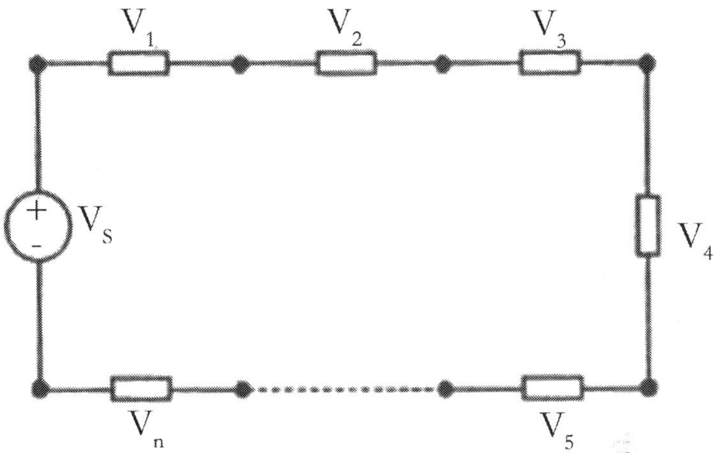

$$\sum_{i=1}^{n} V_i(t)=0 \;..$$

in a closed loop where V_i is the voltage drop or voltage gain across i^{th} element

Kirchhoff's Current Law (KCL)

KCL is a fundamental law, as fundamental as Conservation of Mass in mechanics, for example, because KCL is really conservation of charge.

- KVL and KCL are the starting point for analysis of any circuit.
- KCL and KVL always hold and are usually the most useful piece of information you will have about a circuit after the circuit itself.
- People and computer programs both use KVL and KCL for circuit analysis. Spice (in all its incarnations) starts with KCL.

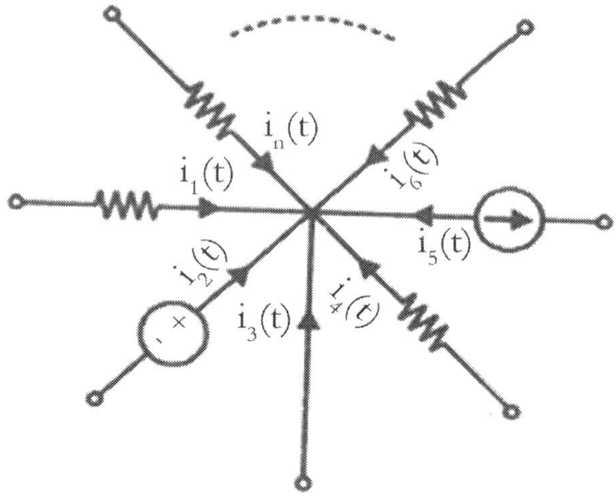

$$\sum_{i=1}^{n} l_i(t) = 0$$

at any node

Where, l_i (t) is the current through i^{th} branch

- KVL is based on the low of conservation of energy.
- KCL is based on the low of conservation of charge.
- KVL and KCL are applicable only to lumped networks.

Ohm's Low

Ohm's law states that the current through a conductor between two points is directly proportional to the potential difference across the two points. Introducing the constant of proportionality, the resistance.

It state's that at constant temperature current density is directly proportional to electric field intensity, that is

$$J = \sigma E$$

Difference form of Ohm's Low

Form	Equation
1st form	$J = \sigma E$
2nd form	$V = IR$
3rd form	$I = GV \qquad (G = 1/R)$
4th form	$V = R\dfrac{dQ}{dt}$
5th form	$v = L\dfrac{di}{dt}$
6th form	$i = \dfrac{1}{L}\displaystyle\int_{-\infty}^{t} v\,dt$
7th form	$i = c\dfrac{dv}{dt}$
8th form	$v = \dfrac{1}{C}\displaystyle\int_{-\infty}^{t} i\,dt$

Ohm's Low is applicable to both lumped and distributed networks till the network is linear.

RESISTOR

A resistor is a passive two-terminal electrical component that implements electrical resistance as a circuit element. Resistors act to reduce current flow, and, at the same time, act to lower voltage levels within circuits.

Combination of Resistors

Resistors do not occur in isolation. They are almost always part of a larger circuit, and frequently that larger circuit contains many resistors. It is often the case that resistors occur in combinations that repeat.

Resistors in series

Resistors are probably the most commonly occurring components in electronic circuits. Practical circuits often contain very complicated combinations of resistors.

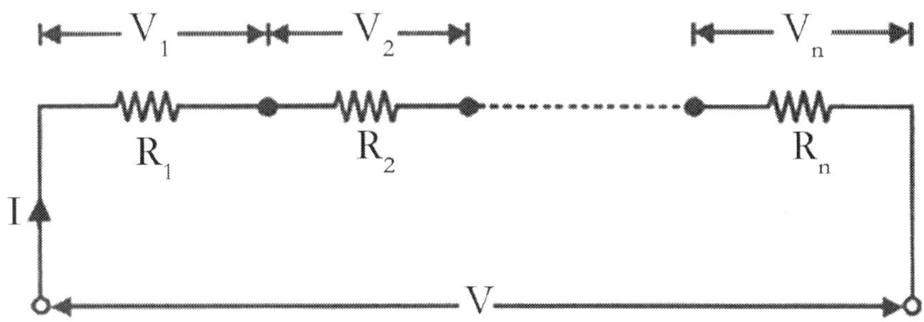

$$R_{eq} = R_1 + R_2 + \ldots\ldots. + R_n$$

Resistors in parallel

Resistors can be connected such that they branch out from a single point, and join up again somewhere else in the circuit. This is known as a parallel connection.

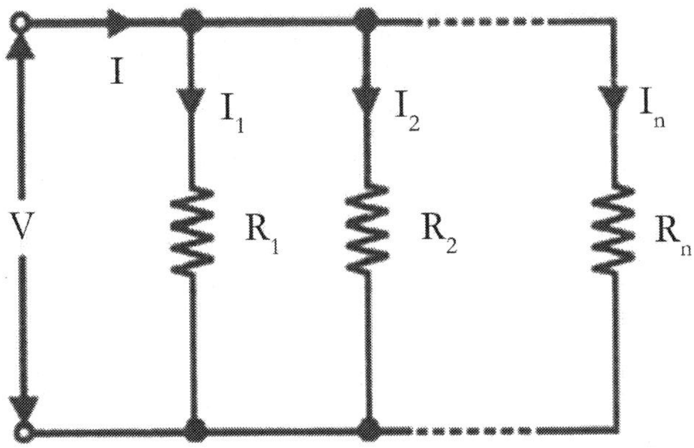

$$\frac{1}{R_{eq}} = \frac{1}{R_1} + \frac{1}{R_2} + ... + \frac{1}{R_n}$$

Voltage Division Equation

A voltage divider referenced to ground is created by connecting two electrical impedances in series

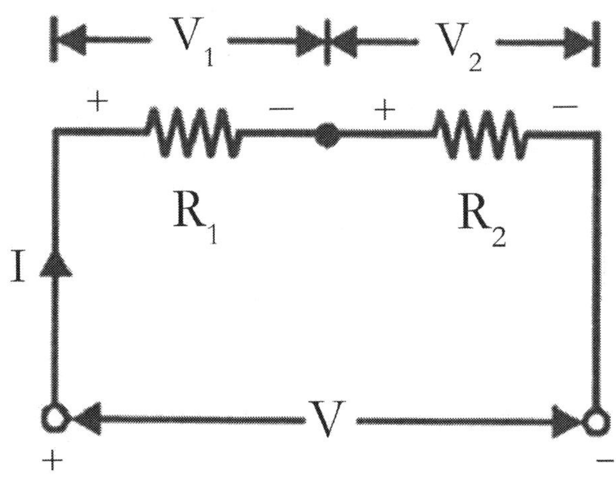

$$V_1 = \frac{R_1}{R_1 + R_2} V$$

$$V_2 = \frac{R_2}{R_1 + R_2} V$$

Current Division Equation

Let's analyze a simple parallel circuit, determining the branch currents through individual resistors:

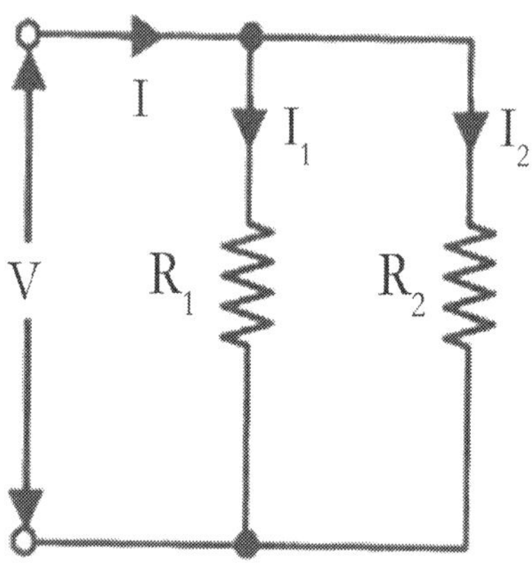

$$1_1 = \frac{R_2}{R_1 + R_2} 1$$

$$1_2 = \frac{R_1 1}{R_1 + R_2}$$

Star to Delta Transformation

Star connection or star network is a network of three electrical components in such a composition such that one point of all the components or resisters are connected at a common point.

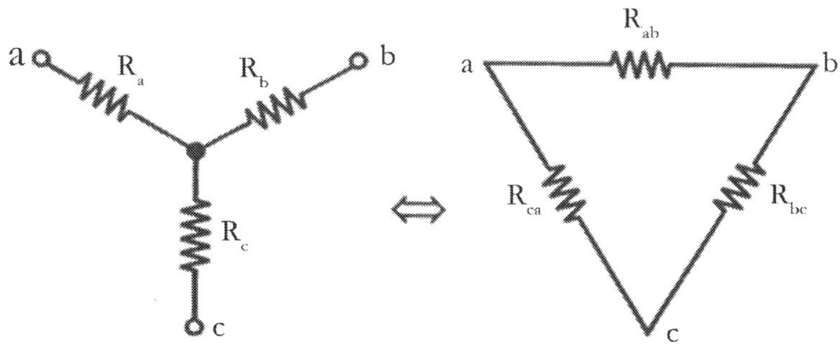

$$R_{ab} = \frac{\Delta}{R_c}, R_{bc} = \frac{\Delta}{R_a}, R_{ca} = \frac{\Delta}{R_b}$$

Where,

$$\Delta = (R_a R_b + R_b R_c + R_c R_a)$$

Delta to Star Transformation

The replacement of delta or mesh by equivalent star connection is known as delta star transformation. The two connections are equivalent or identical to each other if the impedance is measured between any pair of lines.

$$R_a = \frac{R_{ca} R_{ab}}{R_{ab} + R_{bc} + R_{ca}};$$

$$R_b = \frac{R_{ab} R_{bc}}{R_{ab} + R_{bc} + R_{ca}};$$

$$R_c = \frac{R_{bc} R_{ca}}{R_{ab} + R_{bc} + R_{ca}}$$

INDUCTOR

An inductor is a passive electronic component that stores energy in the form of a magnetic field. In its simplest form, an inductor consists of a wire loop or coil. The inductance is directly proportional to the number of turns in the coil.

Combination of Inductor

Combine inductors in series or parallel within an electronic circuit. Then you can use simple equations to calculate the total inductance of the circuit.

- Inductors in series
- Inductors in Parallel

Inductors in series

Inductance for series inductors is more than any one of the individual inductors' inductances. The formula for calculating the series total inductance is the same form as for calculating series resistances:

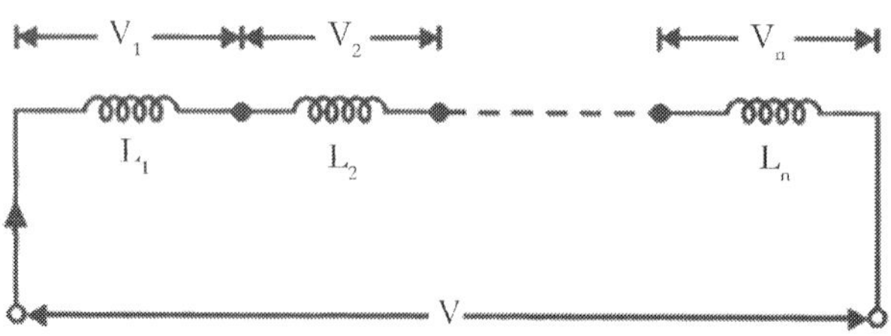

$$L_{eq} = L_1 + L_2 + L_3 + \dots\dots\dots + L_n$$

Inductors in parallel

Inductance is less than any one of the individual inductors' inductances. The formula for calculating the parallel total inductance is the same form as for calculating parallel resistances:

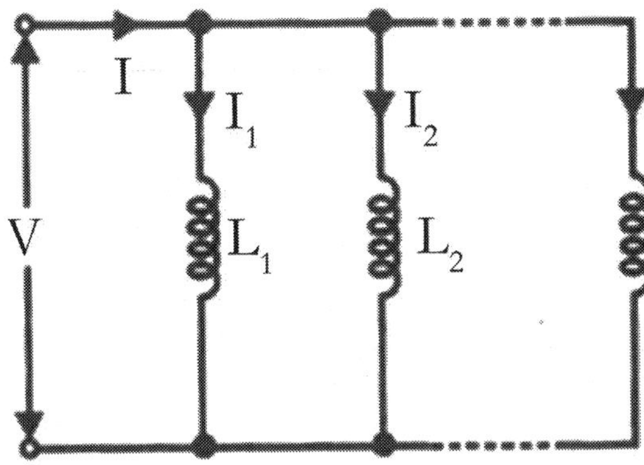

$$\frac{1}{L_{eq}} = \frac{1}{L_1} + \frac{1}{L_2} + ... + \frac{1}{L_n}$$

CAPACITORS

A capacitor is a passive two-terminal electrical component used to store energy electrostatically in an electric field.

Combination of Capacitors

Some circuits have capacitors connected in series and parallel combinations. To determine, for instance, the total charge stored by the set of capacitors, we have to find the single equivalent capacitance of the set

- Capacitors in Series

The case of resistors in parallel, the reciprocal of the circuit's total capacitance is equal to the sum of the reciprocals of the capacitance of each individual capacitor.

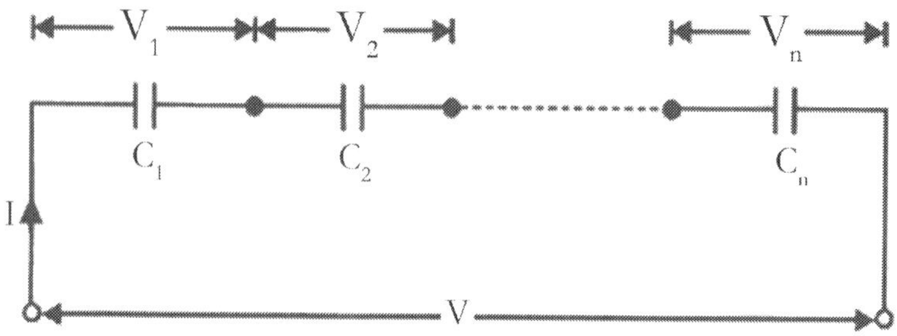

$$V = \frac{1}{C_{eq}} \int_0^t i dt$$

$$\frac{1}{C_{eq}} = \frac{1}{C_1} + \frac{1}{C_2} + \frac{1}{C_3} ..... + \frac{1}{C_n}$$

- Capacitors in Parallel

Total capacitance for a circuit involving several capacitors in parallel (and none in series) can be found by simply summing the individual capacitances of each individual capacitor.

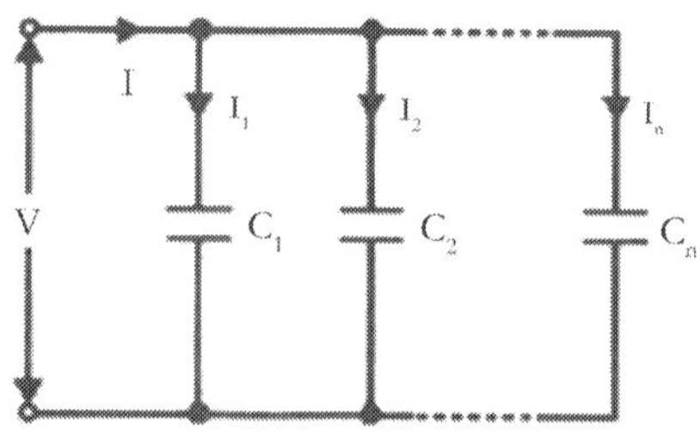

$$1 = C_{eq} \frac{dv}{dt}$$

$C_{eq} = C_1 + C_2 + C_3 + ... + C_n$

SOURCE TRANSFORMATION

The source transformation of a circuit is the transformation of a power source from a voltage source to a current source, or a current source to a voltage source.

Transformation of a resistive voltage source to a resistive current source

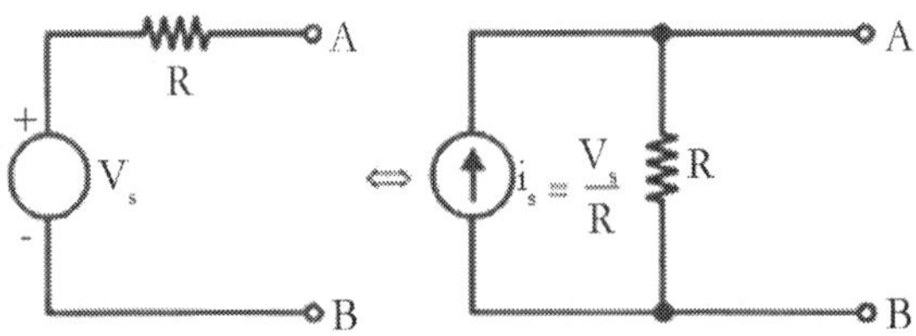

Network Theorems

Electric circuit theorems are always beneficial to help find voltage and currents in multi loop circuits. These theorems use fundamental rules or formulas and basic equations of mathematics to analyze basic components of electrical or electronics parameters such as voltages, currents, resistance, and so on.

Superposition Theorem

The response in any element of a linear, bilateral, R, L, C, network containing more than one independent voltage or current sources is the algebraic sum of the response produced by various sources each acting alone.

- Not applicable to the Network containing
  - a. Unilateral elements
  - b. Non-linear elements

This theorem is used in order to check whether a given element is or nonlinear.

Maximum Power Transfer Theorem

The Maximum Power Transfer Theorem is another useful Circuit Analysis method to ensure that the maximum amount of power will be dissipated in the load resistance when the value of the load resistance is exactly equal to the resistance of the power source.

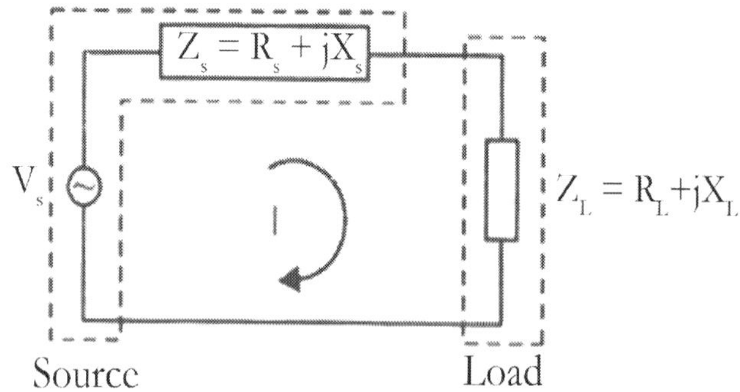

$$Z_L = Z^*_s$$

for a maximum power transfer

Case-1: if Z_L=R_L and Z_s=R_s

$$R_L = R_s$$

Case 2: If Z_s=R_s+R_s+jx_s and Z_L=R_L

$$RL = \sqrt{R^2_s + X^2_s}$$

Tellegan's Theorem

Tellegen's theorem is an important theorem on electrical network analysis. Tellegen's theorem is one of the fundamental theorems among electrical network analysis theorems and most of the network analysis theorems can be easily derived from this theorem.

In any network the sum of instantaneous power consumed by various elements of the branches is always equal to zero

Total power given out by different voltage sources = Total power consumed by various passive elements in various branches of the network.

$$\sum_{k=1}^{b} v_k . i_k = 0$$

Where, b= Number of branches

Tellegan's theorem works based on the principle of law of conservation of energy.

Millman's Theorem

Millman's Theorem is a theorem which helps in simplifying electrical networks with a bunch of parallel branches. Millman's Theorem can be used to find the potential difference between two points of a network which contains only parallel branches.

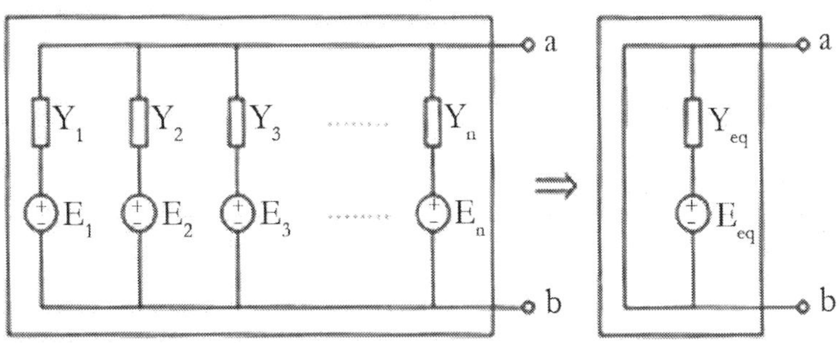

$$E_{eq} = \frac{\sum_{i=1}^{n} E_i Y_i}{\sum_{i=1}^{n} Y_i} \qquad Y_{eq} = \sum_{i=1}^{n} Y_i$$

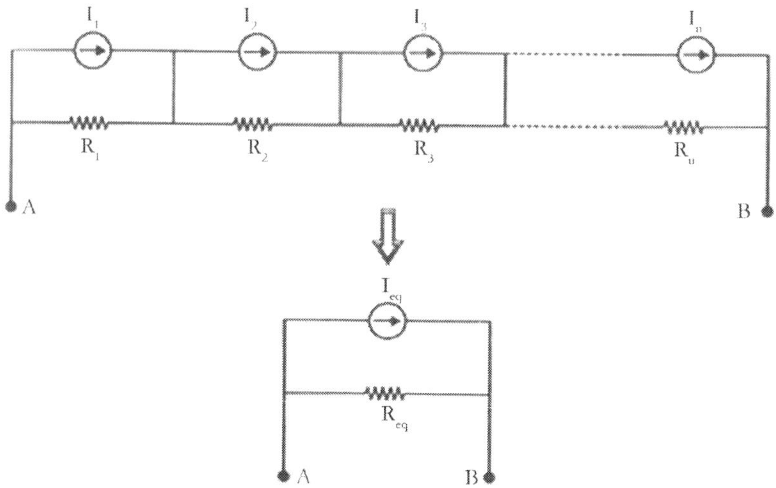

Where, $R_{eq} = R_1 + R_2 + R_3 + \ldots\ldots + R_n$

$$I_{eq} = \frac{I_1R_1 + I_2R_2 + I_3R_3 + \ldots\ldots I_nR_n}{R_1 + R_2 + R_3 + \ldots\ldots + R_n}$$

Thevenin's Theorem

It is used to represent a linear bilateral network by an equivalent circuit having a voltage source and resistance in series with it.

- Any linear electrical network with voltage and current sources and only resistances can be replaced at terminals A-B by an equivalent voltage source V_{th} in series connection with an equivalent resistance R_{th}.
- This equivalent voltage V_{th} is the voltage obtained at terminals A-B of the network with terminals A-B open circuited.
- This equivalent resistance R_{th} is the resistance obtained at terminals A-B of the network with all its independent current sources open circuited and all its independent voltage sources short circuited.

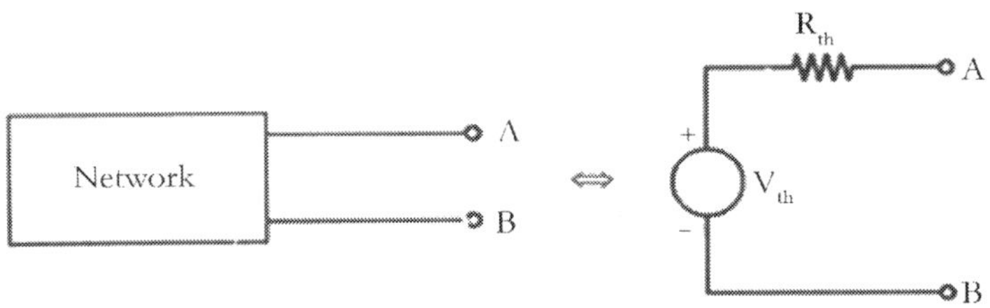

Where V_{th} open circuit voltage across AB.

R_{th} = equivalent impedance seen from terminal AB by replacing all the sources by their internal impedances.

Norton's Theorem

This theorem is just alternative of Thevenin theorem. In Norton theorem, we just replace the circuit connected to a particular branch by equivalent current source. In this theorem, the circuit network is reduced into a single constant current source in which, the equivalent internal resistance is connected in parallel with it. Every voltage source can be converted into equivalent current source

- It is used to represent a linear bilateral network by an equivalent circuit having a current source and an impedance in parallel with it.

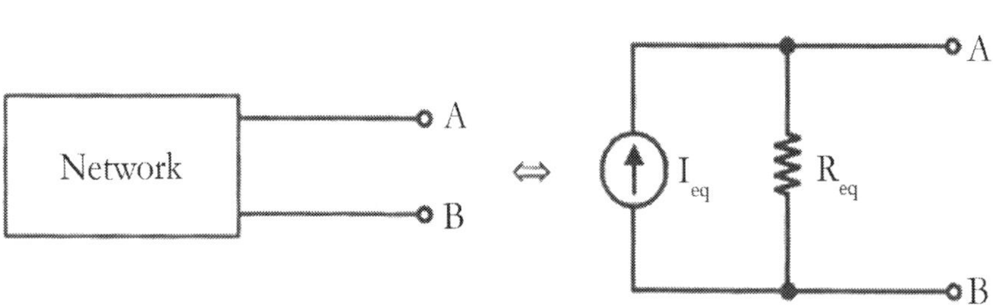

Where l_{eg} = short circuit current though terminal AB.

R_{eq} = equivalent impedance seen from terminal AB by replacing all the sources by their internal impedance

Reciprocity Theorem

Reciprocity theorem is one of the DC network analysis and AC network analysis technique and deals with the relationship between impressed source in a part of the circuit and it is response at some other part of the circuit.

- A linear network is said to be reciprocal or bilateral if it remains invariant due to the interchange of position of cause and effect in the network
- It is applicable to resistor, capacitor, inductor (with and without coupling) and transformers.
- Dependent sources are not permissible.
- All the initial conditions should be zero.
- The particular circuit that has this reciprocal property, is called reciprocal circuit.

3 Graph theory

A circuit graph is a description of the just the topology of the circuit, with details of the circuit elements suppressed. The graph contains branches and nodes. A branch is a curve drawn between two nodes to indicate an electrical connection between the nodes.

- Graph theory concerns the relationship among lines and points.
- A graph consists of some points and some lines between them.
- No attention is paid to the position of points and the length of the lines.

Rank of Graph

$$\text{Rank} = (n-1)$$

Where, n=Number of nodes.

- Tree is a collection of minimum number of branches connecting all the nodes of a graph without making a loop.
- Number of KCL equation =(n=1)
- The number of trees for a given graph=$(n)^{(n-2)}$
- Tree is connected graph.
- Branch of tree is called a twig.
- Remaining part of a graph after removal of twigs is called cotree. It is collection of links.
- Links are the branches removed from the graph to make a tree.

Total number of branch of a graph.

$$b=(n-1)+L$$

Where, L=Number of links

- Number of twigs=(n-1)=Number of KCL equation
- Number of loops=Number of links
- Every connected graph has atleast one tree.
- Rank of a tree is same as that of graph from which it is made.

MATRIXES

In mathematics, a matrix is a rectangular array of numbers, symbols, or expressions, arranged in rows and columns that is treated in certain prescribed ways. The individual items in a matrix are called its elements or entries. An example of a matrix with 2 rows and 3 columns is

$$\begin{bmatrix} 1 & 9 & -13 \\ 20 & 5 & -6 \end{bmatrix}.$$

Incidence Matrix

Let G be a graph with n vertices, m edges and without self-loops. The incidence matrix A of G is an n×m matrix $A = [a_{i\,j}]$ whose n rows correspond to the n vertices and the m columns correspond to m edges such that $a_{ij} = \begin{cases} 1, \\ 0, \end{cases}$

- It is a mathematical representation of a graph in form of a matrix.
- Number of rows of a matrix = Number of nodes in the graph.
- Number of columns of the matrix = Number of branch.

Matrix entry

-1 = if branch is incident on a node.

+1 = if branch is pointing away from node.

0 = if branch is not connected to node.

- Sum of elements of each column is zero.
- Determinant of incidence matrix of a closed loop is always zero.
- Order of incidence matrix = (n x b)

Two graphs having some incidence matrix are called isomorphic. Since rank of a graph cannot be greater than (n - 1) that is why at least one of row can be eliminated from the incidence matrix. Rank of incidence matrix is equal to rank of graph (n - 1).

Reduced Incident Matrix

Obtained after the removal of one line (row) from the Incident called reduced modem matrix. Order of Reduced incidence matrix = (n - 1) × b

Number of tree for a graph

$$T = \det\left\{[A][A]^t\right\}$$

Where

[A] = Reduced incidence matrix

$[A]^t$ =Transposed RIM

Cut Set Matrix

A minimal set of branches of a connected graph such that the removal of these branches causes the graph to be cut into exactly two parts. Fundamental cut set matrix can be used to write KCL equation given network.

- Number of fundamental cut sets of a graph = N twigs (n - 1)
- Number of rows of the matrix = Number of fundamental cut
- Number of columns = Number of branches
- Number of KCL equation = Number of fundamental cut set.

From the cut set matrix we can write equations giving the branch voltages in terms of tree branch voltages by simply reading the columns cut set matrix.

Entry in the matrix

+1=> if orientation of branch is same as the orientation of cut sets related to it

-1 => if orientation of branch is opposite to orientation of cut set related it

0 => if a cut set is not related to the branch.

Circuit Matrix

For a graph having n nodes and b branches, the circuit matrix Ba is rectangular matrix having b columns and as many as rows as there a loops. Its entries are as

b_{hk} = +1=> If branch k is in loop h and there orientation coincide

-1 => If branch k is in loop h and there orientation do not coincide

0 => If branch k is not in loop h

The rank of circuit matrix will be b-(n-1).

Tie set Matrix

A tie-set is a set of branches contained in a loop such that each loop contains one link or chord and the remainder are tree branches.

Number of fundamental tie sets for a graph.

= Number of links

= Number of KVL (mesh equation)

$L = (b - n + 1)$

4 Laplace Transform Analysis and Circuit Transients

The Laplace transform is a frequency domain approach for continuous time signals irrespective of whether the system is stable or unstable.

One sided Laplace transform of the function f(t)

$$L[f(t)] = \int_{0}^{\infty} f(t)e^{-st}dt = F(s)$$

For the two sided Laplace transformation, the limits on the integration are from ∞ to ∞.

Properties of Laplace Transform

The Laplace transform has a number of properties that make it useful for analyzing linear dynamical systems.

£

Linearity

$$L\{af_1(t) + bf_2(t)\} = aF_1(s) + bF_2(s)$$

Differentiation

$$\left\{\frac{d^n f(t)}{dt^n}\right\} = s^n F(s) - s^{n-1} f(0) - s^{n-2} f'(0) - s^{n-3} f''(0)$$

Integration

$$L\left\{\int_{-\infty}^{t} f(t)dt\right\} = \frac{F(s)}{s} + \frac{f(0)}{s}$$

Scaling

$$L\{f(at)\} = \frac{1}{|a|} F\left(\frac{s}{a}\right)$$

Time shifting

$$L\{f(t-a)\} = e^{-as} F(s)$$

Shifting in s-domain

$$L\{e^{-at} f(t)\} = F(s+a)$$

Convolution

$$L\{f_1(t) * f_2(t)\} = F_1(s)F_2(s)$$

Initial Value Theorem

$$\lim_{t\to 0} f(t) = \lim_{s\to\infty} sF(s)$$

Final Value Theorem

$$\lim_{t\to\infty} f(t) = \lim_{s\to 0} sF(s)\ F(s) = L[F(t)] = \frac{L[F_1(t)]}{1 - e^{-st}}$$

Where,

f(t) = Periodic function with time period T

f_1 (t) = Function over one time period

Some Laplace Transform Pairs

f(t)	Laplace ⇌ Inverse Laplace	F(s)
$\delta(t)$	⇌	1
u(t)	⇌	1/s

$r(t) = tu(t)$	$\rightleftharpoons$	$1/s^2$
$t^n\, u(t)$	$\rightleftharpoons$	$\frac{n!}{s^{n+1}}$
$e^{-\alpha t}u(t)$	$\rightleftharpoons$	$\frac{1}{s+\alpha}$
$(\sin \omega t)\, u(t)$	$\rightleftharpoons$	$\frac{\omega}{s^2+\omega^2}$
$(\cos \omega t)\, u(t)$	$\rightleftharpoons$	$\frac{s}{s^2+\omega^2}$

$[(\sin h\, \omega t)](t)$	$\rightleftharpoons$	$\left[\frac{\omega}{s^2-\omega^2}\right]$
$(\cos h\, \omega t)u(t)$	$\rightleftharpoons$	$\left[\frac{s}{s^2-\omega^2}\right]$
$(e^{-\alpha t} t^n)u(t)$	$\rightleftharpoons$	$\frac{n!}{(s+\alpha)^{n+1}}$
$(t \sin \omega t)iu(t)$	$\rightleftharpoons$	$\frac{s\omega s}{(s^2+\omega^2)^2}$

$(t\cos\omega t)u(t)$	$\rightleftharpoons$	$\dfrac{s^2-\omega^2}{(s^2+\omega^2)^2}$

$\left(e^{-\alpha t}\sin\omega t\right)u(t)$	$\rightleftharpoons$	$\dfrac{\omega}{(s+\alpha)^2+\omega^2}$
$\left(e^{-\alpha t}\cos\omega t\right)u(t)$	$\rightleftharpoons$	$\dfrac{s+\alpha}{(s+\alpha)^2+\omega^2}$

Representation of Circuit Elements in S- Domain

Representation of Circuit Elements in S- Domain as follows:

Resistor

Resistors are time and frequency invariant. The resistor is exactly the same in the time domain and the s-Domain.

Time Domain

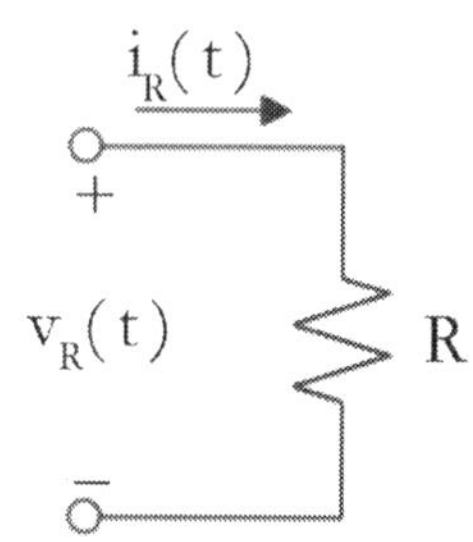

In time domain

$$v_R(t)=Ri_R(t)$$

S-Domain

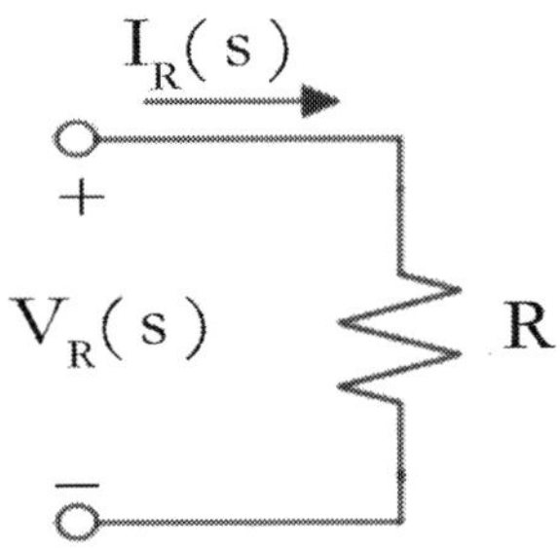

In S-domain

$$V_R(s)=RI_R(s)$$

Inductor

Energy storing element

Time Domain

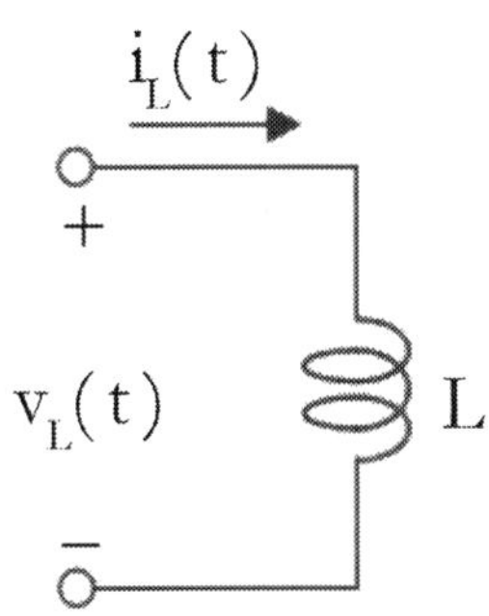

In time domain

$$v_L(t)=L\frac{di_L(t)}{dt}$$

S-Domain

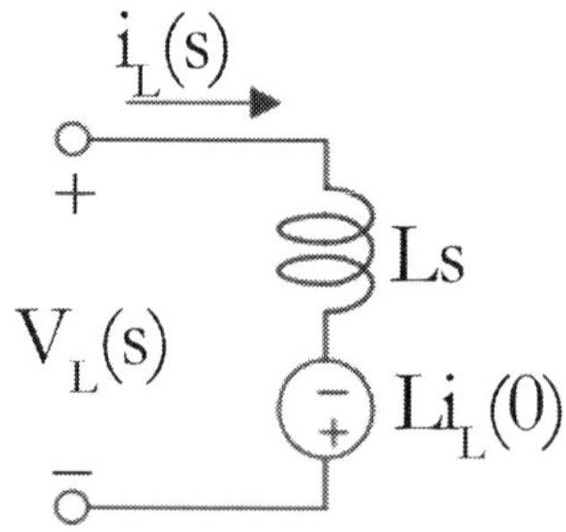

In S- domain

$$V_L(s)=LsI_L(s)-Li_L(0)$$

Capacitor

Time Domain

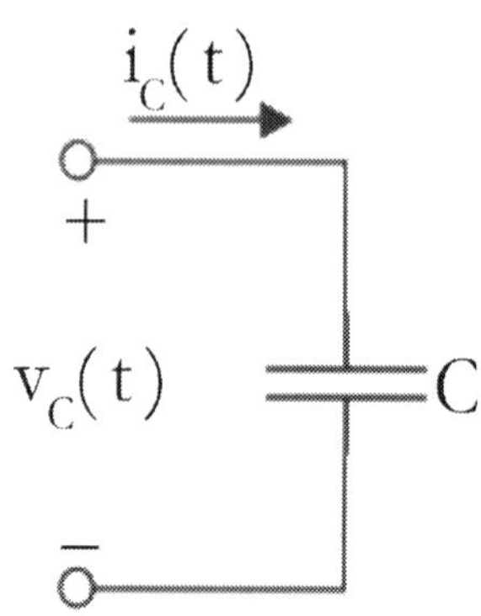

In time domain

$$v_C(t)=\frac{1}{C}\int_0^t i_c(\tau)\,d\tau+v_C(0)$$

S-Domain

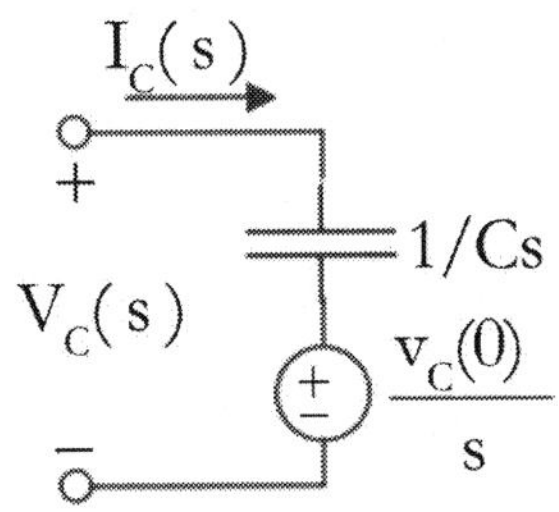

In S- domain

$$V_C(s)=\frac{1}{Cs}I_C(s)+\frac{v_C(0)}{s}$$

Gyrator

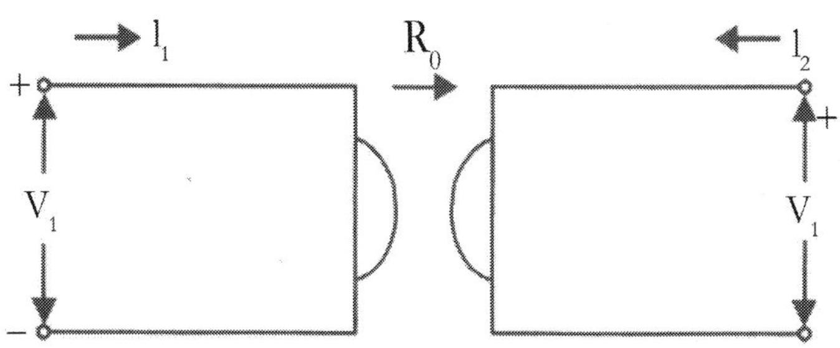

$$\frac{V_1 = +R_0 I_2}{V_2 = -R_0 I_1}$$

Where,

R_0 = Coefficient of gyrator (Depends upon op-AMP parameters and R, C)

Transient Response

Transient response is the response of a system to a change from equilibrium. The transient response is not necessarily tied to "on/off" events but to any event that affects the equilibrium of the system.

Transients are present in circuit, when circuit is subjected to any changes either by changing source magnitude or by changing any circuit elements provided circuit should consists of any energy storage elements. When circuit is having only resistive elements, no transients are present because it allows sudden change of voltage and current.

RL Circuit

The transient current through the inductor L at any time t

$$i_L(t) = i_\infty - \left[i_\infty - i_{0-}\right] e^{-R_{eq}\, t/L}$$

Where

i_∞ = Current through L at t → ∞ i.e. steady state current through L

i_{0^-} = Current through L at t = 0^-

R_{eq} = Thevenin's equivalent resistance seen across L for t > 0.

RC Circuit

The transient voltage across capacitor C at any time t

$$v_C(t) = v_\infty - \left[v_\infty - v_{0-}\right] e^{-t/R_{eq}C}$$

Where

v_∞ = Voltage across capacitor $t \rightarrow \infty$ i.e. steady state voltage across C

v_{0^-} = Voltage across C at $t = 0^-$

R_{eq} = Thevenin's equivalent resistance seen across C for $t > 0$.

AC Transients

- For transient free response θ must be equal to

$$\theta = \tan^{-1}\left(\frac{\omega L}{R}\right)$$

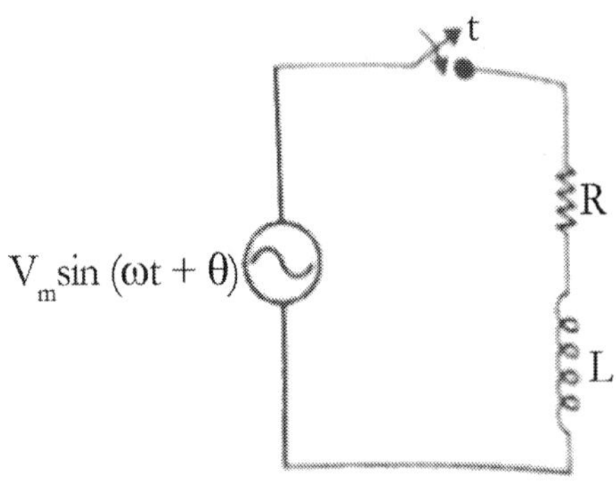

If switch is closed at t=0

$$\theta = \tan^{-1}\left(\frac{\omega L}{R}\right) - \omega t_0$$

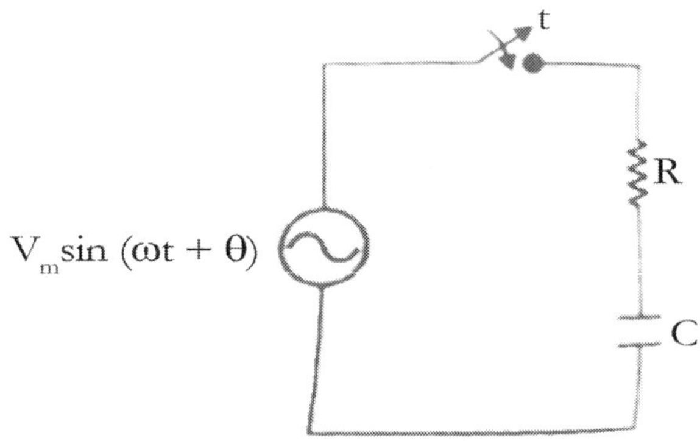

If switch is closed at $t = t_0$

- For transient free response θ must be equal to

$$\theta = \tan^{-1}(\omega RC)$$

If switch is closed at t=0

$$\theta = \tan^{-1}(\omega RC) - \omega t_0$$

Condition for Transient Free Response for Different Inputs for RL, RC series and Parallel Circuits

Time at which switch is closed	Input	RL series or parallel	RC series or parallel
t=0	$V(t)=V_m \sin(\omega t+\theta)$	$\theta=\tan^{-1}(\omega\tau)$ $\left(\tau=\frac{L}{R}=\text{time const.}\right)$	$\theta=\tan^{-1}(\omega\tau)$ $(\tau=RC=\text{time const.})$
t=0	$V(t)=V_m \cos(\omega t+\theta)$	$\theta=\tan^{-1}(\omega\tau)+\frac{\pi}{2}$	$\theta=\tan^{-1}(\omega\tau)+\frac{\pi}{2}$
$t=t_0$	$V(t)=V_m \sin(\omega t+\theta)$	$\theta=\tan^{-1}\omega\tau-\omega_0 t$	$\theta=\tan^{-1}\omega\tau-\omega t_0$
$t=t_0$	$V(t)=V_m \cos(\omega t+\theta)$	$\theta=\tan^{-1}\omega\tau-\omega t_0+\frac{\pi}{2}$	$\theta=\tan^{-1}\omega\tau-\omega t_0+\frac{\pi}{2}$

5 Resonance

Resonance is a phenomenon that consists of a given system being driven by another vibrating system or by external forces to oscillate with greater amplitude at some preferential frequencies. Frequencies at which the response amplitude is a relative maximum are known as the system's resonant frequencies, or resonance frequencies. At resonant frequencies, small periodic driving forces have the ability to produce large amplitude oscillations. This is because the system stores vibrational energy.

Series Resonance

Series Resonance circuits are one of the most important circuits used electrical and electronic circuits. They can be found in various forms such as in AC mains filters, noise filters and also in radio and television tuning circuits producing a very selective tuning circuit for the receiving of the different frequency channels.

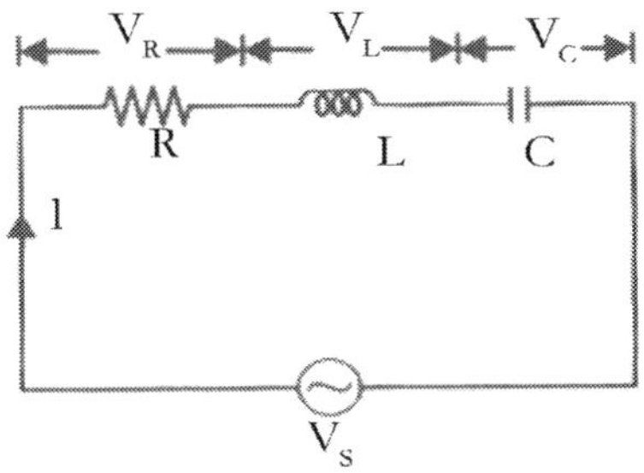

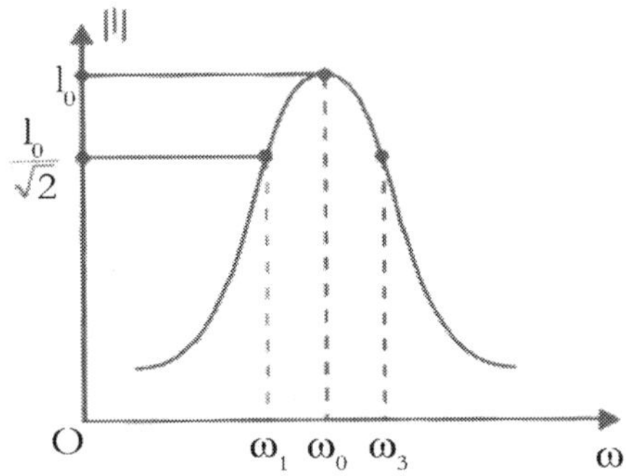

$$Q_0 = \frac{\omega_0 L}{R} = \frac{1}{\omega_0 RC}$$

Where

$\omega_0 \rightarrow$ is resonant frequency

Band Width

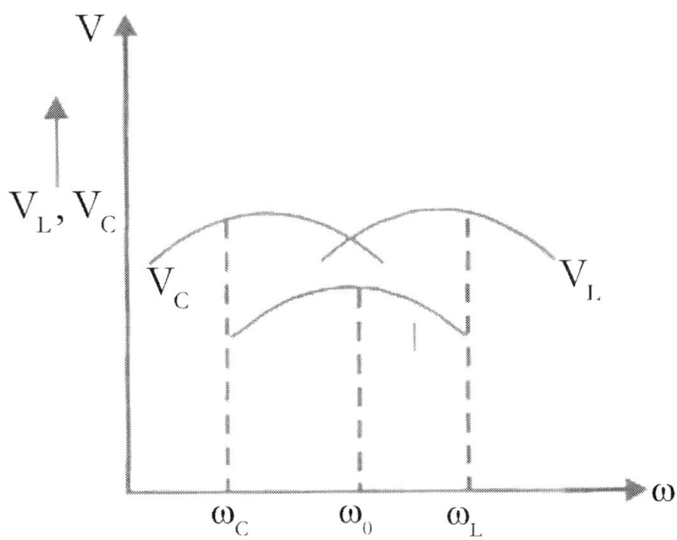

$$\Delta\omega = (\omega_2 - \omega_1) = \frac{R}{L}$$

$$\omega_2 = \omega_0 + \frac{1}{2}\Delta\omega$$

$$\omega_1 = \omega_0 - \frac{1}{2}\Delta\omega$$

$$\omega_0 = \sqrt{\omega_1 . \omega_2}$$

Where

ω_1, ω_2 =Cut off frequency

Frequency at which voltage across capacitor is maximum

$$f_C = \frac{1}{2\pi}\sqrt{\frac{1}{LC} - \frac{R^2}{2L^2}}$$

Frequency at which voltage across inductor is maximum

$$f_L = \frac{1}{2\pi\sqrt{LC - \frac{C^2R^2}{2}}}$$

Selectivity

$$\text{Selectivity} = \frac{\text{Resonance frequency}}{\text{Bandwidth}} = \frac{f_0}{f_2 - f_1}$$

Where

$$Z_{11} = \left.\frac{V_1}{I_1}\right|_{I_2=0} \qquad Z_{12} = \left.\frac{V_1}{I_2}\right|_{I_1=0}$$

$$Z_{21} = \left.\frac{V_2}{I_1}\right|_{I_2=0} \qquad Z_{22} = \left.\frac{V_2}{I_2}\right|_{I_1=0}$$

Admittance or Y-parameters

$$\begin{pmatrix} I_1 \\ I_2 \end{pmatrix} = \begin{pmatrix} Y_{11} & Y_{12} \\ Y_{21} & Y_{22} \end{pmatrix} \begin{pmatrix} V_1 \\ V_2 \end{pmatrix}$$

Where

$$Y_{11} = \left.\frac{I_1}{V_1}\right|_{V_2=0} \qquad Y_{12} = \left.\frac{I_1}{V_2}\right|_{V_1=0}$$

$$Y_{21} = \left.\frac{I_2}{V_1}\right|_{V_2=0} \qquad Y_{22} = \left.\frac{I_2}{V_2}\right|_{V_1=0}$$

h-parameters

$$\begin{pmatrix} V_1 \\ I_2 \end{pmatrix} = \begin{pmatrix} h_{11} & h_{12} \\ h_{21} & h_{22} \end{pmatrix} \begin{pmatrix} I_1 \\ V_2 \end{pmatrix}$$

Where

$$h_{11} = \left.\frac{V_1}{I_1}\right|_{V_2=0} \qquad h_{12} = \left.\frac{V_1}{V_2}\right|_{I_1=0}$$

$$h_{21} = \left.\frac{I_2}{I_1}\right|_{V_2=0} \qquad h_{22} = \left.\frac{I_2}{V_2}\right|_{I_1=0}$$

g-parameters

$$\begin{pmatrix} I_1 \\ V_2 \end{pmatrix} = \begin{pmatrix} g_{11} & g_{12} \\ g_{21} & g_{22} \end{pmatrix} \begin{pmatrix} V_1 \\ I_2 \end{pmatrix}$$

Where

$$g_{11} = \left.\frac{I_1}{V_1}\right|_{I_2=0} \qquad g_{12} = \left.\frac{I_1}{I_2}\right|_{V_1=0}$$

$$g_{21} = \left.\frac{V_2}{V_1}\right|_{I_2=0} \qquad g_{22} = \left.\frac{V_2}{I_2}\right|_{V_1=0}$$

ABCD or T parameters

$$\begin{pmatrix} V_2 \\ I_2 \end{pmatrix} = \begin{pmatrix} A & B \\ C & D \end{pmatrix} \begin{pmatrix} V_1 \\ I_1 \end{pmatrix}$$

Where

$$A = \left.\frac{V_2}{V_1}\right|_{I_1=0} \qquad B = \left.\frac{V_2}{I_1}\right|_{V_1=0}$$

$$C = -\left.\frac{I_2}{V_1}\right|_{I_1=0} \qquad D = -\left.\frac{I_2}{I_1}\right|_{V_1=0}$$

Condition for a network to be Symmetrical and Reciprocal

Reciprocal	Symmetrical
$Z_{12}=Z_{21}$	$Z_{11}=Z_{22}$
$Y_{12}=Y_{21}$	$Y_{11}=Y_{22}$
$\begin{vmatrix} A & B \\ C & D \end{vmatrix} = 1$	$A=D$
$h_{12}=-h_{21}$	$\begin{vmatrix} h_{11} & h_{12} \\ h_{21} & h_{22} \end{vmatrix} = 1$
$g_{12}=-g_{21}$	$\begin{vmatrix} g_{11} & g_{12} \\ g_{21} & g_{22} \end{vmatrix} = 1$

Interconnection between Two port Network

Two-port networks can be connected in series, parallel or cascade.

Cascade Connection of Two-port Network

A cascade connection is defined when the output of one network becomes the input to the next network.

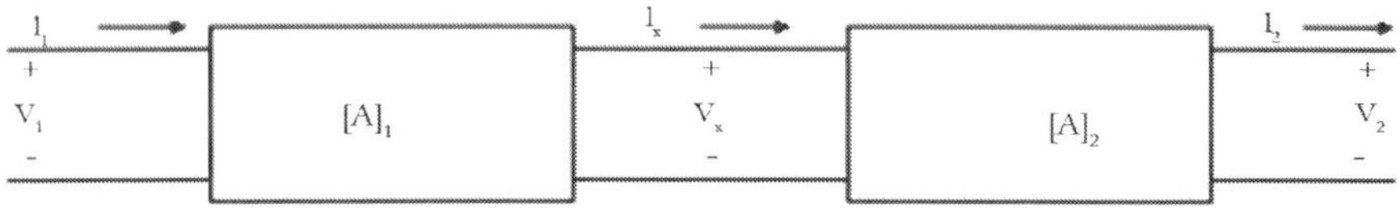

$$\begin{bmatrix} A & B \\ C & D \end{bmatrix} = \begin{bmatrix} A' & B' \\ C' & D' \end{bmatrix} \begin{bmatrix} A'' & B'' \\ C'' & D'' \end{bmatrix}$$

Series-connected Two-port Network

A series connection is defined when the currents in the series elements are equal and the voltages add up to give the resultant voltage.

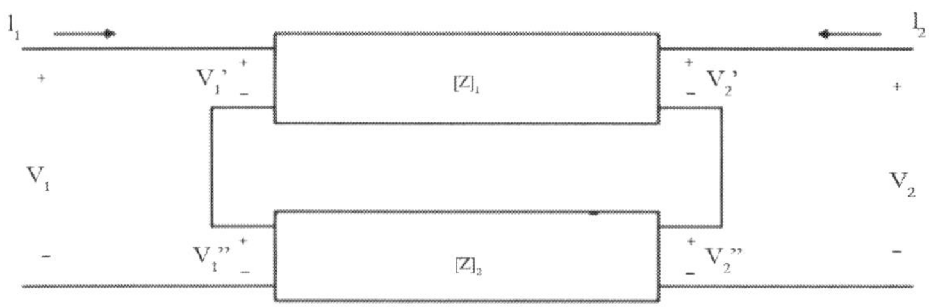

$$\begin{bmatrix} Z_{11} & Z_{12} \\ Z_{21} & Z_{22} \end{bmatrix} = \begin{bmatrix} Z'_{11} + Z''_{11} & Z'_{12} + Z''_{12} \\ Z'_{21} + Z''_{21} & Z'_{22} + Z''_{22} \end{bmatrix}$$

Parallel-connected Two-port Network

A parallel connection is defined when the voltages in the parallel elements are equal and the currents add up to give the resultant current.

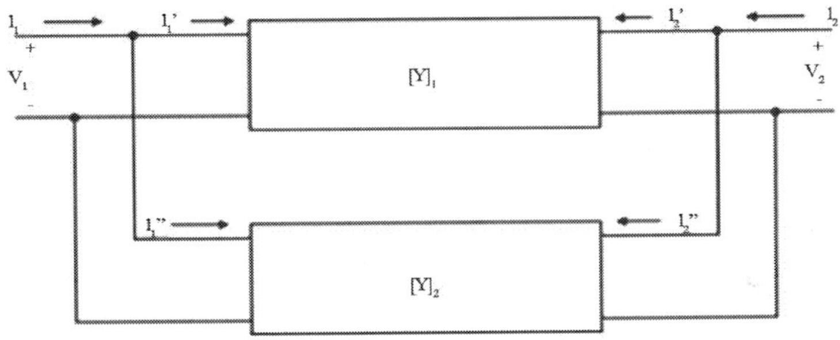

$$\begin{bmatrix} Y_{11} & Y_{12} \\ Y_{21} & Z_{22} \end{bmatrix} = \begin{bmatrix} Y'_{11} + Y''_{11} & Y'_{12} + Y''_{12} \\ Y'_{21} + Z''_{21} & Y'_{22} + Y''_{22} \end{bmatrix}$$

Input Impedance

The input impedance of the network being connected and the output impedance of the source determines the transfer function from the source to the input terminals of the circuit.

If load impedance Z_L is connected to port 2 then input impedance is

$$Z_{in} = \frac{V_1}{I_1}$$

Input Impedance in terms of Z- parameters

$$Z_{in} = \frac{Z_{11}Z_{22} - Z_{12}Z_{21} + Z_{11}Z_L}{Z_{22} + Z_L}$$

Input Impedance in terms of Transmission parameters

$$Z_{in} = \frac{AZ_L + B}{CZ_L + D}$$

Input Impedance in terms of hybrid parameters

$$Z_{in} = h_{11} - \frac{h_{12}h_{21}Z_L}{1 + h_{22}Z_L}$$

Output Impedance

If Z_L is connected to input port 1 then output impedance is

$$Z_0 = V_2/I_2$$

Output Impedance in terms of Z- parameters

$$Z_0 = \frac{Z_{11}Z_{22} - Z_{12}Z_{21} + Z_{22}Z_L}{Z_{11} + Z_L}$$

Output Impedance in terms of Transmission parameters

$$Z_0 = \frac{DZ_L + B}{CZ_L + A}$$

7 Magnetic Coupled Circuit

Two or more electric circuits so arranged that energy can transfer electrically or magnetically from one to another.

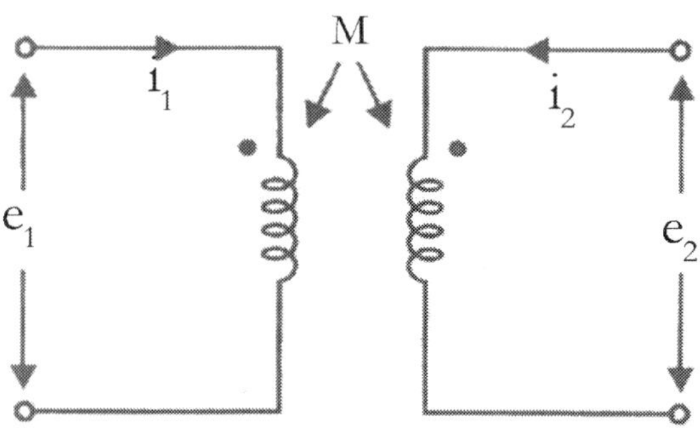

The amount of magnetic coupling is expressed by coefficient coupling (k).

$$k = \frac{\text{useful flux}}{\text{total flux}}$$

- Mutual inductance between two coils is

$$M = k\sqrt{L_1 L_2}$$

L_1 is self-inductance of first coil

L_2 is self-inductance of second coil

Dot Conventions

The dot convention is a convention used to denote the voltage polarity of the mutual inductance of two components. Two conventions are:

- If a current enters the dotted terminal of one coil, then the polarity of the e.m.f. induced in the second coil will be positive at the dotted terminal of the second coil.
- If a current leaves the dotted terminal of one coil, then the polarity of the e.m.f. induced in the second coil will be negative at the dotted terminal of the second coil.

A Handbook on Electronics and Communications Engineering

6

Control System

CONTENTS

1 Mathematical Modelling

A mathematical model is a description of a system using mathematical concepts and language. The process of developing a mathematical model is termed mathematical modelling.

Mechanical System

A mechanical system manages power to accomplish a task that involves forces and movement. A mechanical system consists of (i) a power source and actuators that generate forces and movement, (ii) a system of mechanisms that shape the actuator input to achieve a specific application of output forces and movement, and (iii) a controller with sensors that compares the output to a performance goal and then directs the actuator input. This can be seen in Watt's steam engine in which the power is provided by steam expanding to drive the piston. The walking beam, coupler and crank transform the linear movement of the piston into rotation of the output pulley. Finally, the pulley rotation drives the flyball governor which controls the valve for the steam input to the piston cylinder.

Translational System

Mass

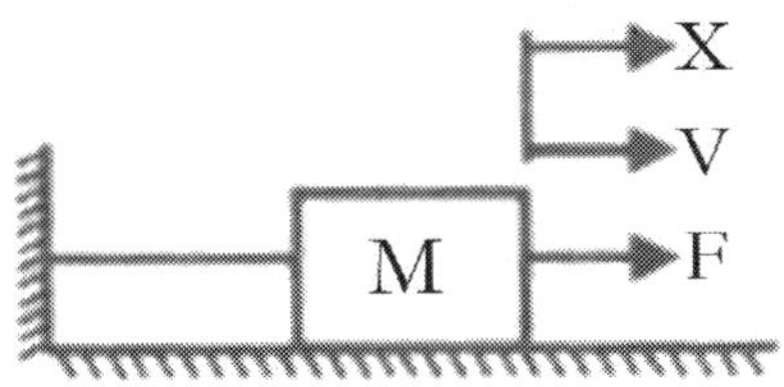

$$F = M\frac{d^2x}{dt^2} = M\frac{dv}{dt}$$

Where,

F=Force on block M,

v=Velocity of block M

x= Displacement of block M

M=Mass of block M

Damper

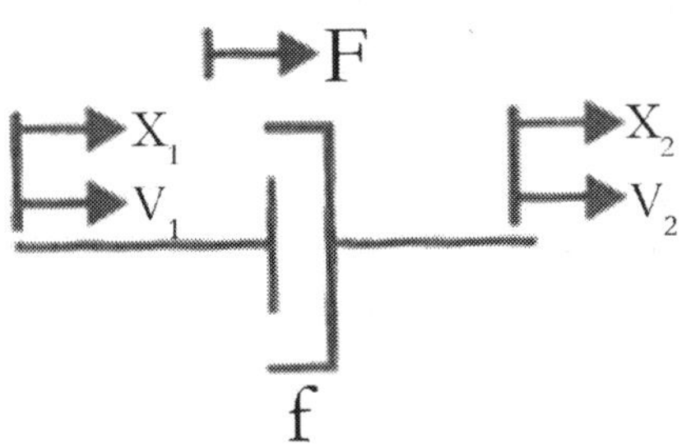

$$F = f\frac{d(x_1 - x_2)}{dt} = f(v_1 - v_2)$$

Spring

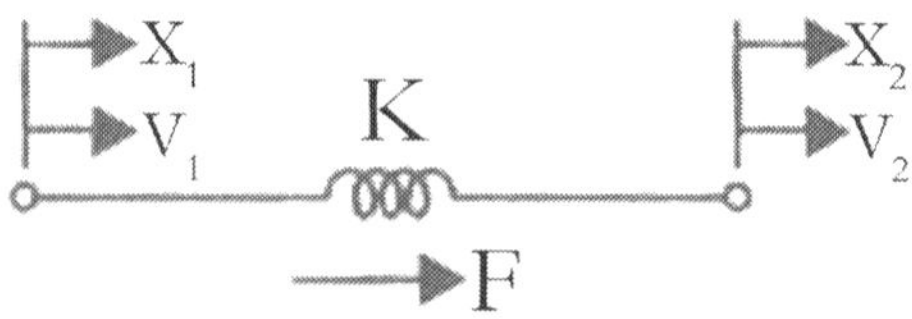

$$F = k(x_1 - x_2) = k\int(v_1 - v_2)\,dt$$

Where,

F=spring force, x_1, x_2 =Compression or expansion of spring v_1, V_2=Velocity at side 1 and 2, K=Spring constant

Rotational System

Inertia

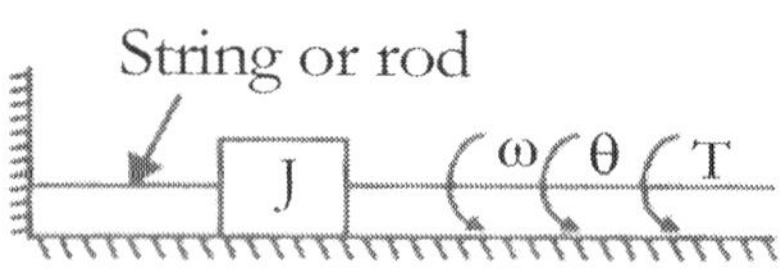

$$T = J\frac{d^2\theta}{dt^2} = J\frac{d\omega}{dt}$$

Damper

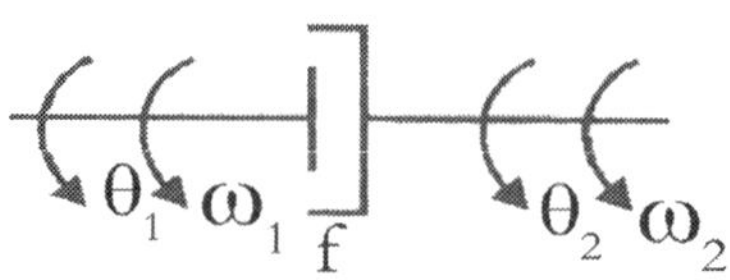

$$T = f\frac{d\theta}{dt} = f(\omega_1 - \omega_2)$$

Spring twisted

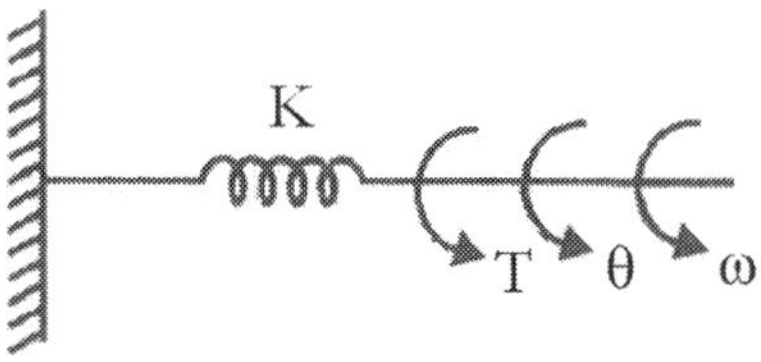

$$T = k\theta = k\int \omega dt$$

Force Voltage Analogy and Force Current Analogy

Voltage Series RLC	Current Parallel RLC	Force Mass Damper	Torque Rotational
V	I	F	T
q	4	x	0
R	1/R	f	**f**
I	V	Linear velocity	Angular velocity
1/C	1/L	K	K
L	C	M	J

Servo Motors

Servo refers to an error sensing feedback control which is used to correct the performance of a system. Servo or RC Servo Motors are DC motors equipped with a servo mechanism for precise control of angular position. Servo motors have been around for a long time and are used in many applications. They are small in size but pack a big punch and are very energy efficient. Because of these features, they can be used to operate remote-controlled or radio-controlled toy cars, robots and airplanes. Servo motors are also used in industrial applications, robotics, in-line manufacturing, pharmaceutics and food services. But how do the little guys work?

Field Controlled DC Motor

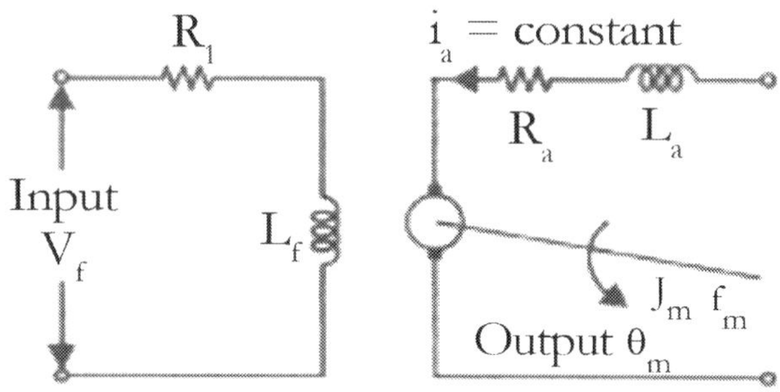

- Block diagram

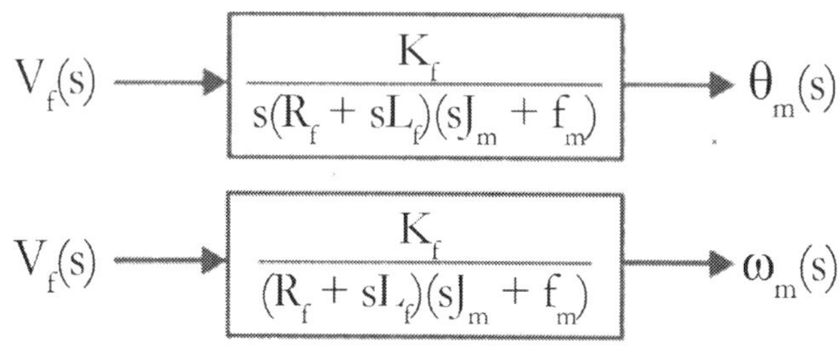

- Dynamic equations

Dynamic equations in time domain	Corresponding equation in s-domain
$V_f = R_f i_f + L_f \dfrac{di_f}{dt}$	$V_f(s) = R_f I_f(s) + sL_f I_f(s)$
$T_M = k_f I_f$	$T_M(s) = k_f I_f(s)$
$T_M = J_m \dfrac{d^2\theta_m}{dt^2} + f_m \dfrac{d\theta_m}{dt}$	$T_M(s) = s^2 J_m \theta_m(s) + sf_m \theta_m(s)$
$T_M = J_m \dfrac{d\omega_m}{dt} + f_m \omega_m$	$T_M(s) = sJ_m \omega_m(s) + f_m \omega_m(s)$

Armature Controlled DC Motor

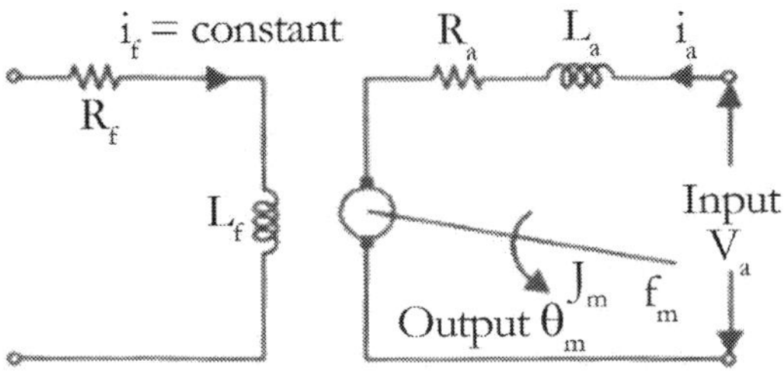

- Block diagram

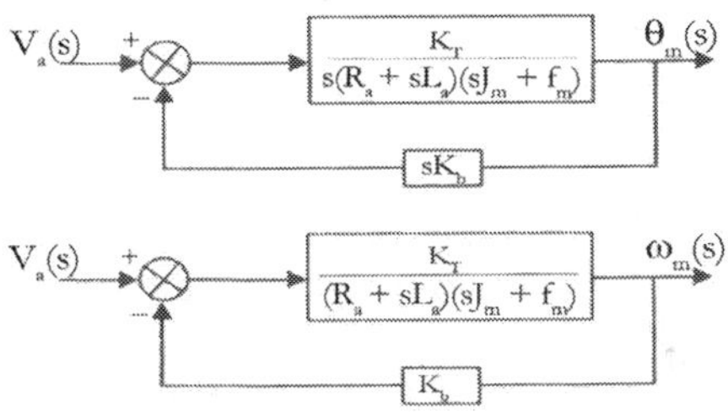

- Dynamic equations

Dynamic equations in time domain	Corresponding equation in s-domain
$V_a - e_b = R_a i_a + L_a \frac{di_a}{dt}$	$V_a(s) - E_b(s) = R_a I_a(s) + sL_a I_a(s)$
$e_b = k_b \frac{d\theta_m}{dt} = k_b \omega_m$	$E_b(s) = sk_b \theta_m(s) = k_b \omega_m$
$T_M = k_T i_a$	$T_M(s) = k_T I_a(s)$
$T_M = J_m \frac{d^2\theta_m}{dt^2} + f_m \frac{d\theta_m}{dt}$	$T_M(s) = s^2 J_m \theta_m(s) + sf_m \theta_m(s)$

$T_M = J_m \frac{d\omega_m}{dt} + f_m \omega_m$	$T_M(s) = sJ_m \omega_m(s) + f_m \omega_m(s)$

AC Servomotor

AC servomotors are AC motors that incorporate encoders are used with controllers for providing feedback and closed-loop control. These motors can be positioned to high accuracy, meaning that they can be controlled exactly as required for the application. Often servomotors have better bearings or higher tolerance designs, and some smaller designs also use higher voltages in order to achieve greater torque. AC servomotors are commonly used in robotics, automation, CNC machinery, and other applications requiring versatility and a high level of precision.

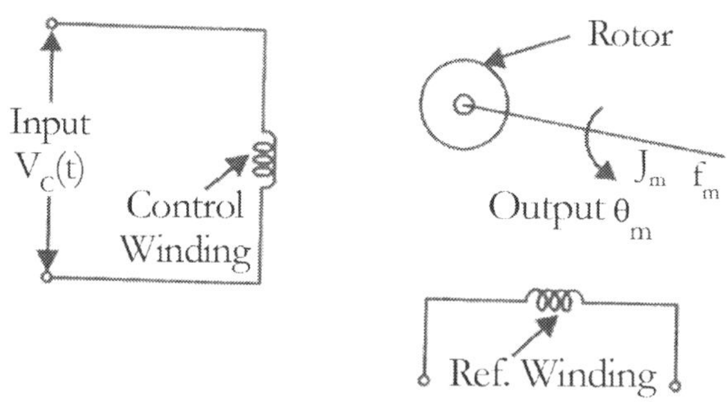

- Block diagram

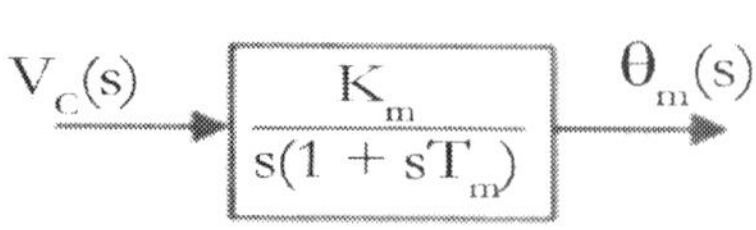

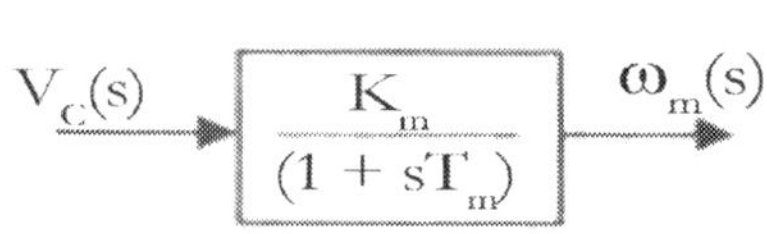

- Dynamic equations

Dynamic equations in time domain	Corresponding equation in s-domain
$T_M = m\frac{d\theta_m}{dt} + KV_c$	$T_M(s) = sm\theta_m(s) + KV_c(s)$
$T_M = J_m\frac{d^2\theta_m}{dt^2} + f_m\frac{d\theta_m}{dt}$	$T_M(s) = s^2 J_m\theta_m(s) + sf_m\theta_m(s)$

Generators

A generator is a device that converts mechanical energy to electrical energy for use in an external circuit. The source of mechanical energy may vary widely from a hand crank to an internal combustion engine. Generators provide nearly all of the power for electric power grids.

Separately Excited DC Generator

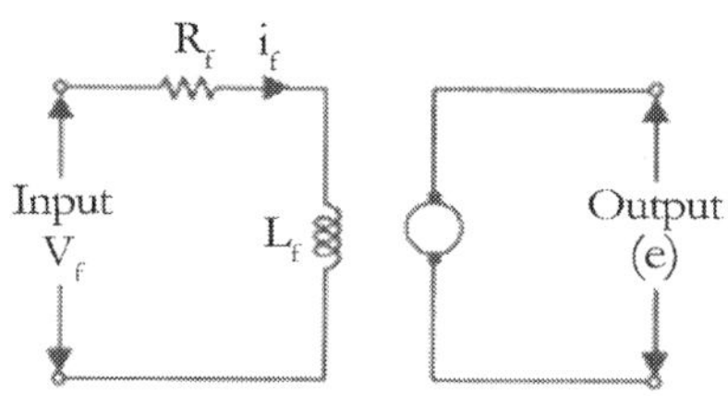

- Block diagram

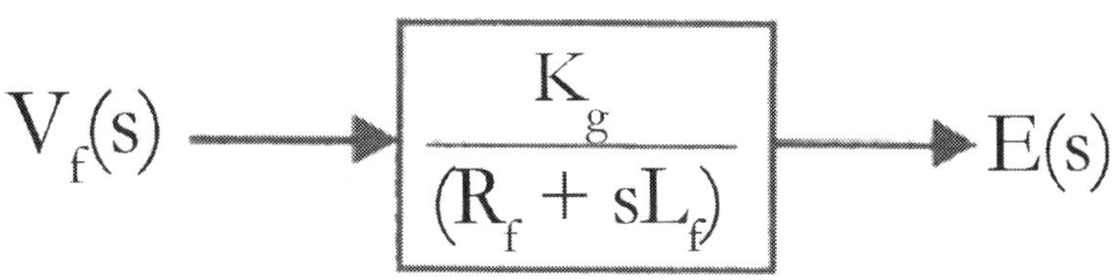

- Dynamic equations

Related equations in time domain	Corresponding equations in s-domain
$V_f = R_f i_f + L_f \dfrac{di_f}{dt}$	$V_f(s) = R_f I_f(s) + sL_f I_f(s)$
$e = k_g i_f$	$E(s) = k_g I_f(s)$

Separately excited DC generator connected to lord

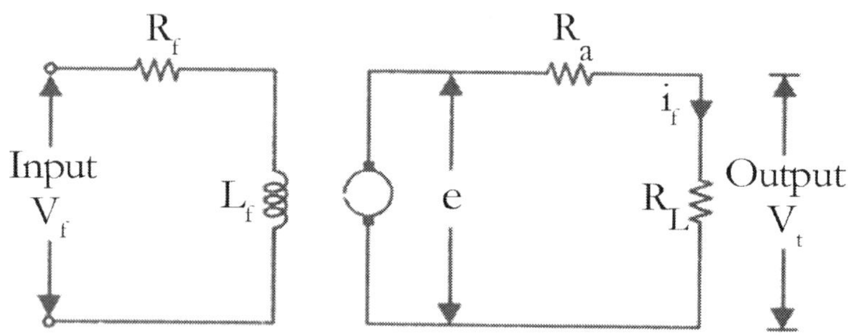

- Block diagram

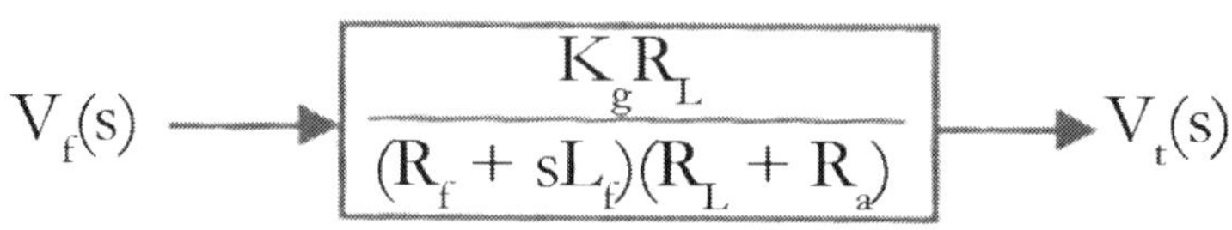

- Dynamic equations

Related equations in time domain	Corresponding equations in s-domain
$V_t = e - R_a i_L$	$V_t(s) = E(s) - R_a I_L(s)$
$I_L = \frac{V_t}{R_L}$	$I_L(s) = \frac{V_t(s)}{R_L}$

2 Block Diagram

A block diagram of a system is a pictorial representation of the functions performed by each component and of the flow of signals. Block Diagrams are a useful and simple method for analysing a system graphically. It is used to represent all types of systems. It can be used, together with transfer functions, to describe the cause and effect relationships throughout the system.

Transform Function

It is defined as the ratio of the Laplace transform of the output variable to the Laplace transform of the input variable, with all zero initial conditions.

Open Loop

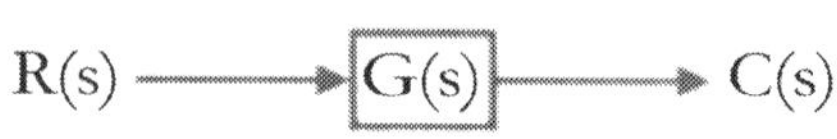

Transfer function

$$G(s) = \frac{C(s)}{R(s)}$$

Where, C(s)=Laplace transform of output

R(s)=Laplace transform of output

Impulse response of a system

$$g(t) = c(t)$$

Where, g(t)=Inverse laplace transform of G(s)

c(t)=Output of the system

Closed Loop transform Function

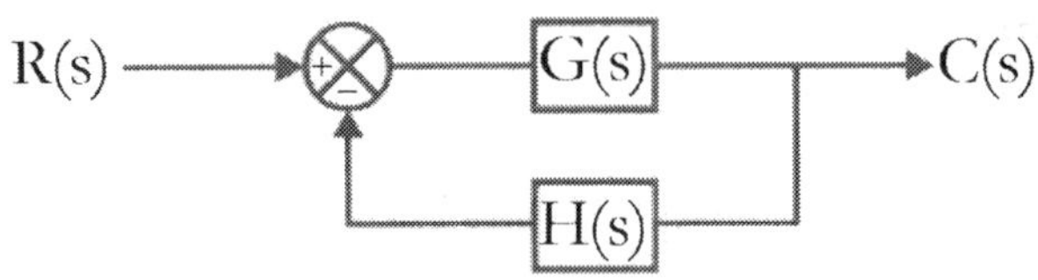

$$\text{T.F.} = \frac{G(s)}{1+G(s)H(s)}$$

Note:

- Transfer function is always calculated with zero initial conditions.
- Zero initial condition ensure the linearity.
- Highest power of S in the denominator of transfer function is called order of the system.
- Transfer function is property of a system itself, independent of magnitude and nature of input or driving function.
- It does not provide any information concerning the physical structure of the system means physically different system can have same transfer functions.
- Transfer function and impulse response function contain the same information about the system dynamics.

Block Diagram Reduction

A general linear control system may be built up from many interconnected subsystems, each of which is a system by itself, and, hence, represented by a transfer function.

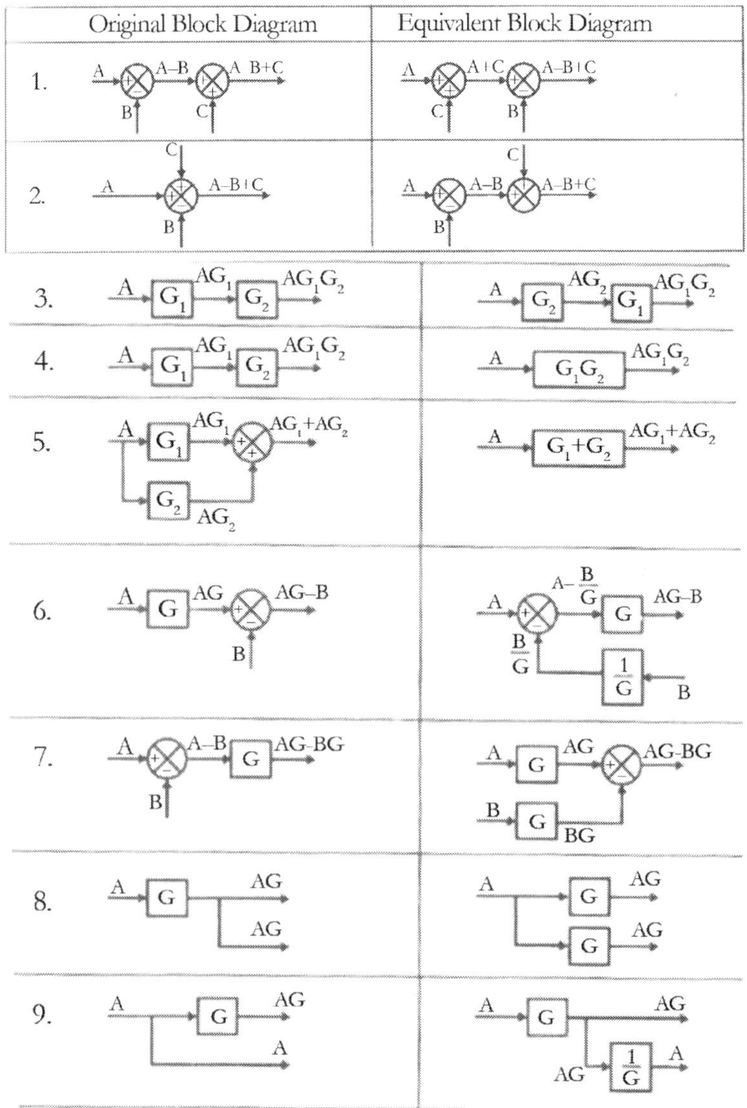

10. 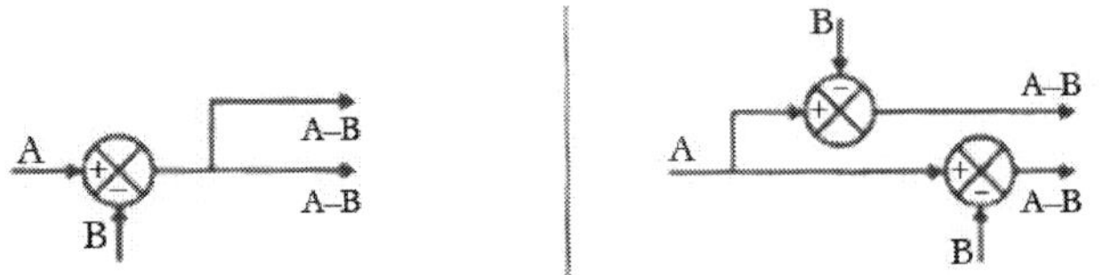

11. 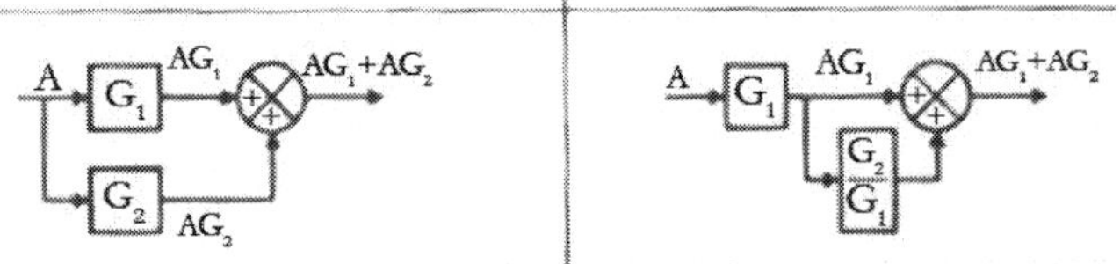

12. 

13. 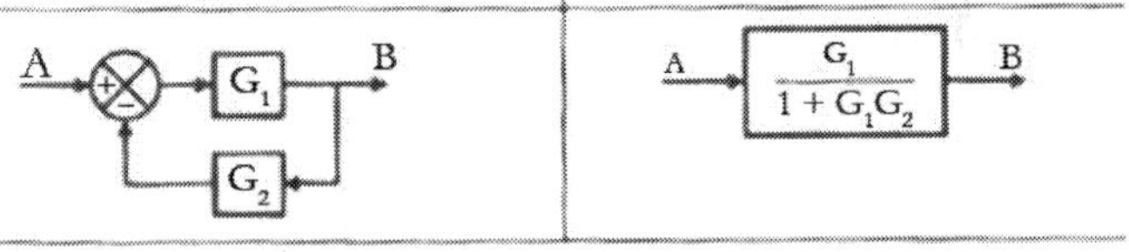

- Drawing signal flow graph from a given block diagram

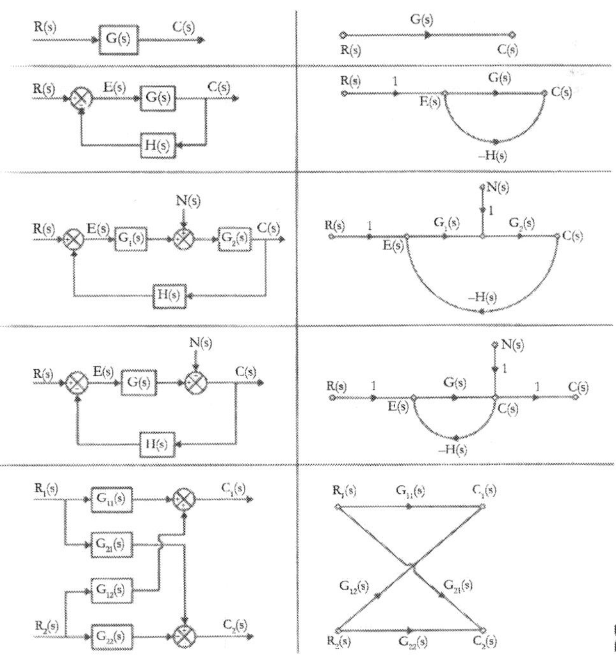

Note:

- A block diagram has the advantage of indicating more realistically the signal flows of the actual system.
- Block diagram of a given system is not unique.
- Feedback element in most cases is a sensor that measures the output of the plant.
- The ratio of output C(s) to the actuating error signal E(s) is called the feed forward transfer function.

Signal Flow Graph

- A signal flow graph represents a set of simultaneous linear algebraic equation.
- A branch represents the functional dependence of one signal on another.
- A node adds the signals of all incoming branches and transmits this sum to all outgoing branches.
- For a given system, a signal flow graph is not unique.

Mason's Gain Formula

$$P = \frac{1}{\Delta} \sum P_k \Delta_k$$

Where

P=Overall gain or transfer function

P_k=Path gain of kth forward path

$\Delta = 1$ -(sum of loop gains of all individual loops)+(sum of gain products of all possible combinations of two nontouching loops)-(sum of gain product of all possible combinations of three nontouching loops)+...

Δ_k =Value of Δ obtained by removing all the loops touching k^{th} forward path

3 Time Response Analysis of Control System

The order of a system is defined as being the highest power of derivative in the differential equation, or being the highest power of s in the denominator of the transfer function. A first-order system only has s to the power one in the denominator, while a second-order system has the highest power of s in the denominator being two.

Standard Test Signals

The dynamic behaviour of a system is manipulated and judged under the application and working of standard and typical test signals. The characteristics of an input signal are constant acceleration, constant velocity, a sudden change or a sudden shock.

Step Function

The step signal defines the sudden change in properties of actual signal. It is being used to see the transient response of system as it gives you the idea about how the system reply to interruption and somehow the system stability.

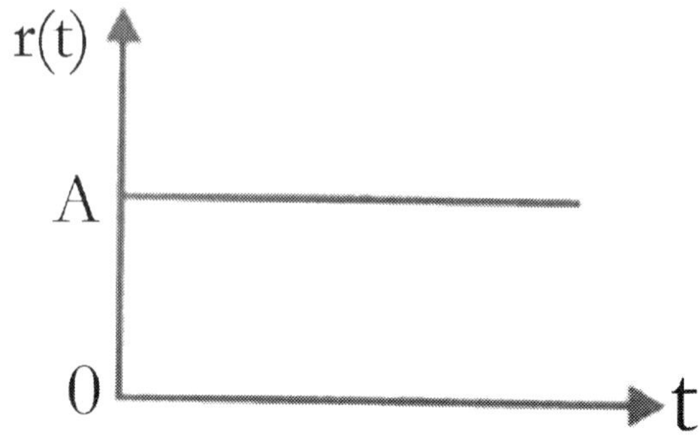

$$r(t) = Au(t)$$
$$\text{where, } u(t) = \begin{cases} 1 & ; t \geq 0 \\ 0 & ; t < 0 \end{cases}$$

Ramp Function

The ramp signal tells you the constant velocity attribute of actual input signal. It is being used to determine the behaviour of system with the velocity factor.

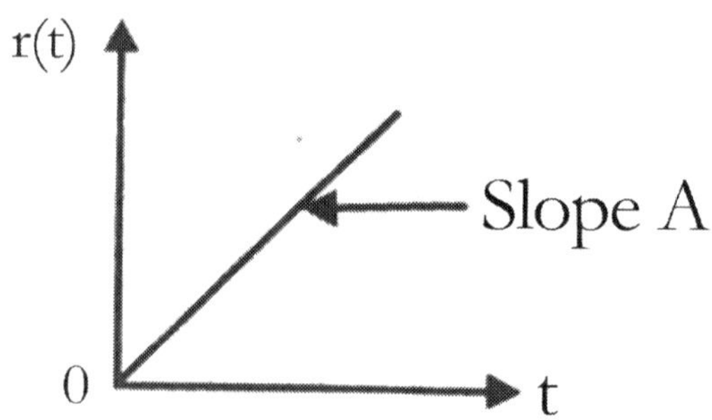

$$r(t) = Atu(t)$$

Parabolic Signal

Parabolic signal gives the constant acceleration distinction of actual input signal. It gives the idea about how the system will respond along with acceleration.

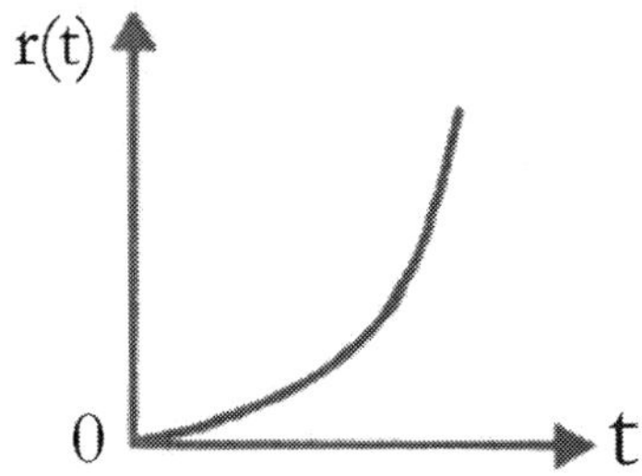

$$r(t) = \frac{At^2}{2} u(t)$$

Impulse Signal

Impulse response in control system imitates sudden shock quality of actual input signal. Impulse is the output of system when given by small input. Impulse response emphasis on change in the system in reaction to some external change. It is the reply of the system to the direct delta input.

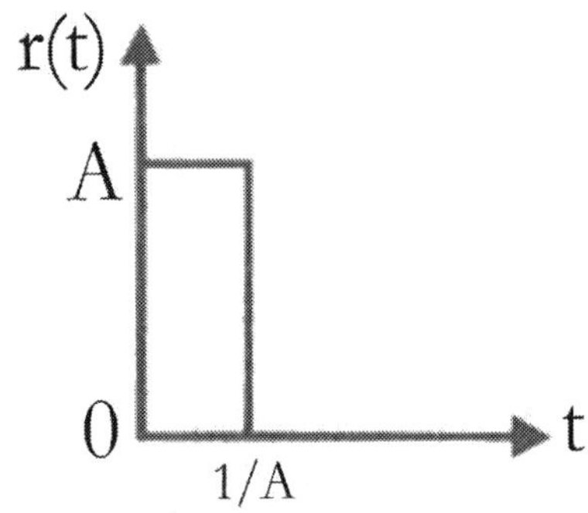

$$\boxed{\delta(t) = \overset{*}{u}(t)}$$

$$\boxed{\delta(t) = 0; t \neq 0}$$

$$\boxed{\lim_{\alpha \to 0} \int_{-\alpha}^{+\alpha} \delta(t)\,dt = 1}$$

Time response Analysis of different systems

Overall response of the system consists of summation of two responses.

$$C(t) = C_{ss}(t) + C_{tr}(t)$$

Where,

C_{ss} (t)=steady state response

C_{tr} (t)=transient sate response

Note:

- Transient response only depends on the system component and their layout not on input.
- Steady state response depends on both input as well as system components and their layout.
- Worse the transient response lesser will be the stability.

Time Response of a First Order Control System

Many systems are approximately first-order. The important feature is that the storage of mass, momentum and energy can be captured by one parameter. Examples of first-order systems are velocity of a car on the road, control of the velocity of a rotating system, electric systems where energy storage is essentially in one capacitor or one inductor, incompressible fluid flow in a pipe, level control of a tank, pressure control in a gas tank, temperature in a body with essentially uniform temperature distribution (e.g. steam filled vessel).

Subject to unit step input function

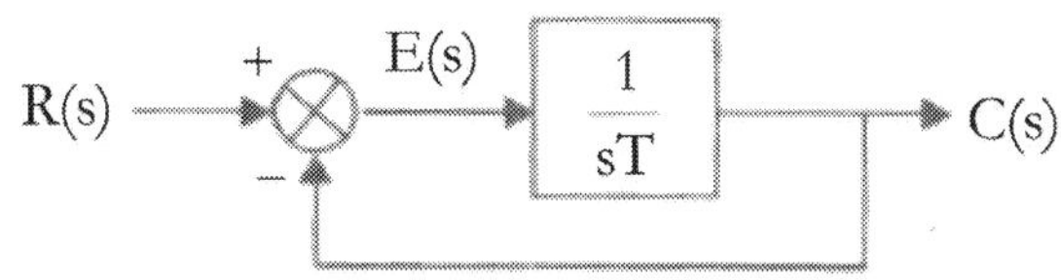

Transfer Function

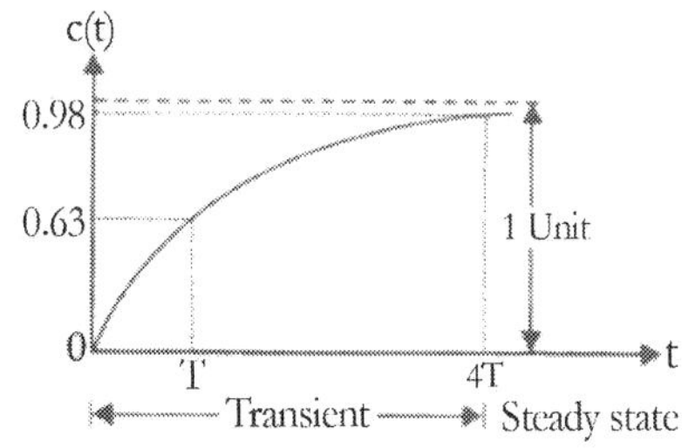

$$\frac{C(s)}{R(s)} = \frac{1}{sT+1}$$

Time response expression $C(t) = L^{-1}C(s) = 1 - e^{-t/T}$

Error $e(t) = r(t) - c(t) = e^{-t/T}$

Subject To Unit Ramp Input Function

Time response expression $C(t) = \left(t - T + Te^{-t/T}\right)$

Error $e(t) = r(t) - c(t) = \left(T - T_e^{-t/T}\right)$

Subject To Unit Impulse Input Function

Time response expression $c(t) = \frac{1}{T}e^{-t/T}$

Note:

Physically first order system may represent a RC circuit, thermal system or the like.

Smaller the time constant T, smaller the steady state error for ramp input.

Time Response of Second Order Control System

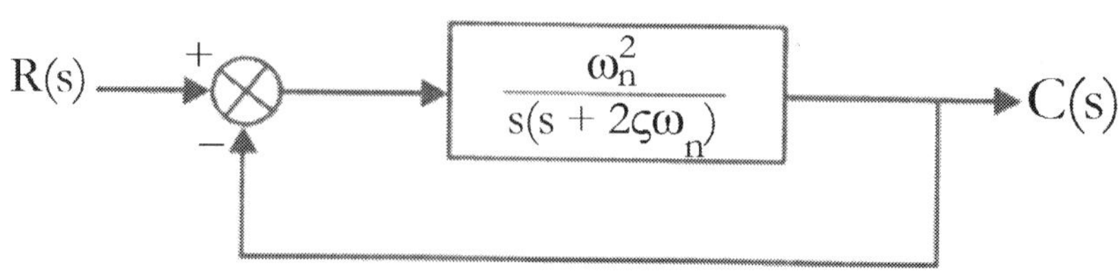

Transfer function $\frac{C(s)}{R(s)} = \frac{\omega_n^2}{s^2 + 2\zeta\omega_n s + \omega_n^2}$

Subject To Unit Step Input Function

Time response expression

Case-1: $\zeta < 1$; Under damped oscillations

$$c(t) = 1 - \frac{e^{-\zeta\omega_n t}}{\sqrt{1-\zeta^2}} \sin(\omega_d t + \phi)$$

$$\phi = \tan^{-1}\left(\frac{\sqrt{1-\zeta^2}}{\zeta}\right)$$

Where,

ω_n =Natural frequency of oscillations

ω_d =Damped frequency of oscillations

ζ =Damping ration

$\zeta\omega_n$ = Damping factor

- Damped frequency $\omega_d = \omega_n\sqrt{1-\zeta^2}$

Note:

- Response settles within 2% of the desired value (1 unit) after damping out the oscillations in a time 4T (or 4/ $\zeta\omega_n$).
- Frequency of transient oscillations is the damped natural frequency ω_d.

Case-2: ζ =0; Sustained (undamped) Oscillations

$$C(t) = (1 - \cos\omega_n t)$$

Note:

- For $\zeta = 0$ the transient response does not die out.
- If the linear system has any amount of clamping then undamped natural frequency cannot be observed experimentally.

Case-3: $\zeta = 1$; Overdamped Oscillations

$$c(t) = \left[1 - e^{-\omega_n t}\left(1 + \omega_n t\right)\right]$$

Case-4: $\zeta > 1$; Overdamped Oscillations

$$c(t) = 1 \frac{e^{-\left(\zeta - \sqrt{\zeta^2 - 1\omega_n t}\right)}}{\left(2\sqrt{\zeta^2 - 1}\right)\left(\zeta - \sqrt{\zeta^2 - 1}\right)}$$

Time constant of the response $T = \dfrac{1}{\left(\zeta - \sqrt{\zeta^2 - 1}\right)\omega_n}$

Note:

For $\zeta > 1$ there will be neither undamped oscillation nor damped oscillations.

Transient Response of Second Order Control System

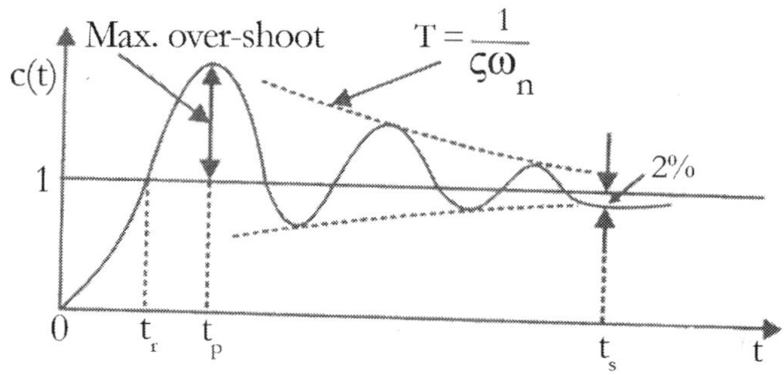

- Rise time $t_r = \dfrac{\pi - \phi}{\omega_n\sqrt{1-\zeta^2}}$
- Peak time $t_p = \dfrac{\pi}{\omega_n\sqrt{1-\zeta^2}}$

Maximum overshoot $M_p = e^{-\left(\frac{\zeta\pi}{\sqrt{1-\zeta^2}}\right)}$

Setting time

on 2% basic $t_s = 4\dfrac{1}{\zeta\omega_n}$

on 5% basic $t_s = 3\dfrac{1}{\zeta\omega_n}$

Two second order system having the same z but different ω_n will exhibit the same over shoot and same oscillatory response, such system are said to have relative stability.

Subjected To Impulse Input Function

- Time response expression

Case-1: $\zeta < 1$

$$c(t) = \frac{\omega_n}{\sqrt{1-\zeta^2}} . e^{-\zeta\omega_n t} \sin\left[\left(\omega_n \sqrt{1-\zeta^2}\right) t\right]$$

Case-2: $\zeta = 1$

$$c(t) = t\omega_n^2 e^{-\omega_n t}$$

Case-3: $\zeta > 1$

$$c(t) = \frac{\omega_n}{\sqrt{\zeta^2 - 1}} e^{-\zeta\omega_n t} \sin h\left[\left(\omega_n \sqrt{\zeta^2 - 1}\right) t\right]$$

- Characteristic roots of second order system

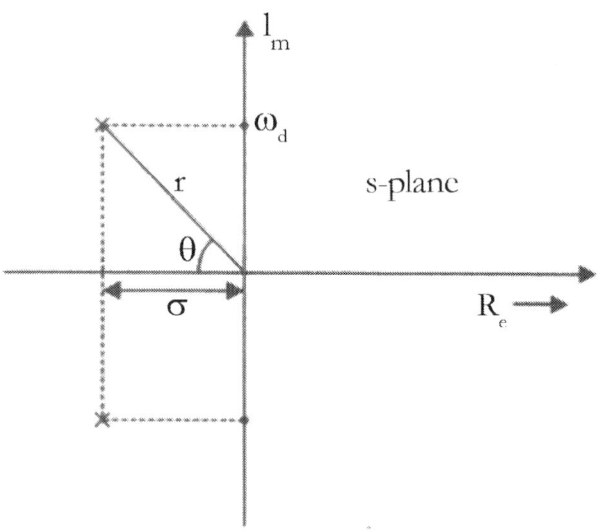

$$S^2 + 2\zeta\omega_n s + \omega_n^{\ 2} = 0$$
$$S_1, S_2 = -\sigma \pm j\omega_d$$
$$\cos\theta = \zeta$$

Radial length

$$r = \sqrt{\sigma^2 + \omega_d^2}$$
$$\Rightarrow r = \omega_n$$

Constant £ Locus

Root Locus Analysis is a graphical method for examining how the roots of a system change with variation of a certain system parameter, commonly again within a feedback system. This is a technique used as a stability criterion in the field of control systems.

Steady State Error

Steady-state error is a measure of the accuracy of a control system. The steady-state error of a control system is judged by the steady-state error due to step, ramp or acceleration input. We shall investigate a type of steady-state error which is caused by the incapability of a system to follow particular types of inputs. Any physical control system inherently suffers steady-state error in response to certain types of inputs. A system may have no steady-state error to a step input but the same system may exhibit non-zero steady-state error to a ramp input. We can, however, eliminate this error by modifying the system structure. Generally, steady state error should be as low as possible.

$$E(s) = \frac{R(s)}{1 + G(s)H(s)}$$

Static Error Coefficients

Static acceleration error coefficient is associated with e_n for unit parabolic input. The steady-state actuating error of the system with unit-parabolic input (acceleration) is given by

$$e_{ss} = \lim_{s\to 0} \frac{sR(s)}{1+G(s)H(s)}$$

$$= \lim_{s\to 0} \frac{s}{1+G(s)H(s)} \frac{1}{s^3}$$

$$= \lim_{s\to 0} \frac{s}{s^2 + s^2G(s)H(s)} = \frac{1}{K_a}$$

$$K_a = \lim_{x\to\infty} s^2G(s)H(s).$$

Sensivity

The controller parameters are typically matched to the process characteristics and since the process may change, it is important that the controller parameters are chosen in such a way that the closed loop system is not sensitive to variations in process dynamics. One way to characterize sensitivity is through the nominal sensitivity peak M_s

$$S_K^A = \frac{\partial A / A}{\partial K / K}$$

Where S_K^A =sensitivity of variable A with respect to parameter K

For Open Loop control system

$$S_G^M = \frac{G(s)}{M(s)} . \frac{\partial M(s)}{\partial G(s)} = 1$$

For Close Loop control System

$$S_G^M = \frac{1}{1+G(s)H(s)}$$

Control Actions

Action by a control mechanism that makes changes to the inputs of a manufacturing process based on the accumulated error over a period of time. Integral action controllers are often used in conjunction with proportional controllers, which make corrective changes in proportion to the amount of error in an input, in order make input adjustments faster and more accurate.

$$\omega_n' = \omega_n\sqrt{K_p}$$

$$\zeta' = \frac{\zeta}{\sqrt{K_p}}$$

W_n Increase by $\sqrt{k_p}$ ξ Decrease $\sqrt{k_p}$ m_p Increases

Derivative Controller

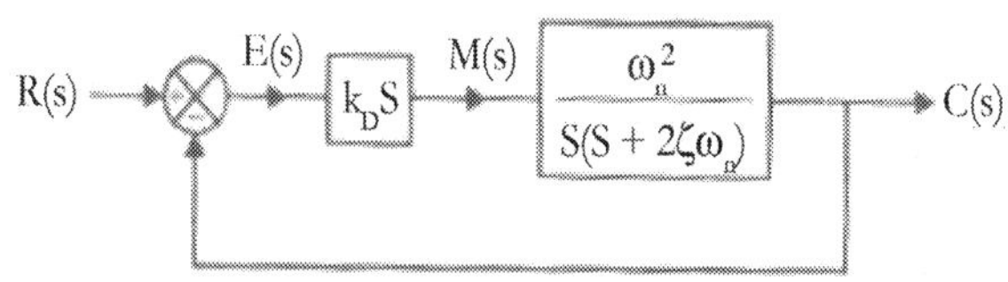

4 Stability Analysis

Stability

The stability of a general dynamical system with no input can be described with Lyapunov stability criteria.

- A linear system is called bounded-input bounded-output (BIBO) stable if its output will stay bounded for any bounded input.
- Stability for nonlinear systems that take an input is input-to-state stability (ISS), which combines Lyapunov stability and a notion similar to BIBO stability.

For simplicity, the following descriptions focus on continuous-time and discrete-time **linear** systems.

Mathematically, this means that for a causal linear system to be stable all of the poles of its transfer function must have negative-real values, i.e. the real part of all the poles are less than zero. Practically speaking, stability requires that the transfer function complex poles reside

- In the open left half of the complex plane for continuous time, when the Laplace transform is used to obtain the transfer function.
- Inside the unit circle for discrete time, when the Z-transform is used.

Root Locus

Root locus analysis is a graphical method for examining how the roots of a system change with variation of a certain system parameter, commonly a gain within a feedback system. This is a technique used as a stability criterion in the field of control systems developed by Walter R. Evans which can determine stability of the system. The root locus plots the poles of the closed loop transfer function in the complex s plane as a function of a gain parameter.

$$1+G(z)H(z)=0$$
$$1+GH(z)=0$$
$$1+L(z)=0$$

Angle of Asymptotes

It is a normal occurrence to have less finite zeros than finite poles, so often a system will have infinite zeros. For each infinite zero, the root locus will have a branch that travels to the infinite zero along an asymptote. So there will be one asymptote for each infinite zero. But where are the infinite zeros? This rule answers that question.

$$\theta = \pm \frac{360^\circ \times q}{(P-Z)}$$

Q=0, 1, 2, 3

Angles of Deparyres

The angle of departure is the angle at which the locus leaves a pole in the s-plane. The angle of arrival is the angle at which the locus arrives at a zero in the s-plane. By convention, both types of angles are measured relative to a ray starting at the origin and extending to the right along the real axis in the s-plane

Angle of arrival

Measurement is a method for determining the direction of propagation of a radio-frequency wave incident on an antenna array. Aoa determines the direction by measuring the time difference of arrival Aoa at individual elements of the array from this delay.

Bode Plot

A Bode plot is a graph of the frequency response of a system. It is usually a combination of a Bode magnitude plot, expressing the magnitude of the frequency response, and a Bode phase plot, expressing the phase shift. Both quantities are plotted against a horizontal axis proportional to the logarithm of frequency. The magnitude axis of the Bode plot is usually expressed as decibels of power, that is by the 20 log rule: 20 times the common (base 10) logarithm of the amplitude gain. With the magnitude gain being logarithmic, Bode plots make multiplication of magnitudes a simple matter of adding distances on the graph (in decibels), since

$$\log(a.b) = \log(a) + \log(b)$$

Phase Margin

The phase margin (pm) is the difference between the phase, measured in degrees, and 180°, for an amplifier's output signal (relative to its input), as a function of frequency. Typically the open-loop phase lag (relative to input) varies with frequency, progressively increasing to exceed 180°, at which frequency the output signal becomes inverted.

Gain Margin

Gain Margin: Greater will the gain margin greater will be the stability of the system. It refers to the amount of gain, which can be increased or decreased without making the system unstable. It is usually expressed in dB.

Gain Margin

Gain Margin: The gain margin is the factor by which the gain |G(jω)| needs to be increased for the closed-loop system to be neutrally stable. The gain margin (GM) is defined as

$$GM = \left|G\left(j\omega_{180^\circ}\right)\right|^{-1}$$

Where, ω180◦ is such that G(jω180◦) = 180◦ The gain margin is normally expressed in dB

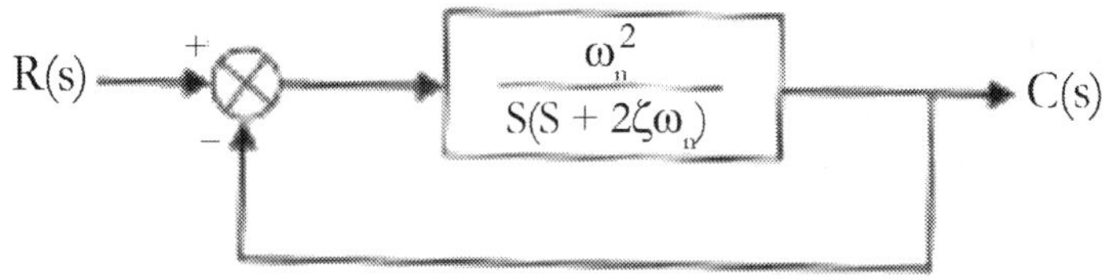

Phase Margin

Phase Margin: Phase margin (PM) is the amount by which the phase of G(jω) exceeds 180◦ when |G(jω)| = 1. It is defined as, PM = G(jω$_c$) − 180◦ where, the frequency ω_c is such that |G(jω$_c$)| = 1. It is also called the phase cross-over frequency.

$$PM = \angle G\left(j\omega_c\right) - 180^\circ$$

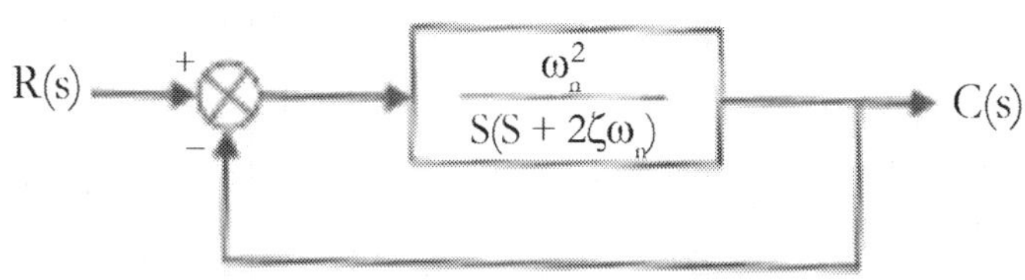

Constant Magnitudes loci(M-Circle)

The method discussed above is complicated to obtain closed loop frequency respon t us study how to obtain constant magnitude loci and constant phase angle loci for sed loop transfer function. Such loci are called m and n circles. These circles are vi ich convenient in determining the dosed loop frequency response from the polar plot.

C(s) G(s)

R(s)=1+G(s)

Now is' is the complex variable hence Gaci is also a complex variable.

G(joi)=X+jY

Radius of M Circles

$$\text{Radius} = \frac{M}{1-M^2}$$

Center of M circles

$$\text{Centre} = \left(\frac{M^2}{1-M^2}, 0\right)$$

Constant of Phase Angles Loci (N- circles)

Radius of n - circles

$$\text{Radius} = \sqrt{\frac{1}{4} + \frac{1}{4N^2}}$$

Centre of n-circles

$$\text{Centre} = \left(-\frac{1}{2}, \frac{1}{2N}\right)$$

Cut Off Frequency and Bandwidth

Closed-loop frequency response showed in figure 6.46 and 6.51*b* so that we can define the closed-loop frequency bandwidth and cut-off frequency. We define the bandwidth of a closed-loop control system in a manner similar to other electronic equipment such as amplifiers. The bandwidth of a closed-loop control system is defined as the frequency range where the magnitude of the closed loop gain does not drop below −3 db as shown in figure 6.54. Therefore, the bandwidth of the control system, ω_b, is defined to be that frequency range in which the magnitude of the closed-loop frequency response is greater than −3 db. The frequency ω_b is defined as the cut off frequency. At frequencies greater than ω_b, the closed-loop frequency response is attenuated by more than−3 db.

Cut off frequency

$$\omega_c = \omega_n \left[1 - 2\zeta^2 + \sqrt{4\zeta^4 - 4\zeta^2 + 2}\right]^{1/2}$$

Bandwidth

$$B.W = \omega_c$$

Second Order Systems

The transfer function of the second order system is $(\omega^2)/(s(s+2\zeta\omega))$. We are going to analyse the **transient state response of control system** for the following standard signal. The standard form of a second-order system is:

$$\frac{C(s)}{R(s)} = \frac{\omega_n^2}{s^2 + 2\zeta\omega_n s + \omega_n^2}$$

Where, ω_n and ζ are called the natural frequency (in rad/s) and damping ratio (dimensionless), respectively. The response will depend on the nature of the poles:

5 State Space Analysis

General Representations of State Model

A state-space representation is a mathematical model of a physical system as a set of input, output and state variables related by first-order differential equations. "State space" refers to the space whose axes are the state variables. The state of the system can be represented as a vector within that space.

State Equation

$$\dot{x} = Ax + Bu$$

Output equation

$$y = Cx + Du$$

Where:

$x(\cdot)$ is called the "state vector", $x(t) \in \mathbb{R}^n$

$y(\cdot)$ is called the "output vector", $y(t) \in \mathbb{R}^q$

$u(\cdot)$ is called the "input (or control) vector", $u(t) \in \mathbb{R}^p$

$A(\cdot)$ is the "state (or system) matrix", $\dim[A(\cdot)] = n \times n$

$B(\cdot)$ is the "input matrix", $\dim\left[B(\cdot)\right] = n \times p$

$C(\cdot)$ is the "output matrix", $\dim\left[C(\cdot)\right] = q \times n$

$D(\cdot)$ is the "feed through (or feed forward) matrix" (in cases where the system model does not have a direct feed through, $D(\cdot)$ is the zero matrix), $\dim\left[D(\cdot)\right] = q \times p$

State Space Representations in Canonical Forms

For any given system, there are essentially an infinite number of possible state space models that will give the identical input/output dynamics. Thus, it is desirable to have certain standardized state space model structures: these are the so-called canonical forms. Given a system transfer function, it is possible to obtain each of the canonical models. And, given any particular canonical form it is possible to transform.

$$\frac{Y(s)}{U(s)} = H(s) = \frac{b_0 S^n + b_1 S^{n-1} + b_2 S^{n-2} + \dots\dots\dots + b_n}{S^n + a_1 S^{n-1} + a_2 S^{n-2} + \dots\dots\dots + a_n}$$

$$\begin{bmatrix} \dot{x}_1 \\ \dot{x}_2 \\ \dot{x}_3 \\ \vdots \\ \dot{x}_n \end{bmatrix} = \begin{bmatrix} 0 & 0 & \dots & 0 & -a_n \\ 1 & 0 & \dots & 0 & -a_{n-1} \\ 0 & 1 & \dots & 0 & -a_{n-2} \\ \vdots & \vdots & \ddots & \vdots & \vdots \\ 0 & 0 & \dots & 1 & -a_1 \end{bmatrix} \begin{bmatrix} x_1 \\ x_2 \\ x_3 \\ \vdots \\ x_n \end{bmatrix} + \begin{bmatrix} b_n - a_n b_0 \\ b_{n-1} - a_{n-1} b_0 \\ b_{n-2} - a_{n-2} b_0 \\ \dots \\ b_1 - a_1 b_0 \end{bmatrix} u$$

$$y = [0\ 0\ 0\ 0 \ldots\ldots..1] \begin{bmatrix} x_1 \\ x_2 \\ x_3 \\ . \\ . \\ . \\ x_n \end{bmatrix} + b_0 u$$

State Transitions Matrix

In control theory, the state-transition matrix is a matrix whose product with the state vector x at an initial time t_0 gives x at a later time t. The state-transition matrix can be used to obtain the general solution of linear dynamical systems.

$$\phi(t) = e^{At} = 1 + At + \frac{A^2t^2}{\underline{|2}} + \frac{A^3t^3}{\underline{|3}} + \ldots\ldots\ldots.$$

$$\phi(t) = L^{-1}\left[(sI - A)\right]^{-1}$$

Lead Compensators

Lead compensators are sometimes the best controller to use to get a system to do what you want it to do. It's as simple as that. They are an option that you may need if you cannot use anything in the PID family to bring a system's performance within specifications. There's no guarantee that a lead compensator will do the trick, but it is another weapon in the arsenal.

$$\frac{V_0(s)}{V_i(s)} = \frac{a(1+Ts)}{1+Ts}$$

$$\alpha = \frac{R_2}{R_1 + R_2}$$

$$T = R_1C$$

Lag Compensators

Lag compensators are sometimes the best controller to use to get a system to do what you want it to do. Like other compensators, lag compensators can be used to adjust frequency response by adding equal numbers of poles and zeroes to systems. Those added singularities may possibly be manipulated to give better stability, better performance and general improvement. Lag compensators are one tool that should be in the toolkit of every control system designer.

$$\frac{V_0(s)}{V_i(s)} = \frac{1+Ts}{1+\beta Ts}$$

$$\alpha = \frac{R_2}{R_1 + R_2}$$

$$T = R_2C$$

Lead and Lag Compensator

Lead–lag compensators influence disciplines as varied as robotics, satellite control, automobile diagnostics, and laser frequency stabilization. They are an important building block in analog control systems, and can also be used in digital control.

$$\frac{V_0(s)}{V_i(s)} = \frac{\alpha(1+T_1 s)(1+T_2 s)}{(1+\alpha T_1 s)(1+\beta T_2 s)}$$

$$\alpha = \frac{R_2}{R_1 + R_2}$$

$$\beta = \frac{R_1 + R_2}{R_2}$$

$$T_1 = R_1 C_1$$

$$T_2 = R_2 C_2$$

A Handbook on Electronics and Communications Engineering

7

Electromagnetic Theory

CONTENTS

1 Co-ordinate System and Transformation

The analysis of most semiconductor devices includes the calculation of the electrostatic potential within the device as a function of the existing charge distribution. Electromagnetic theory and more specifically electrostatic theory are used to obtain the potential.

CO-ORDINATE SYSTEM AND TRANSFORMATION

In geometry, a coordinate system is a system which uses one or more numbers, or coordinates, to uniquely determine the position of a point or other geometric element on a manifold such as Euclidean space. The order of the coordinates is significant and they are sometimes identified by their position in an ordered tuple and sometimes by a letter, as in "the x-coordinate". The coordinates are taken to be real numbers in elementary mathematics, but may be complex numbers or elements of a more abstract system such as a commutative ring. The use of a coordinate system allows problems in geometry to be translated into problems about numbers and vice versa; this is the basis of analytic geometry.

Cartesian Co-ordinate System(X, Y, Z)

Cartesian coordinates, also called rectangular coordinates, provide a method of rendering graphs and indicating the positions of points on a two-dimensional (2D) surface or in three-dimensional (3D) space. The Cartesian coordinate system is used to define positions on computer displays and in virtual reality (VR) renderings.

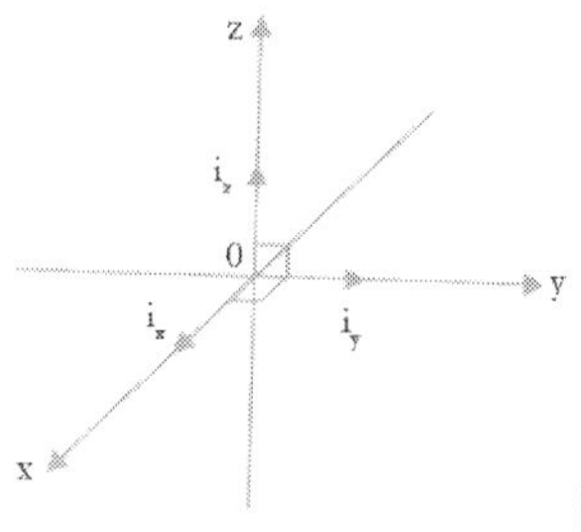

$$\begin{cases} -\infty < x < \infty \\ -\infty < y < \infty \\ -\infty < z < \infty \end{cases}$$

A vector drawn from the origin to an arbitrary point P(x, y, z) is called the position vector of point P.

Position vector of any arbitrary point(x, y, z)

$$\overline{r} = xi_x + yi_y + zi_z$$

Displacement vector

$$dl = dxi_x + dyi_y + dzi_z; \ \left|\overline{dl}\right| = \sqrt{(dx)^2 + (dy)^2 + (dz)^2}$$

Unit Vector

$$\vec{r} = xi_x + yi_y + zi_z$$
$$|\vec{r}| = \sqrt{x^2 + y^2 + z^2}$$

Differential length elements

$$dx\, i_x, dy\, i_y, dz\, i_z$$

Differential normal surface areas

$$d\vec{S} = dy\, dz\, i_x$$
$$dz\, dx\, i_y$$
$$dx\, dy\, i_z$$

Differential volume

$$dv = dx\, dy\, dz$$

Cylindrical Co-ordinate System(r, ϕ, z)

Cylindrical coordinates are obtained by replacing the x and y coordinates with the polar coordinates r and theta (and leaving the z coordinate unchanged).

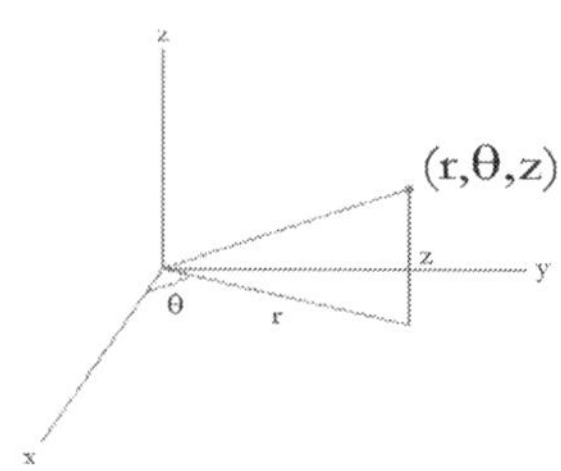

Thus, we have the following relations between Cartesian and cylindrical coordinates:

From cylindrical to Cartesian:

$$x = r\cos\theta,\ y = r\sin\theta,\ z = z$$

From Cartesian to cylindrical:

$$r = \sqrt{x^2 + y^2},\ \tan\theta = \frac{y}{x},\ z = z$$

As an example, the point (3,4,-1) in Cartesian coordinates would have polar coordinates of (5,0.927,-1).Similar conversions can be done for functions. Using the first row of conversions, the function

$$x^2 + y^2 + z^2 = 9$$

in Cartesian coordinates would have a cylindrical coordinate representation of

$$r^2 + z^2 = 9$$

Cylindrical coordinates are most convenient when some type of cylindrical symmetry is present. The surfaces r=constant, theta=constant, and z=constant are a cylinder, a vertical plane, and a horizontal plane, respectively.

Spherical Coordinates

The coordinates used in spherical coordinates are rho, theta, and phi. Rho is the distance from the origin to the point. Theta is the same as the angle used in polar coordinates. Phi is the angle between the z-axis and the line connecting the origin and the point.

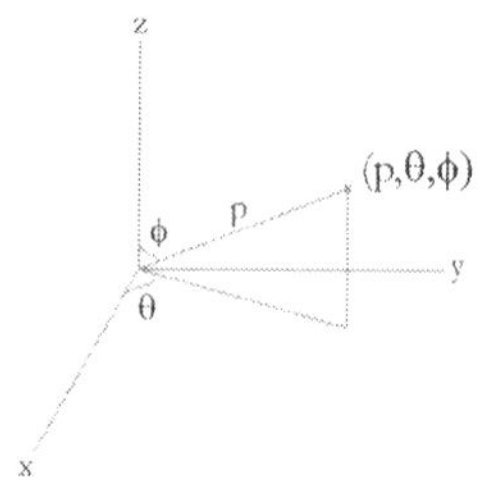

The following are the relations between Cartesian and spherical coordinates:

From spherical to Cartesian:

$$x = \rho \sin\phi \cos\theta,\ y = \rho \sin\phi \sin\theta,\ z = \rho \cos\phi$$

From Cartesian to spherical:

$$\rho = \sqrt{x^2 + y^2 + z^2},\ \tan\theta = \frac{y}{x},\ \tan\phi = \sqrt{\frac{x^2 + y^2}{z}}$$

Relations between cylindrical and spherical coordinates also exist:

From spherical to cylindrical:

$$r = \rho \sin\phi,\ \theta = \theta,\ z = \rho \cos\phi$$

From cylindrical to spherical:

$$\rho = \sqrt{r^2 + z^2},\ \theta = \theta,\ \tan\phi = \frac{r}{z}$$

The point (5, 0, 0) in Cartesian coordinates has spherical coordinates of (5, 0, 1.57). The surfaces pho=constant, theta=constant, and phi=constant are a sphere, a vertical plane, and a cone (or horizontal plane), respectively. Spherical coordinates are of course very useful when any type of spherical symmetry is present.

It should be noted that formulas for objects like the gradient, curl, and divergence have a different form in different coordinate systems. Always be careful to use the proper formula when dealing with these objects.

2 Vector Calculus

Vector calculus (or vector analysis) is a branch of mathematics concerned with differentiation and integration of vector fields, primarily in 3-dimensional Euclidean space R^3. The term "vector calculus" is sometimes used as a synonym for the broader subject of multivariable calculus, which includes vector calculus as well as partial differentiation and multiple integration. Vector calculus plays an important role in differential geometry and in the study of partial differential equations.

Vector Products

The vector product and the scalar product are the two ways of multiplying vectors which see the most application in physics and astronomy. The magnitude of the vector product of two vectors can be constructed by taking the product of the magnitudes of the vectors times the sine of the angle (<180 degrees) between them. The magnitude of the vector product can be expressed in the form:

$$\vec{A} \times \vec{B}_{\text{magunitude}} = AB \sin\theta$$

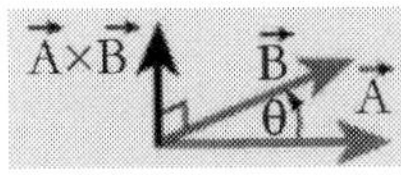

$\vec{A}\times\vec{B}$ is perpendicular to both A and B

And the direction is given by the right-hand rule. If the vectors are expressed in terms of unit vectors i, j, and k in the x, y, and z directions, then the vector product can be expressed in the rather cumbersome form:

$$\vec{A}\times\vec{B}=\vec{i}\left(A_yB_z-A_zB_y\right)-\vec{j}\left(A_xB_z-A_zB_x\right)+\vec{k}\left(A_xB_y-A_yB_x\right)$$

Which may be stated somewhat more compactly in the form of a determinant?

Spherical Coordinate System (r,θ,ϕ)

Spherical coordinates can be a little challenging to understand at first. Spherical coordinates determine the position of a point in three-dimensional space based on the distance ρ from the origin and two angles θ and ϕ. If one is familiar with polar coordinates, then the angle θ is not too difficult to understand as it is essentially the same as the angle θ from polar coordinates. But some people have trouble grasping what the angle ϕ is all about.

Relationship between spherical and Cartesian coordinates

Spherical coordinates are defined as indicated in the following figure, which illustrates the spherical coordinates of the point P.

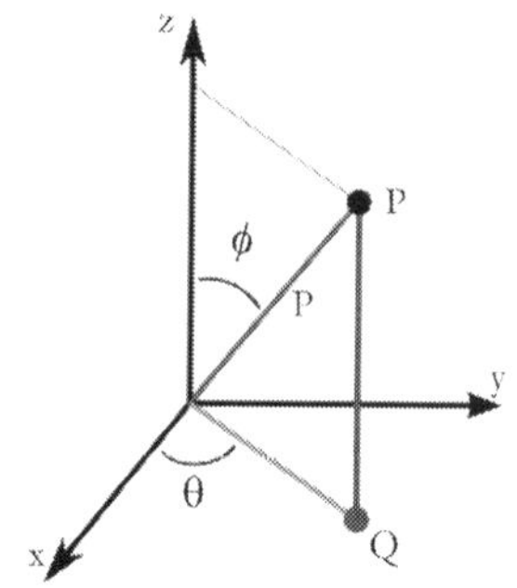

The coordinate ρ is the distance from P to the origin. If the point Q is the projection of P to the xy-plane, then θ is the angle between the positive x-axis and the line segment from the origin to Q. Lastly, ϕ is the angle between the positive z-axis and the line segment from the origin to P.

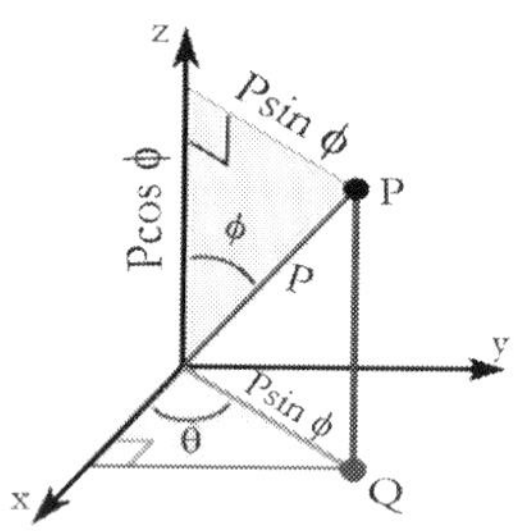

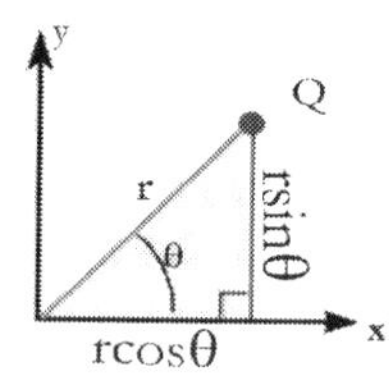

Position vector of an arbitrary point P $\vec{r} = r i_r$

Displacement vector

$$d\vec{l} = dr\, i_r + r d\theta\, i_\theta + r \sin\theta\, d\phi\, i_\phi$$

$$\left|\vec{dl}\right| = \sqrt{(dr)^2 + (rd\theta)^2 + (r\sin\theta d\phi)^2}$$

Differential length elements

$$dr\, i_r,\quad rd\theta\, i_\theta,\quad r \sin\theta\, d\phi\, i_\phi$$

Differential normal surface areas

$$d\vec{S} = r^2 \sin\theta\, d\theta\, d\phi\, i_r$$
$$r \sin\theta\, dr\, d\phi\ \ i_\theta$$
$$r\, dr\, d\theta\quad i_z$$

Differential volume

$$dv = r^2 \sin\theta\, dr\, d\theta\, d\phi$$

Cylindrical Co-Ordinate System

A cylindrical coordinate system is a three-dimensional coordinate system that specifies point positions by the distance from a chosen reference axis, the direction from the axis relative to a chosen reference direction, and the distance from a chosen reference plane perpendicular to the axis. The latter distance is given as a positive or negative number depending on which side of the reference plane faces the point.

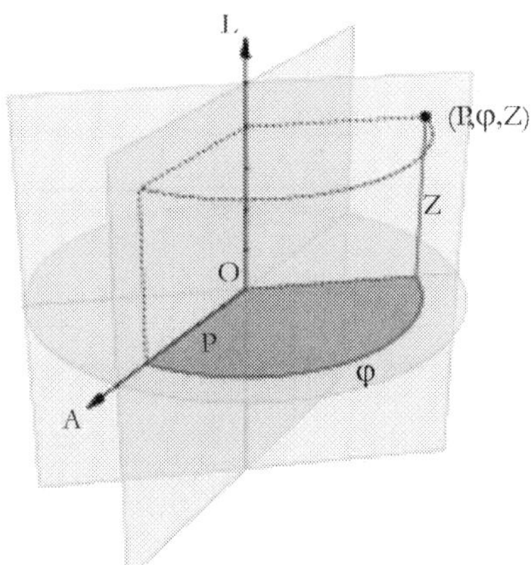

Postion Vector an Orbitary point $P(r, \phi, z)$

$$\vec{r} = ri_r + zi_z$$

Displacement Vector

$$d\vec{l} = dri_r + rd\phi i_\phi + dzi_z$$

$$\left|d\vec{l}\right| = \sqrt{(dr)^2 + (rd\phi)^2 + (dz)^2}$$

Differential Length Elements

$$dr\, i_r,\ rd\phi\, i_\phi,\ dz\, i_z$$

Differential Normal Surface Area

$$\begin{aligned} d\vec{S} = r^2\ d\phi\ dz\ i_r \\ dr\ dz\quad i_\phi \\ r\ d\phi\ dr\quad i_z \end{aligned}$$

Differential Volume

$$dv = rdr\ d\phi\ dz$$

Spherical Coordinate Systems (r, θ, ϕ)

A spherical coordinate system is a coordinate system for three-dimensional space where the position of a point is specified by three numbers: the radial distance of that point from a fixed origin, its polar angle measured from a fixed zenith direction, and the azimuth angle of its orthogonal projection on a reference plane that passes through the origin and is orthogonal to the zenith, measured from a fixed reference direction on that plane. The radial distance is also called the radius or radial coordinate. The polar angle may be called co-latitude, zenith angle, normal angle, or inclination angle.

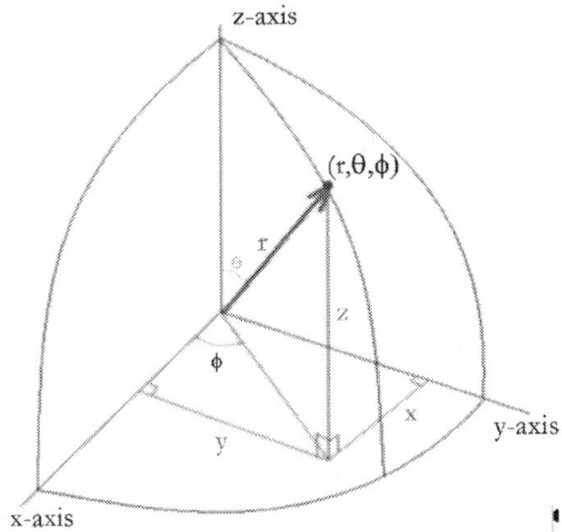

$$\begin{cases} 0 & \leq r & < \infty \\ 0 & \leq \theta & \leq \pi \\ \theta & \leq \phi & < 2\pi \end{cases}$$

Cross Product

The cross product or vector product is a binary operation on two vectors in three-dimensional space and is denoted by the symbol ×. The cross product a × b of two linearly independent vectors a and b is a vector that is perpendicular to both and therefore normal to the plane containing them. It has many applications in mathematics, physics, engineering, and computer programming. The cross product is ant commutative (a × b = −b × a) and is distributive over addition (a × (b + c) = a × b + a × c). The space R^3 together with the cross product is an algebra over the real numbers,

$$\vec{A} \times \vec{B} = |\vec{A}||\vec{B}| \sin\theta.\hat{i}_n$$

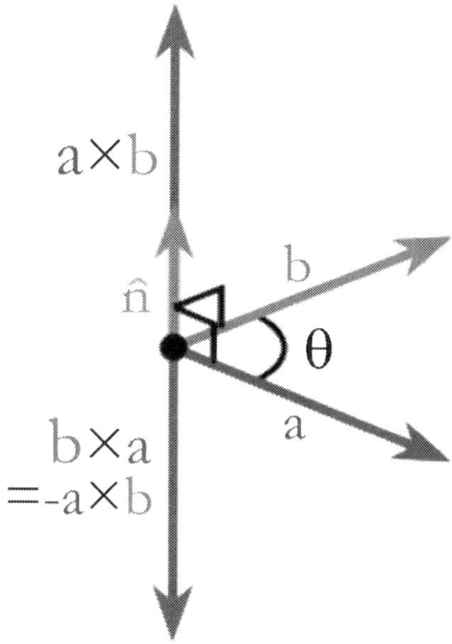

Scalar Triple Products

The scalar triple product of three vectors a, b, and c is (a×b) c. It is a scalar product because, just like the dot product, it evaluates to a single number. (In this way, it is unlike the cross product, which is a vector.) The scalar triple product is important because its absolute value | (a×b)·c| is the volume of the parallelepiped spanned by a, b, and c (i.e., the parallelepiped whose adjacent sides are the vectors a, b, and c).

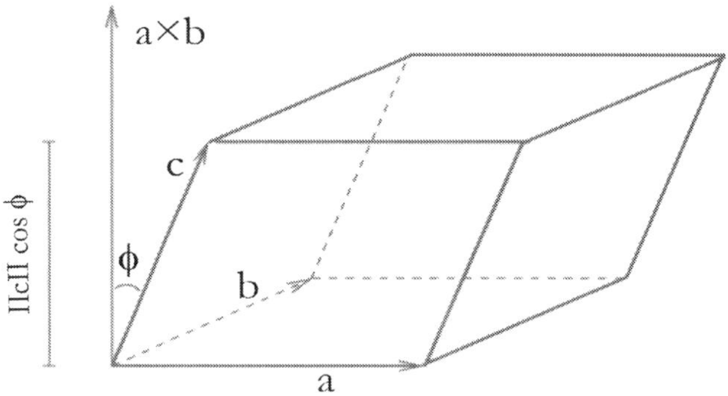

$$\vec{A}\times\left(\vec{B}\times\vec{C}\right)=\vec{B}.\left(\vec{C}\times\vec{A}\right)-\vec{C}.\left(\vec{A}\times\vec{B}\right)$$

Differentiation of Vectors

A curve C is defined by r = r (t), a vector-valued function of one (scalar) variable. Let us imagine that C is the path taken by a particle and t is time. The vector r (t) is the position vector of the particle at time t and r (t + h) is the position vector at a later time t + h. The average velocity of the particle in the time interval [t, t + h] is then

$$\vec{A}=A_x i_x + A_y i_y + A_z i_z$$

$$d\vec{A}=dA_x i_x + dA_y i_y + dA_z i_z$$

Gradient of Scalar Field V

The gradient of a scalar field is a vector field and whose magnitude is the rate of change and which points in the direction of the greatest rate of increase of the scalar field. If the vector is resolved, its components represent

the rate of change of the scalar field with respect to each directional component. Hence for a two-dimensional scalar field Ø (x, y). For a three dimensional scalar, its gradient is given by: $\text{grad}(V) = \underline{a}_n \frac{dV}{dn} = \nabla V$
Gradient is a vector that represents both the magnitude and the direction of the maximum space rate of increase of a scalar. dV = (∇V) $d\underline{l}$, where $d\underline{l} = \underline{a}_i \, d\underline{l}$

$$\text{grad}\phi(x,y) = \nabla\phi(x,y) = \left(\frac{\partial}{\partial x}, \frac{\partial}{\partial y}\right)\phi = \left(\frac{\partial \phi}{\partial x}, \frac{\partial \phi}{\partial y}\right)$$

3 Electrostatic Field

Electric field is defined as the electric force per unit charge. The direction of the field is taken to be the direction of the force it would exert on a positive test charge. The electric field is radially outward from a positive charge and radially in toward a negative point charge.

Coulomb's Law

Consider two point charges q_1 and q_2 placed at a distance of r from each other. Let the electrostaticforce between them is F

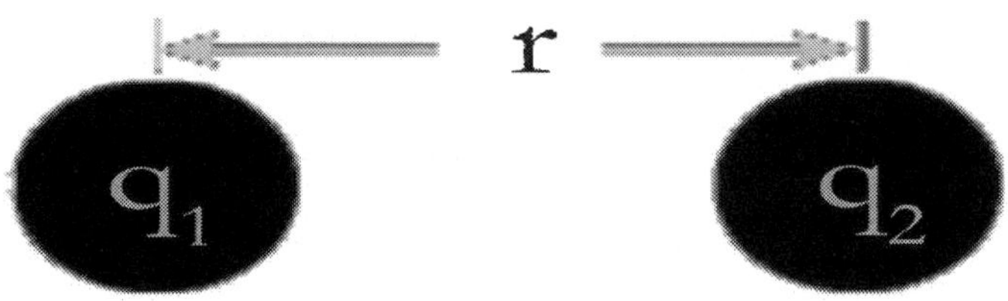

$$F \,\alpha\, q_1 q_2$$

$$F \,\alpha\, \frac{1}{r^2}$$

$$F \,\alpha\, \frac{q_1 q_2}{r^2}$$

or

$$F = K \frac{q_1 q_2}{r^2}$$

$$F = \frac{1}{4\pi\varepsilon_o} \frac{q_1 q_2}{r^2}$$

Electric Field Intensity

Electric Field Intensity E(r) The static Electric Field Intensity (or Electrostatic Field Intensity) E(r) = E(x, y, z), at any fixed point $r = \hat{1}_x x + \hat{1}_y y + \hat{1}_z z$ in space is defined as the limiting force per unit charge exerted on a test charge q at that point as the magnitude of the test charge goes to zero: E(r) ≡ lim q→0 F(r) q (N/C = V/m) (5) The limit q → 0 is introduced in order that the test charge does not influence the charge sources that produce the electrostatic field

$$\boxed{E(r) \equiv \lim_{q \to 0} \frac{F(r)}{q}} \left(N/C = V/m\right)$$

Electric Flux Density

A measure of the strength of an electric field generated by a free electric-charge, corresponding to the number of electric lines of force passing through a given area. It is equal to the electric field strength multipliedby the permittivity of the material through which the electric field extends. It is measured in coulombs per square meter. Also called

$$D = \varepsilon E$$

$$E = \frac{q_1}{4\pi\varepsilon R^2}$$

$$D = \varepsilon E = \frac{q_1}{4\pi\varepsilon R^2}$$

Gauss Law

The electric flux out of any closed surface is proportional to the total charge enclosed within the surface. The integral form of Gauss' Law finds application in calculating electric fields around charged objects. In applying Gauss' law to the electric field of a point charge, one can show that it is consistent with Coulomb's law. While the area integral of the electric field gives a measure of the net charge enclosed, the divergence of the electric field gives a measure of the density of sources. It also has implications for the conservation of charge.

Intgral from

$$\oint \vec{E}.\vec{dA} = \frac{q}{\varepsilon_0} = 4\pi kq$$

Differential from

$$\nabla.E = \frac{\rho}{\varepsilon_0} = 4\pi kq$$

Electric Potential

Electric potential, the amount of work needed to move a unit charge from a reference point to a specific point against an electric field. Typically, the reference point is the Earth, although any point beyond the influence of the electric field charge can be used. The forces acting on a positive charge q located between two plates, A and B, of an electric field E. The electric force F exerted by the field on the positive charge is F= qE to move the charge from plate A to plate B, an equal and opposite force (F′= qE) must then be applied. The work W done in moving the positive charge through a distance d is W= F′d= qEd.

$$V = \frac{kQ}{r} = \frac{Q}{4\pi\varepsilon_0 r}$$

$$V_a - V_b = kQ\left[\frac{1}{r_a} - \frac{1}{r_b}\right]$$

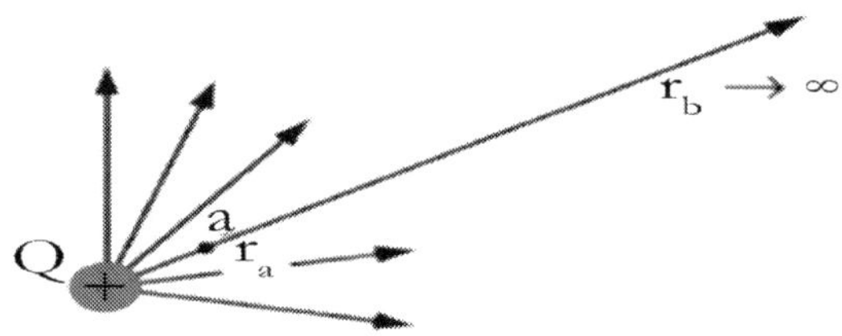

Electric dipole

Electric dipole moment is a measure of the separation of positive and negative electrical charges in a system of electric charges, that is, a measure of the charge system's overall polarity. The SI units are Coulomb-meter (C m). This article is limited to static phenomena, and does not describe time-dependent or dynamic polarization. The magnitude of dipole moment determines the electric field strength.

$$p = qd$$

Energy Density in Electrostatic Field

Energy density is the amount of energy stored in a given system or region of space per unit volume or mass, though the latter is more accurately termed specific energy. Often only the useful or extractable energy is measured, which is to say that chemically inaccessible energy such as rest mass energy is ignored. In cosmological and other general relativistic contexts, however, the energy densities considered are those that correspond to the elements of the stress–energy tensor and therefore do include mass energy as well as energy densities

$$w_E = \frac{1}{2}\varepsilon_0 E^2 = (0.5)(8.85\times10^{-12})(2.0\times10^6)^2 = 18\ \mathrm{Jm}^{-3}$$

$$w_B = \frac{1}{2}\frac{B^2}{\mu_0} = \frac{(0.5)(1.0\times10^{-2})^2}{(4\pi\times10^{-7})} = 40\ \mathrm{Jm}^{-3}$$

$$w = w_E + w_B = (18+40) = 58\ \mathrm{Jm}^{-3}$$

Conductor

Any charged object - positive or negative, conductor or insulator - creates an electric field that permeates the space surrounding it. In the case of conductors there are a variety of unusual characteristics about which we could elaborate. That a conductor is material that allows electrons to move relatively freely from atom to atom. It was emphasized that when a conductor acquires an excess charge, the excess charge moves about and distributes itself about the conductor in such a manner as to reduce the total amount of repulsive forces within the conductor.

Boundary Condition

General boundary conditions satisfied by the electric field at the interface between two media: *e.g.*, the interface between a vacuum and a conductor? Consider an interface P between two media A and B. Let us, first of all, apply Gauss' law,

$$\oint_S E.ds = \frac{1}{\varepsilon_0}\int_V \rho \, dV,$$

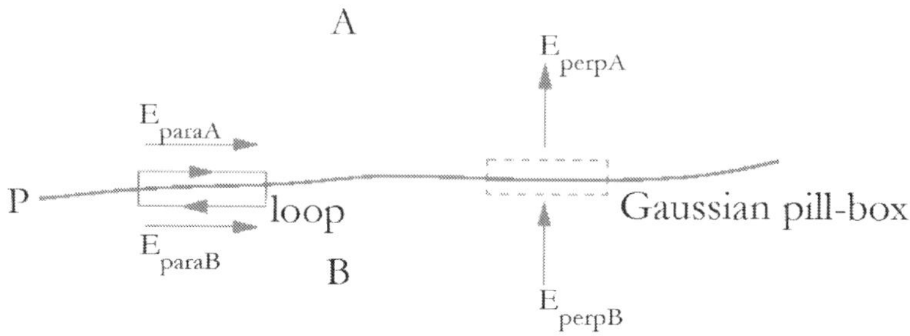

Tangential Field

Tangential component of E by applying Faraday's law to a small rectangular loop positioned across the boundary, and in the plane of E_1 and E_2.

$$\oint E.dI \rightarrow \int_a^b E_1.dI + \int_c^d E_2.dI = 0$$

Normal Field

Normal field or Normal electric field is the vector component of an electric field which is normal or perpendicular to the surface. It is denoted by the symbol: En if E is the electric field across the surface.

$$\oint B.dS = 0$$

$$\oint B.dS \rightarrow B_1.\hat{n}\ \Delta s + B_2.(-\hat{n})\Delta s = B_{n1}\Delta s - B_{n2}\Delta s$$

4 Magnetostaic Field

Biot Sevart's Law

The Biot-Savart Law relates magnetic fields to the currents which are their sources. In a similar manner, Coulomb's law relates electric fields to the point charges which are their sources. Finding the magnetic field resulting from a current distribution involves the vector product, and is inherently a calculus problem when the distance from the current to the field point is continuously changing.

$$\overrightarrow{dB} = \frac{\mu_0 I \overrightarrow{dL} x \vec{1}_r}{4\pi r^2}$$

Where $\overrightarrow{DL}$ = Infinitesimal length of conductor carrying electric current I

$\vec{1}_r$ = Unit vector to specify the direction of the vector distance r from the current to the field point

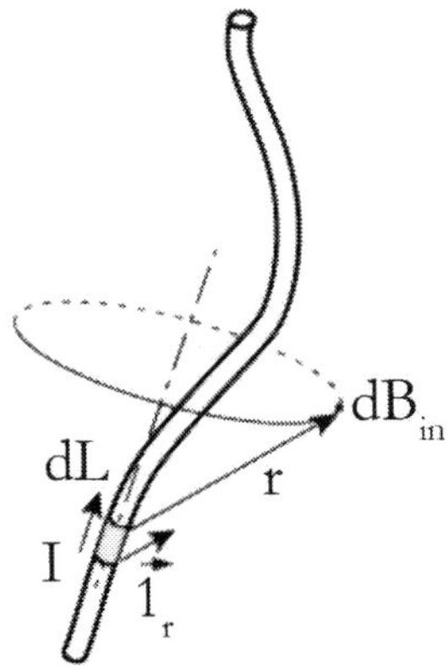

The magnetic field in space around an electric current is proportional to the electric current which serves as its source, just as the electric field in space is proportional to the charge which serves as its source. Ampere's Law states that for any closed loop path, the sum of the length elements times the magnetic field in the direction of the length element is equal to the permeability times the electric current enclosed in the loop.

$$\oint H.dL = I_{enc}$$

$$\oint H.dL = 2\pi r H = I_{enc}$$

$$\Rightarrow H = \frac{I_{enc}}{2\pi r}$$

Magnetic Flux Density

The magnetic flux per unit area perpendicular to the direction of the magnetic force: it is measured in teslas or gausses: symbol, B

$$B = \mu H$$

Where μ Permeability of free space.

Magnetic scalar and vector potential

Magnetic Scalar quantity and easier to handle than E which is a vector quantity. B is derivable from the gradient of a potential. Therefore B can be expressed as the gradient of a scalar quantity φ_m. φ_m is called as the magnetic scalar potential

$$\nabla \times E = 0$$

$$\vec{E} = -\vec{\nabla} V$$

Force Due To Magnetic Field

The force on a charged particle due to an electric field is directed parallel to the electric field vector in the case of a positive charge, and anti-parallel in the case of a negative charge. It does not depend on the velocity of the particle. In contrast, the magnetic force on a charge particle is orthogonal to the magnetic field vector, and depends on the velocity of the particle. The right hand rule can be used to determine the direction of the force.an electric field may do work on a charged particle, while a magnetic field does no work. The Lorentz force is the combination of the electric and magnetic force, which are often considered together for practical applications. Electric field lines are generated on positive charges and terminate on negative ones. The field lines of an isolated charge are directly radially outward. The electric field is tangent to these lines. Magnetic field lines, in the case of a magnet, are generated at the North Pole and terminate on a south pole. Magnetic poles do not exist in isolation. Like in the case of electric field lines, the magnetic field is tangent to the field lines. Charged particles will spiral around these field lines.

Magnatic Tarque and Magnatic Moment

The magnetic moment of a magnet is a quantity that determines the force that the magnet can exert on electric currents and the torque that amagnetic field will exert on it. A loop of electric current, a bar magnet, an electron, a molecule, and a planet all have magnetic moments.

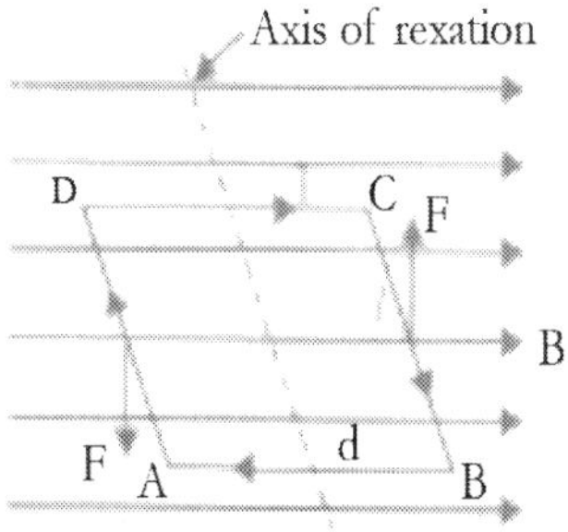

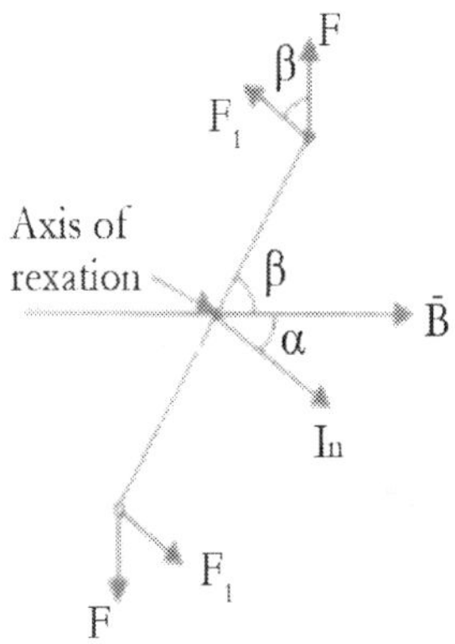

$$\boxed{T = BIS\sin\alpha}$$

Magnetic torque $\boxed{\vec{T} = \vec{m}\times\vec{B}}$ where, S= area of the loop

Magnetic dipole moment $\vec{m} = ISi_n$

Magnetic Boundary Conditions

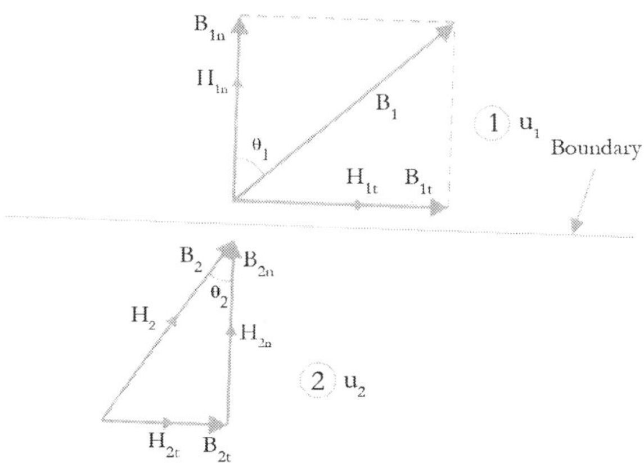

Tangential component relation $H_{1t} = H_{2t}$ or $\frac{B_{1t}}{\mu_1} = \frac{B_{2t}}{\mu_2}$

Normal component relation $B_{1n} = B_{2n}$ or $\mu_1 H_{1n} = \mu_2 H_{2n}$

- The tangential component of H is continuous while that of B is discontinuous at boundary.

The normal component of B is continuous while that of H is discontinuous at boundary. $\frac{\tan\theta_1}{\tan\theta_2} = \frac{\mu_1}{\mu_2}$

Magnetic Energy

Magnetic energy and electric energy are related by Maxwell's equations. The potential energy of a magnet of magnetic moment m in a magnetic field B is defined as the work of magnetic force (actually of magnetic torque) on re-alignment of the vector of the magnetic dipole moment, and is equal:

$$E_{p,m} = -m.B$$

while the energy stored in an inductor (of inductance L) when current I is passing via it is

$$E_{p,m} = \frac{1}{2} LI^2$$

This second expression forms the basis for superconducting magnetic energy storage.

Energy is also stored in a magnetic field. The energy per unit volume in a region of space of permeability μ mu containing magnetic field B is:

$$u = \frac{1}{2}\frac{B^2}{\mu}$$

5 Maxwell's Equations

Maxwell's equations are a set of partial differential equations that, together with the Lorentz force law, form the foundation of classical electrodynamics, classical optics, and electric circuits. These fields in turn underlie modern electrical and communications technologies. Maxwell's equations describe how electric and magnetic fields are generated and altered by each other and by charges and currents.

Faraday's law

The Maxwell-Faraday's equation version of Faraday's law describes how a time varying magnetic field creates ("induces") an electric field. This dynamically induced electric field has closed field lines just as the magnetic field, if not superposed by a static (charge induced) electric field. This aspect of electromagnetic induction is the operating principle behind many electric generators: for example, a rotating bar magnet creates a changing magnetic field, which in turn generates an electric field in a nearby wire.

TRANSFORMER AND MOTIONAL ELECTROMOTIVE FORCES

Electromotive force, also called emf is the voltage developed by any source of electrical energy such as a battery or dynamo. It is generally defined as the potential for a source in a circuit.

Sationary loop in time-varying magnetic field

Transformer emf

$$V_{emf} = \int_L \vec{E}.d\vec{l} = -\frac{d}{dt}\int_S \vec{B}.d\vec{S}$$

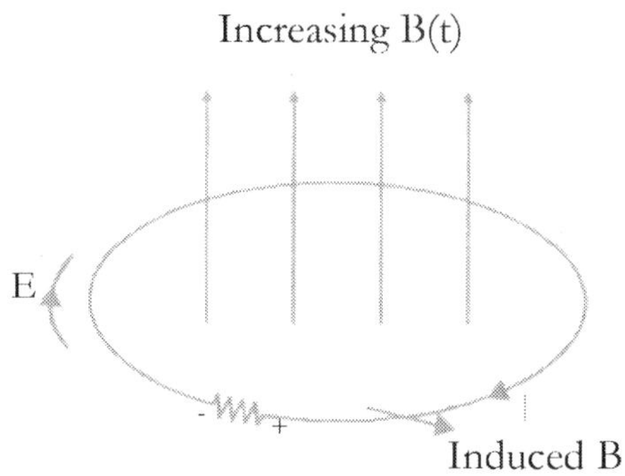

Maxwell's equations for time-varying field $\nabla \times \vec{E} = -\frac{\partial \vec{B}}{\partial t}$

Time varying electric field E is not conservative $(\nabla \times E \neq 0)$

Moving Loop in Static Magnetic Field

Motional electric field

$$\vec{E}_m = \frac{\vec{F}_m}{Q} = \vec{u} \times \vec{B}$$

Where, F_m =Force on charge moving with uniform Velocity u, I a magnetic field

Motional emf or flux-cutting emf

$$V_{emf} = \oint \vec{E}_m.d\vec{l} = \oint_L \left(\vec{u} \times \vec{B}\right).dl$$

Moving loop in time-varying field

$$V_{emf} = \oint_L \vec{E}.d\vec{l} = -\int_S \frac{\partial \vec{B}}{\partial t}.d\vec{S} + \oint_L \left(\vec{u} \times \vec{B}\right).d\vec{l}$$

$$\nabla \times \vec{E} = -\frac{\partial \vec{B}}{\partial t} + \nabla \times \left(\vec{u} \times \vec{B}\right)$$

MAXWELL'S EQUATIONS IN FINAL FORMS

Maxwell's equations serve many purposes and take many forms. On the one hand, they are used in the solution of actual real-world problems of electromagnetic fields and radiation. On the other hand, they are the subject of admiration for their elegance. There are many T-shirts, typically obtainable on college campuses, sporting various forms of these equations.

Maxwell's equations for static electric and magnetic fields

Differential (or Point from)	Integral From	Remarks
$\nabla.D = \rho_v$	$\oint_S D.dS = \int_v \rho_v dv$	Gauss's law
$\nabla.B = 0$	$\oint_S B.dS = 0$	Nonexistence of magnetic monopole

$\nabla \times E = 0$	$\oint_L E.dl = 0$	Conservative of electrostatic field
$\nabla \times H = J$	$\oint_L H.dI = \int_S J.dS$	Ampere's Law

Maxwell's equations for time-varying conditions

Differential from	Integral From	Remarks
$\nabla.D = \rho_v$	$\oint_S D.dS = \int_v \rho_v dv$	Gauss's law
$\nabla.B = 0$	$\oint_S B.dS = 0$	Nonexistence of magnetic monopole
$\nabla \times E = -\frac{\partial B}{dt}$	$\oint_L E.dl = -\frac{\partial}{\partial t}\int_S B.dS$	Faradays law
$\nabla \times H = J + \frac{\partial D}{\partial t}$	$\oint_L H.dI = \int_S \left(J + \frac{\partial D}{\partial t}\right) dS$	Ampere's Circuit Law

6 Electromagnetic Wave Propagation

Electromagnetic waves are waves which can travel through the vacuum of outer space. Mechanical waves, unlike electromagnetic waves, require the presence of a material medium in order to transport their energy from one location to another. Sound waves are examples of mechanical waves while light waves are examples of electromagnetic waves.

Wave Velocity

Wave velocity is a general concept, of various kinds of wave velocities, for a wave's phase and speed concerning energy (and information) propagation. The phase velocity is given as:

$$v_p = \frac{\omega}{k}$$

Where:

v_p is the phase velocity (in meters per second, m/s),

ω is the angular frequency (in radians per second, rad/s),

k is the wavenumber (in radians per meter, rad/m).

Wave Propagation in Lossy Dielectrics

ave propagation in lossy dielectrics is a general case from which wave propagation in other types of media can be derived as special cases.

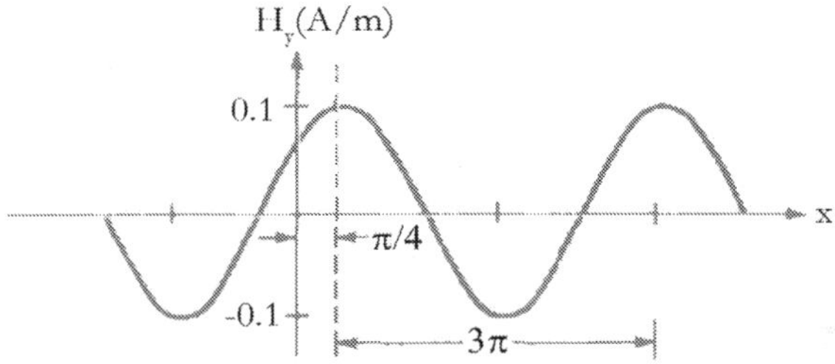

A lossy dielectric is a medium in which an EM wave loses power as it propagates due to poor conduction.

In other words, a lossy dielectric is a partially conducting medium (imperfect dielectric or imperfect conductor) with $\sigma \neq 0$, as distinct from a lossless dielectric (perfect or good dielectric) in which $\sigma = 0$.

Consider a linear, isotropic, homogeneous, lossy dielectric medium that is charge free ($\rho_v = 0$). Assuming and suppressing the time factor $e^{j\omega t}$, Maxwell's equations

$$\nabla \cdot E_s = 0$$
$$\nabla \cdot H_s = 0$$
$$\nabla \times E_s = -j\omega\mu H_s$$
$$\nabla \times H_s = (\sigma + j\omega\varepsilon) E_s$$

Taking the curl of both sides

$$\nabla \times \nabla \times E_s = -j\omega\mu \, \nabla \times H_s$$

Applying the vector identity

$$\nabla \times \nabla \times A = \nabla(\nabla.A) - \nabla^2 A$$

to the left-hand side of eq.

$$\nabla^2 E_s - \gamma^2 E_s = 0$$

Plane Waves in Lossless Dielectrics

In a lossless dielectric, $\sigma << \omega\varepsilon$ we. It is a special case of except that

$$\boxed{\sigma \simeq 0, \quad \varepsilon = \varepsilon_0 \varepsilon_r, \quad \mu = \mu_o \mu_r}$$

Substituting these into eqs and gives

$$\alpha = 0, \quad \beta = \omega\sqrt{\mu\varepsilon}$$

Also

$$u = \frac{\omega}{\beta} = \frac{1}{\sqrt{\mu\varepsilon}}, \quad \lambda = \frac{2\pi}{\beta}$$

$$\eta = \sqrt{\frac{\mu}{\varepsilon}} \angle 0^\circ$$

and thus E and H are in time phase with each other.

Plane Waves in Free Space

In free space $\sigma = 0, \in = \in_0, \mu = \mu_0$

$$\alpha = 0, \beta = \omega\sqrt{\mu_0 \in_0} = \frac{\omega}{c};$$

$$u = \frac{1}{\sqrt{\mu_0 \in_0}} = c; \lambda = \frac{2\pi}{\beta}$$

$$\eta_0 = \sqrt{\frac{\mu_0}{\in_0}} \simeq 377\Omega; \;\; i_E \times i_H = i_k$$

Where, i_k=direction of wave propagation

Skin Depth

$$E_o e^{-\alpha\delta} = E_o e^{-1}$$

through which the wave amplitude decreases by a factor e^{-I} (about 37%) is called skin depth or penetration depth

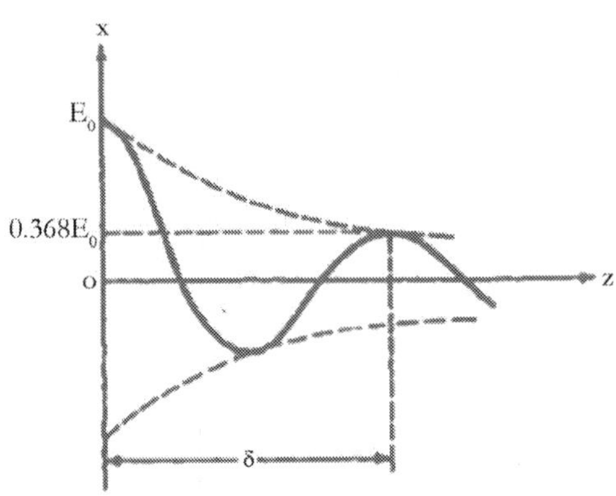

$$\delta = \frac{1}{\alpha}$$

For good conductors

$$\delta = \frac{1}{\sqrt{\pi f \mu \sigma}} \qquad \eta = \frac{1}{\sigma\delta}\sqrt{2}e^{j\pi/4} = \frac{1+j}{\sigma\delta}$$

$$E = E_o e^{-z/\delta} \cos\left(\omega t - \frac{z}{\delta}\right) a_x$$

Skin Depth in Copper*

Frequency (Hz)	10	60	100	500	10^4	10^8	10^{10}
Skin depth (mm)	20.8	8.6	6.6	2.99	0.66	6.6×10^{-3}	6.6×10^{-4}

Poynting's Theorem

Poynting's theorem is a statement of conservation of energy for the electromagnetic field, in the form of a partial differential equation. Poynting's theorem is ANALOGOUS to the work-energy theorem in classical mechanics, and mathematically similar to the continuity equation, because it relates the energy stored in the electromagnetic field to the work done on a charge distribution, through energy flux.

$$\nabla \times \underline{E} = -\frac{\partial B}{\partial t}$$

$$\nabla \times \underline{H} = \underline{J} + \frac{\partial \underline{D}}{\partial t}$$

From these we obtain

$$\underline{H} \cdot (\nabla \times \underline{E}) = -\underline{H} \cdot \frac{\partial \underline{B}}{\partial t}$$

$$\underline{E} \cdot (\nabla \times \underline{H}) = \underline{J} \cdot \underline{E} + \underline{E} \cdot \frac{\partial \underline{D}}{\partial t}$$

7 Transmission Line

A transmission line is a specialized cable or other structure designed to carry alternating current of radio frequency, that is, currents with a frequency high enough that their wave nature must be taken into account. Transmission lines are used for purposes such as connecting radio transmitters and receivers with their antennas, distributing cable television signals, trunklines routing calls between telephone switching centers, computer network connections and high speed computer data buses.

Transmission Line Equations

Short section of line δ x long. v(x, t) and i(x, t) depend on both position and time.

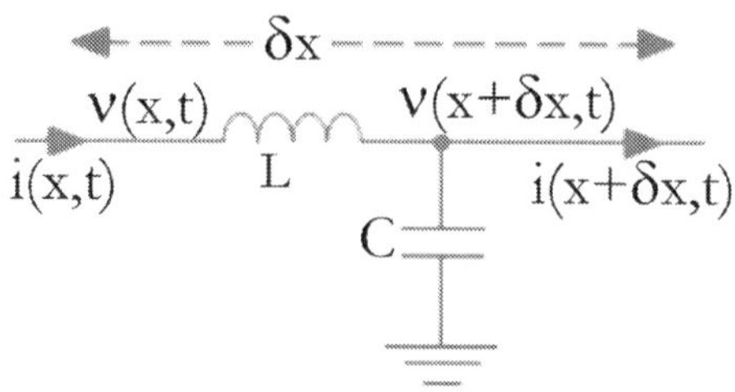

Small δ x $\Rightarrow$ignore 2nd order derivatives:

$$\frac{\partial v(x,t)}{\partial t} = \frac{\partial v(x+\delta x,t)}{\partial t} = \frac{\partial v}{\partial t}$$

Capacitor equation: $C\frac{\partial v}{\partial t} = i(x,t) - i(x+\delta x,t) = -\frac{\partial i}{\partial x}\delta x$

Inductor equation: $L\frac{\partial i}{\partial t} = v(x,t) - v(x+\delta x,t) = -\frac{\partial v}{\partial x}\delta x$

Lossless Transmission Line Model

As a signal propagates down the pair of conductors, each new section acts electrically as a small lumped circuit element. In its simplest form, called the lossless model, the equivalent circuit of a transmission line has just inductance and capacitance. These elements are distributed uniformly down the length of the line, as shown in Equivalent Circuit Model of a Lossless Transmission Line.

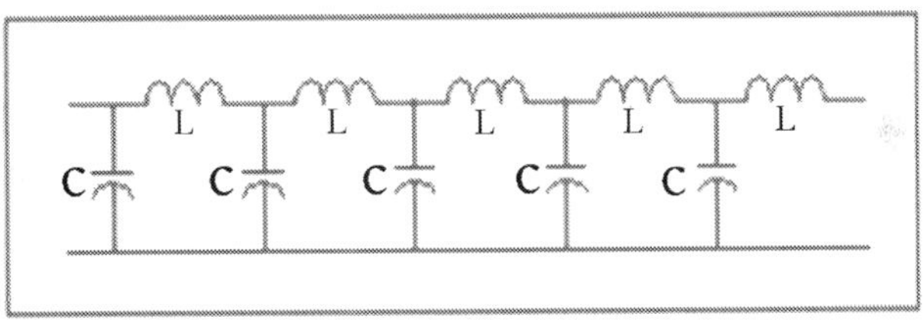

From this electrical circuit model, the two important terms that characterize a transmission line can be derived: the velocity of a signal (v) and the characteristic impedance (Z0).

$$v = \frac{1}{\sqrt{L_L C_L}}$$

and

$$z_0 = \sqrt{\frac{L_L}{C_L}}$$

LL = inductance per length

CL = capacitance per length

This is the basis for the T Element used in Star-Hspice. It accounts for a characteristic impedance (Z0) and a time delay (TD). The time delay depends on the distance (d) between the two ends of the transmission line:

$$TD = \frac{d}{v}$$

Impedance

The impedance of a device (Z) is defined as the instantaneous ratio of the voltage across the device (V) to the current through it:

$$Z = \frac{V}{I}$$

Impedance of Simple Lumped Elements

The impedance of a device can be thought of as the quality of the device that causes it to transform a current through it into a voltage across it:

$$V = ZI$$

The admittance (Y) is less often used to characterize a device. It is the inverse of the impedance:

$$Y = \frac{1}{Z} = \frac{1}{V}$$

Inductance

Mutual Inductance and Self Inductance

The most confusing, subtle and important parameter in high-speed packaging and interconnect design is inductance. It plays a key role in the origin of simultaneous switching noise, also called common ground inductance, and a key role in crosstalk between transmission line structures.

Operational Definition of Inductance

Consider an inductor to be any section of circuit element which carries current: an interconnect trace, a ground plane, a TAB lead frame, a lead in a DIP package, the lead of a resistor or a pin in a connector. An inductor does not have to be a closed circuit path, but can be a small section of a circuit path.

A changing current passing through an inductor generates a voltage drop. The magnitude of the voltage drop (Δ V) for an inductance (L) and change in current (dI/dt) is:

$$\Delta V = L\frac{dI}{dt}$$

This definition can always be used to evaluate the inductance of a section of a circuit.

$$V_L = \frac{\mu_0}{2\pi} Ln\left(\frac{s}{r} - \frac{r}{s}\right)\frac{dI}{dt}$$

The effective inductance per length of one wire is found to be:

$$L_L = \frac{\mu_0}{2\pi} Ln\left(\frac{s}{r} - \frac{r}{s}\right) \approx 5Ln\left(\frac{s}{r}\right)(s >> r)$$

Transmission Matrices

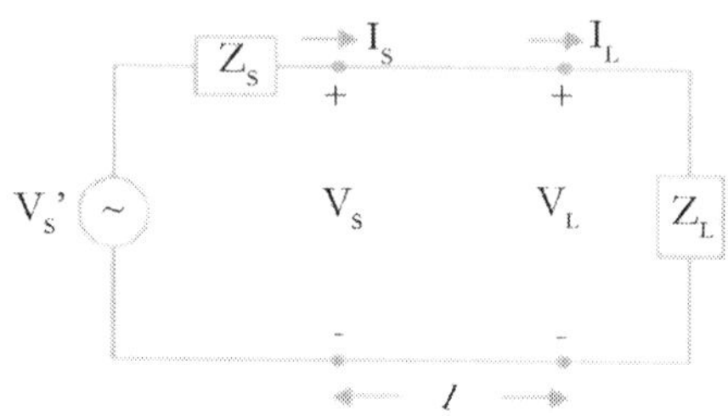

$$\begin{bmatrix} V_S \\ I_S \end{bmatrix} = \begin{bmatrix} \cos\beta l & jZ_0 \sin\beta l \\ \frac{j}{Z_0}\sin\beta l & \cos\beta l \end{bmatrix} \begin{bmatrix} V_L \\ I_L \end{bmatrix}$$

$$\begin{bmatrix} V_L \\ I_L \end{bmatrix} = \begin{bmatrix} \cos\beta l & -jZ_0 \sin\beta l \\ \frac{-j}{Z_0}\sin\beta l & \cos\beta l \end{bmatrix} \begin{bmatrix} V_S \\ I_S \end{bmatrix}$$

Smith chart

Smith chart is useful in calculating transmission line parameters graphically.

8 Antennas

Antennas couple propagating electromagnetic waves to and from circuits and devices, typically using wires or apertures.

- An antenna is an electrical conductor or system of conductors
  a. Transmission - radiates electromagnetic energy into space
  b. Reception - collects electromagnetic energy from space
- In two-way communication, the same antenna can be used for transmission and reception

Half Wave Dipole

The half wave dipole is formed from a conducting element which is wire or metal tube which is an electrical half wavelength long. It is typically fed in the center where the impedance falls to its lowest. In this way, the antenna consists of the feeder connected to two quarter wavelength elements in line with each other.

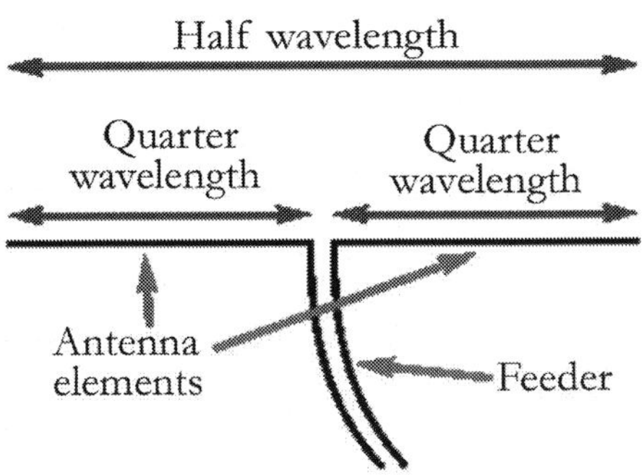

Hertzian Dipole

The Hertzian dipole or Elementary doublet refers to a theoretical construction, rather than a physical antenna design. It may be defined as a finite oscillating current (in a specified direction) of $Ie^{j\omega t}$ over a tiny or infinitesimal length $\delta\ell$ at a specified position. The solution of the fields from a Hertzian dipole can be used as the basis for analytical or numerical calculation of the radiation from more complex antenna geometries (such as practical dipoles) by forming the superposition of fields from a large number Hertzian dipoles comprising the current pattern of the actual antenna.

Also called oscillatory current element of small length.

$$E(r,\theta,\phi,t) = \left(\frac{I_m dl \sin\theta.\omega}{4\pi \in rc}\right) \sin\omega t\, e^{-j\beta r}\hat{a}_\phi$$

$$H(r,\theta,\phi,t) = \left(\frac{I_m dl \sin\theta.\omega}{4\pi \in rc}\right) \sin\omega t\, e^{-j\beta r}\hat{a}_\phi$$

Total power radiated is

$$W_r = I_{rms}^2 \times 80\pi^2 \left(\frac{dl}{\lambda}\right)^2$$

Where, dl-length of antenna

Radiation Resistance

One knows the far field radiation pattern due to a given antenna current, then it is possible to compute the radiation resistance directly. For the above fields due to the Hertzian dipole, we can compute the power flux according to the Poynting vector, resulting in a power (as averaged over one cycle) of:

$$\langle S \rangle = \frac{1}{2}\mathrm{Re}\left(E \times H^*\right)$$

Although not required, it is simplest to do the following exercise at a large r where the far field expressions for E and H apply. Consider a large sphere surrounding the source with a radius r. We find the power per unit area crossing the surface of that sphere to be in the $\hat{r}$ direction according to:

$$\langle S_r \rangle = \frac{Z_0}{2}\frac{k^2 |I|^2 (\delta\ell)^2}{(4\pi r)^2}\sin^2\theta$$

Integration of this flux over the complete sphere results in:

$$P_{net} = \int_0^{2\pi}\int_0^{\pi}\langle S_r \rangle r^2 d\phi d\theta$$

$$= \frac{Z_0}{12\pi}k^2 |I|^2 (\delta\ell)^2 = \frac{\pi Z_0}{3}|I|^2\left(\frac{\delta\ell}{\lambda}\right)^2$$

Directive Gain

Using the above expression for the radiated flux given by the Poynting vector, it is also possible to compute the directive gain of the Hertzian dipole. Dividing the total power computed above by $4\pi r^2$ we can find the flux averaged over all directions P_{avg} as

$$P_{avg} = \frac{P_{net}}{4\pi r^2} = \frac{Z_0}{48\pi^2 r^2}k^2 |I|^2 (\delta\ell)^2$$

Dividing the flux radiated in a particular direction by P_{avg} we obtain the directive gain $G(\theta)$:

$$G(\theta) = \frac{\langle S_r \rangle}{P_{avg}} = \frac{3}{2}\sin^2\theta$$

Antenna Gain

Power output, in a particular direction, compared to that produced in any direction by a perfect omnidirectional antenna (isotropic antenna)

Expressed in terms of effective area

- Related to physical size and shape of antenna
- Relationship between antenna gain and effective area

$$G = \frac{4\pi A_e}{\lambda^2} = \frac{4\pi f^2 A_e}{c^2}$$

- G = antenna gain
- Ae = effective area
- f = carrier frequency
- c = speed of light ($\approx 3 \times 10^8$ m/s)
- λ = carrier wavelength

Line-of-Sight Equations

Optical line of sight

$$d = 3.57\sqrt{h}$$

Effective, or radio, line of sight

$$d = 3.57\sqrt{Kh}$$

- d = distance between antenna and horizon (km)
- h = antenna height (m)
- K = adjustment factor to account for refraction, rule of thumb K = 4/3

Maximum distance between two antennas for LOS propagation:

$$3.57\left(\sqrt{Kh_1}+\sqrt{Kh_2}\right)$$

- h_1 = height of antenna one
- h_2 = height of antenna two

Antenna arrays

If there are N-antennas having successive phase difference α and placed at a distance d successively, then over all pjase difference ata appoint P at angle θ is

$$\Psi = \beta d \cos\theta + \alpha$$

Where, $\beta = \frac{2\pi}{\lambda}$

Maximum radiation gets for Ψ =0.

Broad side array

If maximum radiation occurs at $\pi/2$ them this is called broad side array, so maximum radiation.

Ψ =0, $\Rightarrow$ $\alpha = 0$ for broad side array (as $\theta=\pi/2$)

Endfire array

If maximum radiation occurs at $\theta=0$ or π then this is called end fire array for maximum radiation

$\Psi=0,\ 0=\beta d(\pm 1)+\alpha;\ \alpha=\pm\beta d$ for endfire arrey.

A Handbook on Electronics and Communications Engineering

8

Digital Electronics

CONTENTS

1 Number System and Codes

The binary number system is the most important one in digital systems, but several others are also important. The decimal system is important because it is universal used to represent quantities outside a digital system. This means that there will be situations where decimal values have to be converted to binary values before they are entered into the digital system.

Digital Number System

A digital system can understand positional number system only where there are only a few symbols called digits and these symbols represent different values depending on the position they occupy in the number.

Number Representation

$$(N)_b = d_{n-1}\, d_{n-2} \cdots d_i \cdots d_1\, d_0 . d_{-1} d_{-2} \cdots d_{-f} \cdots d_{-m}$$

Where N= number, b= base or radix

Weighted Number System

Weighted binary codes are those which obey the positional weighting principles, each position of the number represents a specific weight.

The BCD (Binary Coded Decimal) is a straight assignment of the binary equivalent. It is possible to assign weights to the binary bits according to their positions. The weights in the BCD code are 8,4,2,1.

Un-weighted Number System

It is the un-weighted code and it is not arithmetic codes. That means there are no specific weights assigned to the bit position. It has a very special feature that has only one bit will change, each time the decimal number is incremented as shown in fig. As only one bit changes at a time, the gray code is called as a unit distance code. The gray code is a cyclic code. Gray code cannot be used for arithmetic operation.

Decimal	BCD	Gray
0	0 0 0 0	0 0 0 0
1	0 0 0 1	0 0 0 1
2	0 0 1 0	0 0 1 1
3	0 0 1 1	0 0 1 0
4	0 1 0 0	0 1 1 0
5	0 1 0 1	0 1 1 1
6	0 1 1 0	0 1 0 1
7	0 1 1 1	0 1 0 0
8	1 0 0 0	1 1 0 0
9	1 0 0 1	1 1 0 1

Codes

In digital electronics, codes are used to communicate the information between computers. These codes represent the information symbolically as a string of bits 0 and 1 and rules defined by the code decide the arrangement of these bits.

Binary Coded Decimal Code

Binary-coded decimal (BCD) is a class of binary encodings of decimal numbers where each decimal digit is represented by a fixed number of bits, usually four or eight. Special bit patterns are sometimes used for a sign or for other indications

Binary coded decimal (BCD) is a system of writing numerals that assigns a four-digit binary code to each digit 0 through 9 in a decimal (base-10) numeral. The four-bit BCD code for any particular single base-10 digit is its representation in binary notation, as follows:

0 = 0000

1 = 0001

2 = 0010

3 = 0011

4 = 0100

5 = 0101

6 = 0110

7 = 0111

8 = 1000

9 = 1001

Excess-3 Code

Excess-3 is a digital code related to BCD that is derived by adding 3 to each decimal digit and then converting the result of that addition to 4-bit binary. Since no de-finite weights can be assigned to the four digit position, excess-3 is and un-weighted code that has advantaged in certain arithmetic operations. The excess-3 code for decimal 2 is 2+3=5= (0101)2. The excess-3 code for each decimal digit is found by the same procedure. The entire code is shown in the following table.

Decimal	Binary	Excess-3
0	0000	0011
1	0001	0100
2	0010	0101
3	0011	0110
4	0100	0111
5	0101	1000
6	0110	1001
7	0111	1010
8	1000	1011
9	1001	1100

Gray Code

The reflected binary code, also known as Gray code after Frank Gray, is a binary numeral system where two successive values differ in only one bit (binary digit). The reflected binary code was originally designed to prevent spurious output from electromechanical switches. Today, Gray codes are widely used to facilitate error correction in digital communications such as digital terrestrial television and some cable TV systems.

2 Logic Gate

A logic gate is an idealized or physical device implementing a Boolean function; that is, it performs a logical operation on one or more logical inputs, and produces a single logical output. Depending on the context, the term may refer to an ideal logic gate, one that has for instance zero rise time and unlimited fan-out, or it may refer to a non-ideal physical device.

Not Gate

In digital logic, an inverter or NOT gate is a logic gate which implements logical negation. The digital Logic NOT Gate is the most basic of all the logical gates and is sometimes referred to as an Inverting Buffer or simply a Digital Inverter. It is a single input device which has an output level that is normally at logic level "1" and goes "LOW" to a logic level "0" when its single input is at logic level "1", in other words it "inverts" (complements) its input signal. The output from a NOT gate only returns "HIGH" again when its input is at logic level "0" giving us the Boolean expression of: $\overline{A} = Q$.

The truth table and symbol are shown in follows:

INPUT	OUTPUT
A	NOT A
0	1
1	0

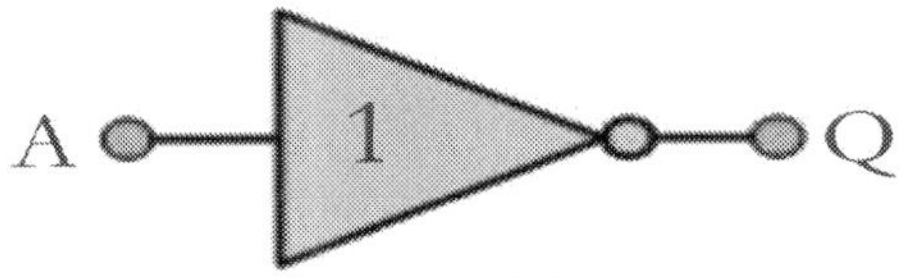

Inverter or NOT Gate

Transistor Circuit

A simple 2-input logic NOT gate can be constructed using a RTL Resistor-transistor switches as shown below with the input connected directly to the transistor base. The transistor must be saturated "ON" for an inverted output "OFF" at Q.

A	OUT
0	1
1	0

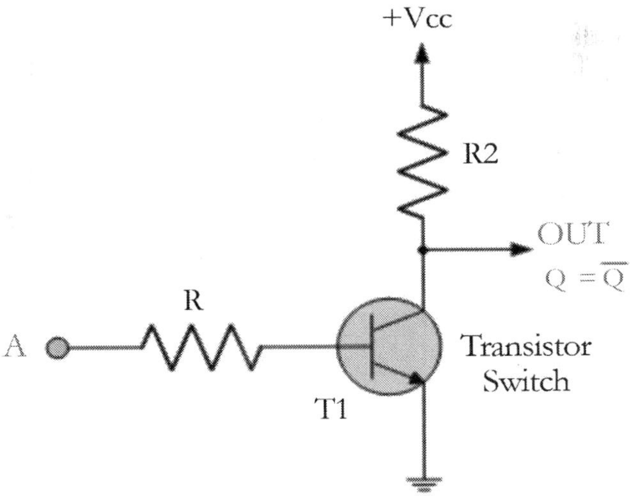

For an A stable Multivibrator (AMV)

Time period of AMV= $T = 2nt_{pd}$

Where

n= number of inverters (Not Gates)

t_{pd} = propagation delay time of each inverter

T= Time period of a square wave generated by AMV or Ring oscillator

Frequency of AMV

$$f = \frac{1}{T} = \frac{1}{2nt_{pd}}$$

AND Gate

The AND gate is a basic digital logic gate that implements logical conjunction - it behaves according to the truth table to the right. A high output (1) results only if both the inputs to the AND gate are high (1). If neither or only one input to the AND gate is high, a low output results. In another sense, the function of AND effectively finds the minimum between two binary digits, just as the OR function finds the maximum. Therefore, the output is always 0 except when all the inputs are 1s.

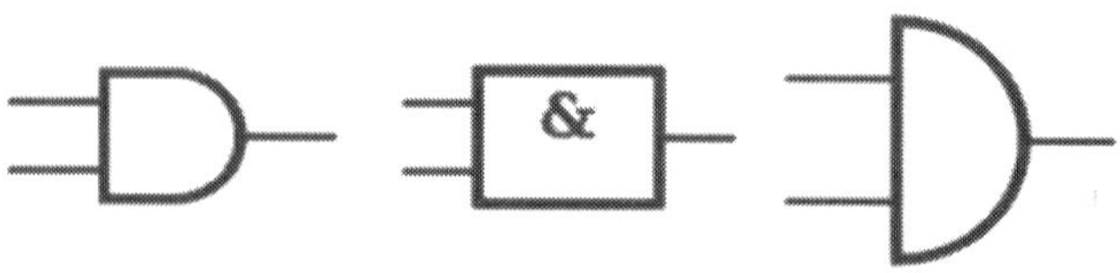

MIL/ANSI Symbol IEC Symbol DIN Symbol

Input		Output
A	B	A AND B
0	0	0
0	1	0
1	0	0
1	1	1

Transistor Circuit

A simple 2-input logic AND gate can be constructed using RTL Resistor-transistor switches connected together as shown below with the inputs connected directly to the transistor bases. Both transistors must be saturated "ON" for an output at Q.

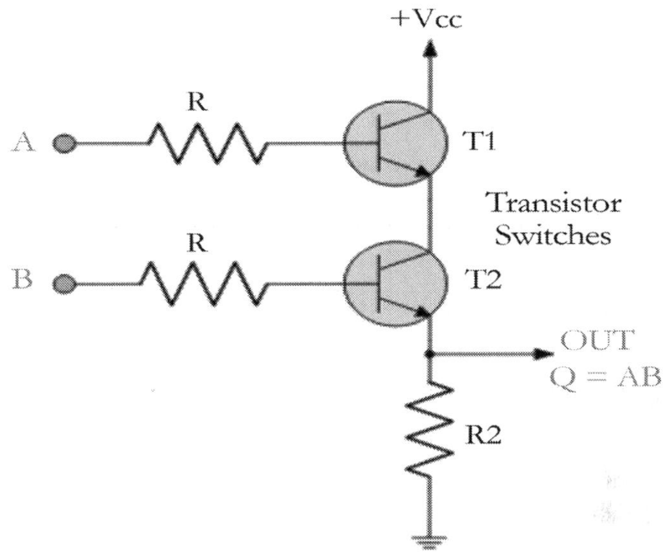

A	B	T_1	T_2	Y
0	0	OFF	OFF	0
0	1	OFF	ON	0
1	0	ON	OFF	0
1	1	ON	ON	1

Diode Circuit Diagram

The diode circuit diagram and truth table are follows:

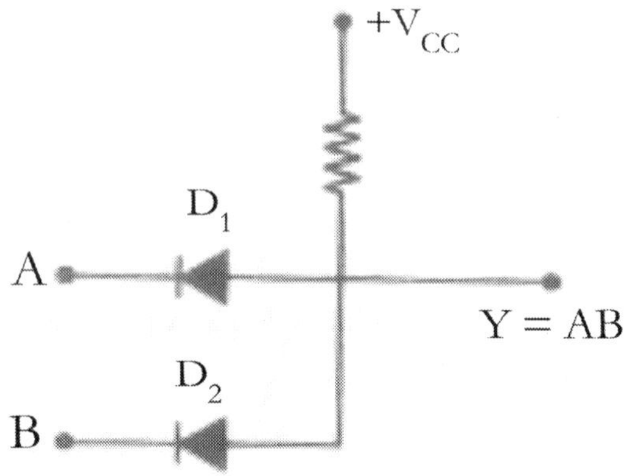

A	B	D1	D2	y
0	0	ON	ON	0
0	1	ON	OFF	0
1	0	OFF	ON	0
1	1	OFF	OFF	1

OR Gate

A Logic OR Gate or Inclusive-OR gate is a type of digital logic gate that has an output which is normally at logic level "0" and only goes "HIGH" to a logic level "1" when one or more of its inputs are at logic level "1". The output, Q of a "Logic OR Gate" only returns "LOW" again when ALL of its inputs are at a logic level "0". In other words for a logic OR gate, any "HIGH" input will give a "HIGH", logic level "1" output.

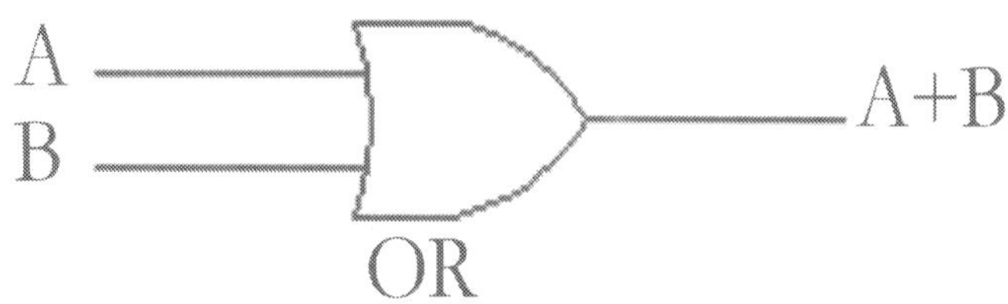

2 Input OR gate		
A	B	A+B
0	0	0
0	1	1
1	0	1
1	1	1

Transistor Circuit

A simple 2-input logic OR gate can be constructed using RTL Resistor-transistor switches connected together as shown below with the inputs connected directly to the transistor bases. Either transistor must be saturated "ON" for an output at Q.

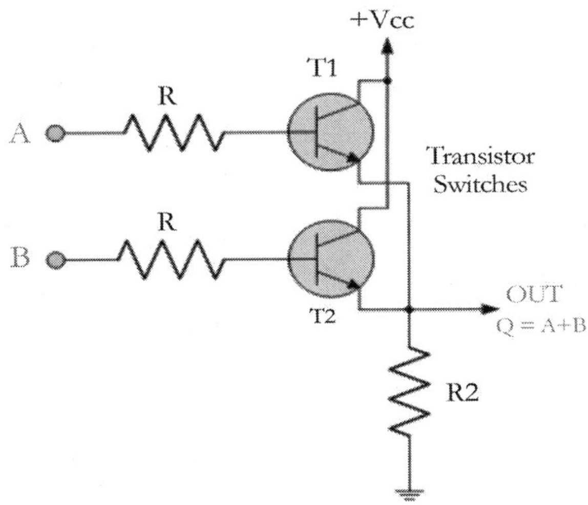

B	A	OUT
0	0	0
0	1	1
1	0	1
1	1	1

Diode circuit diagram:

The diode circuit diagram and truth table are follows:

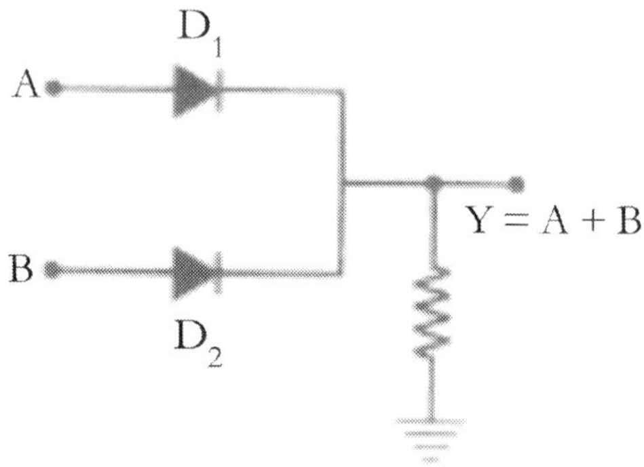

A	B	D_1	D_2	Y
0	0	ON	ON	1
0	1	ON	OFF	1
1	0	OFF	ON	1
1	1	OFF	OFF	0

NAND Gate

The Logic NAND Gate is a combination of the digital logic AND gate with that of an inverter or NOT gate connected together in series. The NAND (Not – AND) gate has an output that is normally at logic level "1" and only goes "LOW" to logic level "0" when ALL of its inputs are at logic level "1". The Logic NAND Gate is the reverse or "Complementary" form of the AND gate we have seen previously.

The digram and truth table of NAND gate are follows:

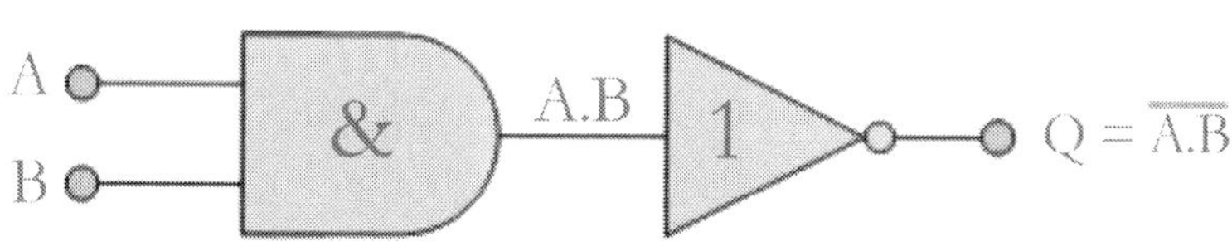

INPUT		OUTPUT
A	B	A NAND B
0	0	1
0	1	1
1	1	1
1	0	0

Transistor Circuit

A simple 2-input logic NAND gate can be constructed using RTL Resistor-transistor switches connected together as shown below with the inputs connected directly to the transistor bases. Either transistor must be cut-off "OFF" for an output at Q.

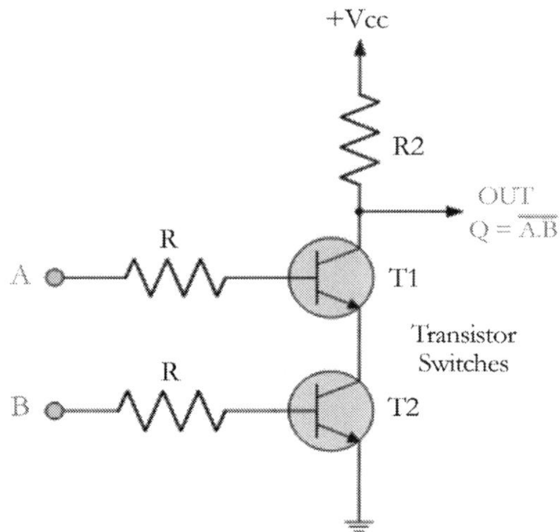

B	A	OUT
0	0	1
0	1	1
1	0	1
1	1	0

Other Gates Using Universal Gate NAND

AND gate

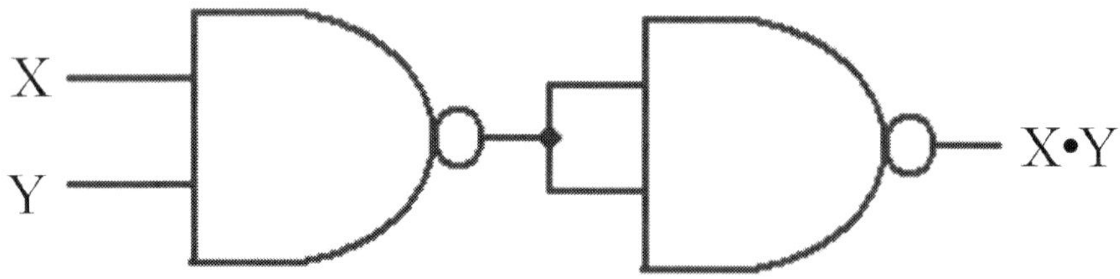

OR Gate

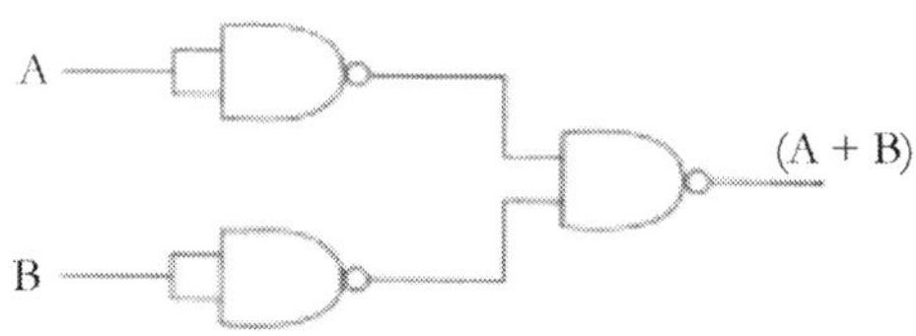

NOT Gate

EXOR Gate

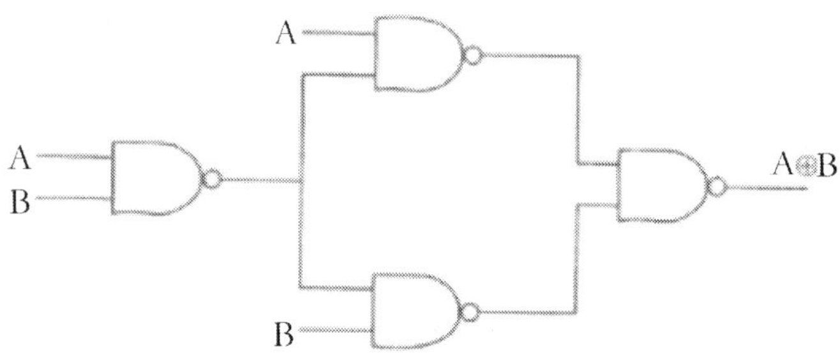

EXNOR Gate

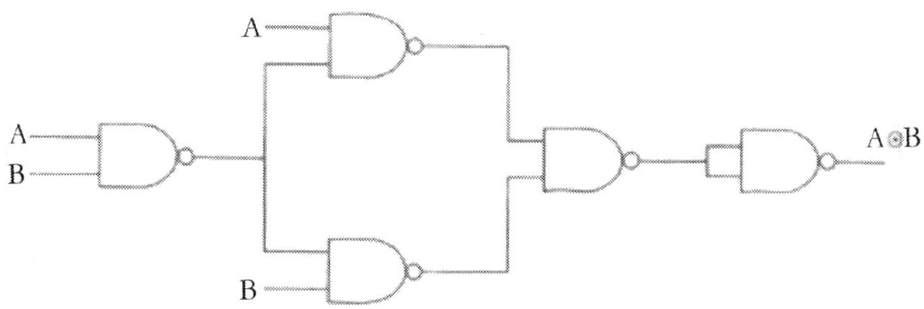

NOR Gate

The Logic NOR Gate or Inclusive-NOR gate is a combination of the digital logic OR gate with that of an inverter or NOT gate connected together in series. The NOR (Not – OR) gate has an output that is normally at logic level "1" and only goes "LOW" to logic level "0" when ANY of its inputs are at logic level "1". The Logic NOR Gate is the reverse or "Complementary" form of the OR gate we have seen previously.

The diagram and the truth table of NOR gate are follows:

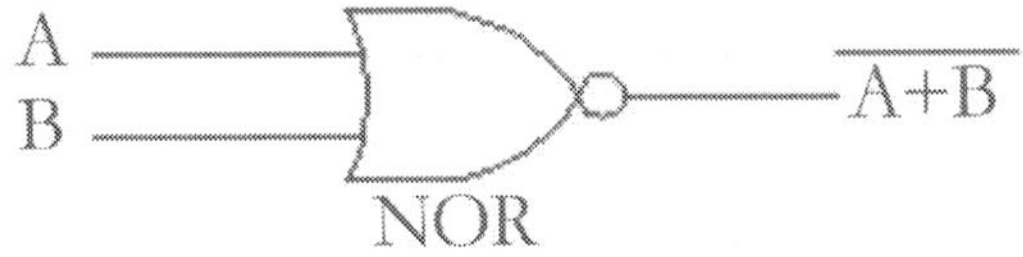

INPUT		OUTPUT
A	B	A NOR B
0	0	1
0	1	0
1	1	0
1	0	0

Tranaisitor Circuit

A simple 2-input logic NOR gate can be constructed using RTL Resistor-transistor switches connected together as shown below with the inputs connected directly to the transistor bases. Both transistors must be cut-off "OFF" for an output at Q.

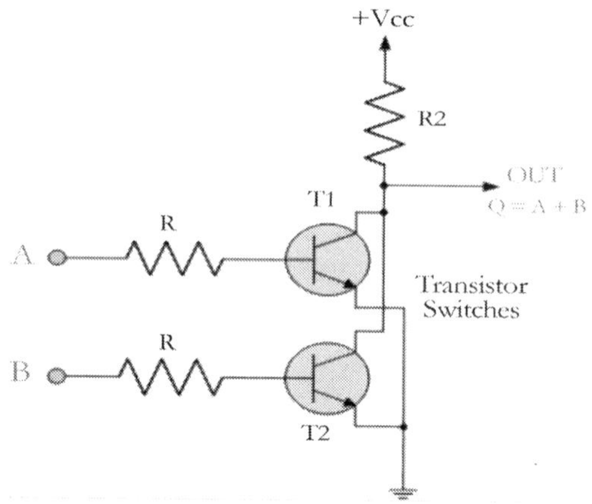

B	A	OUT
0	0	1
0	1	0
1	0	0
1	1	0

Other Gates Using Universal NOR Gate

AND Gate

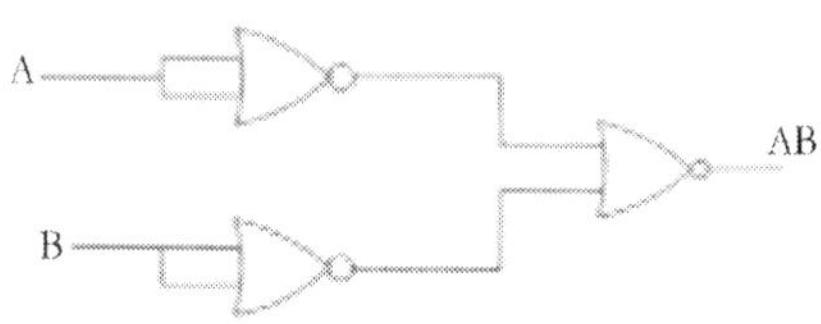

OR Gate

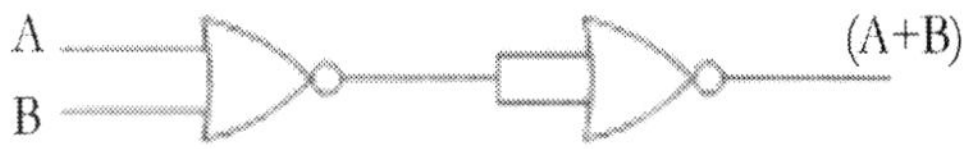

NOT Gate

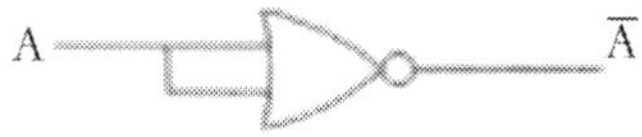

EXOR Gate

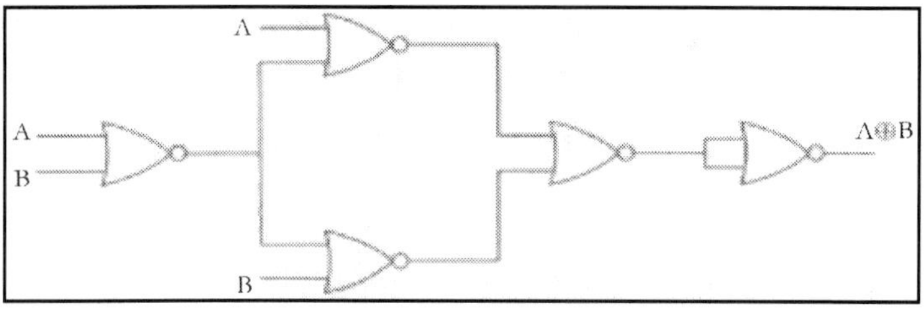

EXNOR Gate

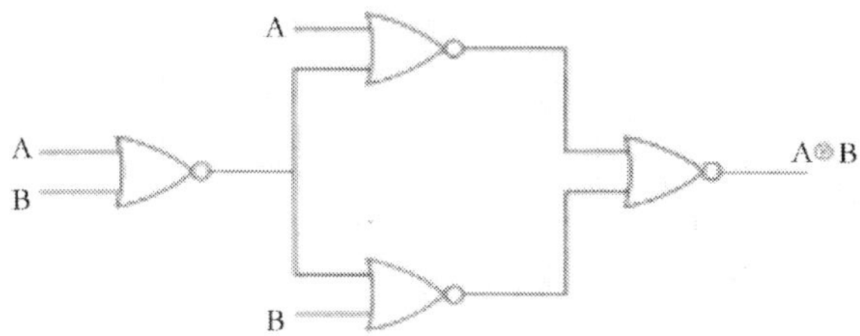

EXOR Gate

In previous, we saw that by using the three principal gates, AND Gate, the OR Gate and the NOT Gate, we can build many other types of logic gate functions, such as a NAND Gate and a NOR Gate or any other type of digital logic function we can imagine.

But there are two other types of digital logic gates which although they are not a basic gate in their own right as they are constructed by combining together other logic gates, their output Boolean function is important enough to be considered as complete logic gates. These two "hybrid" logic gates are called the Exclusive-OR (Ex-OR) Gate and its complement the Exclusive-NOR (Ex-NOR) Gate.

The symbol and the truth table are follows:

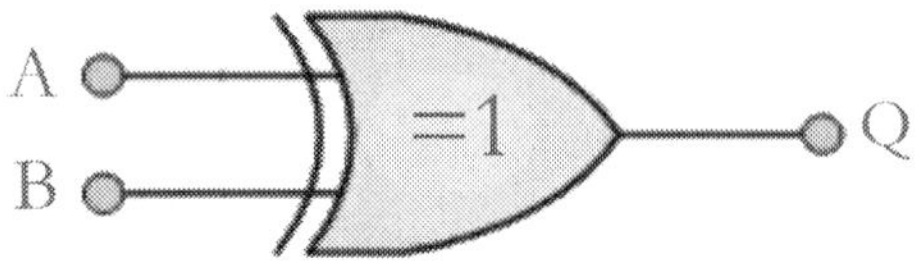

INPUT		OUTPUT
A	B	A Ex-OR B
0	0	0
0	1	1
1	1	1
1	0	0

Boolean function of 2-input E-XOR operation

$$Y = A \oplus B = \overline{A}B + A\overline{B}$$

XNOR Gate

The Exclusive-NOR Gate function or Ex-NOR for short, is a digital logic gate that is the reverse or complementary form of the Exclusive-OR function we look at in the previous tutorial. Basically the "Exclusive-NOR Gate" is a combination of the Exclusive-OR gate and the NOT gate but has a truth table similar to the standard NOR gate in that it has an output that is normally at logic level "1" and goes "LOW" to logic level "0" when ANY of its inputs are at logic level "1".

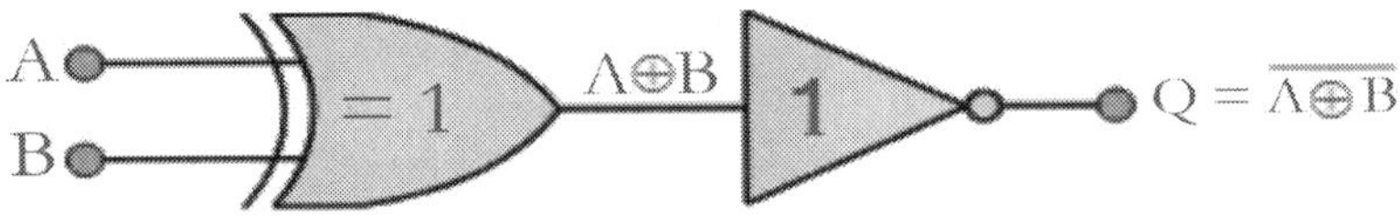

A	B	Q
0	0	1
0	1	0
1	1	0
1	0	1

Boolean function of 2-input X-NOR operation

$$Q = \overline{A \oplus B}$$

Alternative Symbols of Gates

The left side of the illustration shows the standard symbol for each logic gate, and the right side shows the alternate symbol. The alternate symbol for each gate is obtained from the standard symbol by doing the following:

1. Invert each input and output of the standard symbol. This is done by adding bubbles (small circles) on input and output lines that do not have bubbles, and by removing bubbles that are already there.
2. Change the operation symbol from AND to OR, or from OR to AND. (In the special case of the INVERTER, the operation symbol is not changed.)

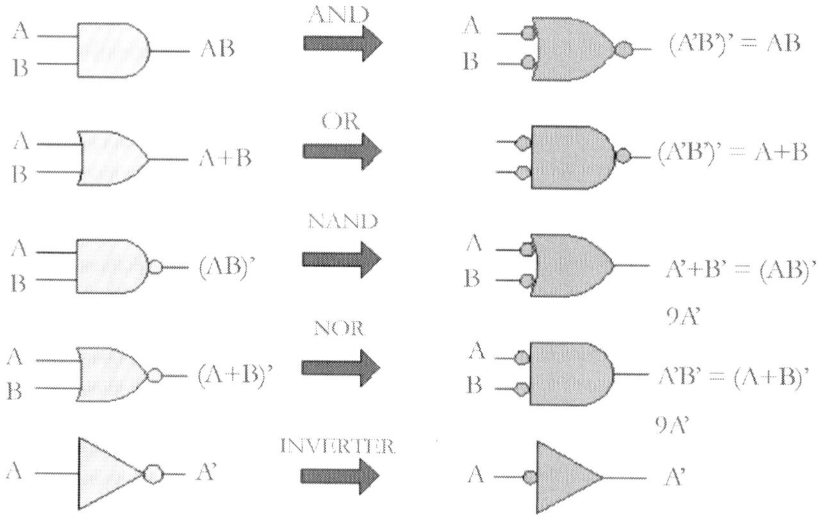
A
B
AB
AND
(A'B')' = AB
A+B
OR
(A'B')' = A+B
(AB)'
NAND
A'+B' = (AB)'
(A+B)'
NOR
A'B' = (A+B)'
A'
INVERTER
A'

3 Boolean Algebra and Reduction Techniques

Boolean algebra finds its most practical use in the simplification of logic circuits. If we translate a logic circuit's function into symbolic (Boolean) form, and apply certain algebraic rules to the resulting equation to reduce the number of terms and/or arithmetic operations, the simplified equation may be translated back into circuit form for a logic circuit performing the same function with fewer components. If equivalent function may be achieved with fewer components, the result will be increased reliability and decreased cost of manufacture.

Boolean algebra Laws

As well as the logic symbols "0" and "1" being used to represent a digital input or output, we can also use them as constants for a permanently "Open" or "Closed" circuit or contact respectively. A set of rules or Laws of Boolean Algebra expressions have been invented to help reduce the number of logic gates needed to perform a particular logic operation resulting in a list of functions or theorems known commonly as the Laws of Boolean Algebra.

Commutative Law

The "Commutative Laws" say we can swap numbers over and still get the same answer when you add. Or when you multiply.

Commutative Law of Addition:

A+B = B+A

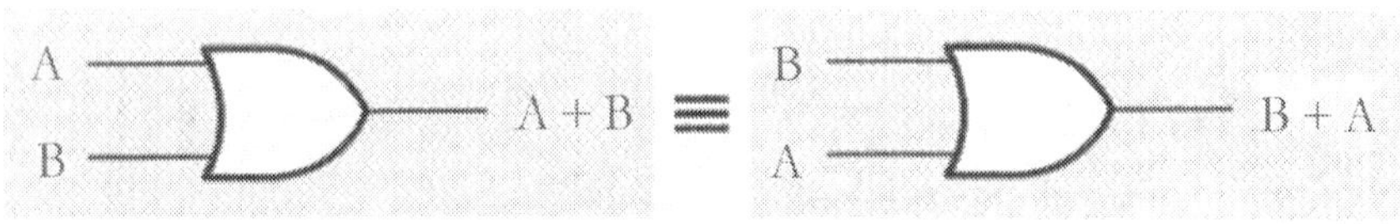

Commutative Law of Multiplication:

A*B = B*A

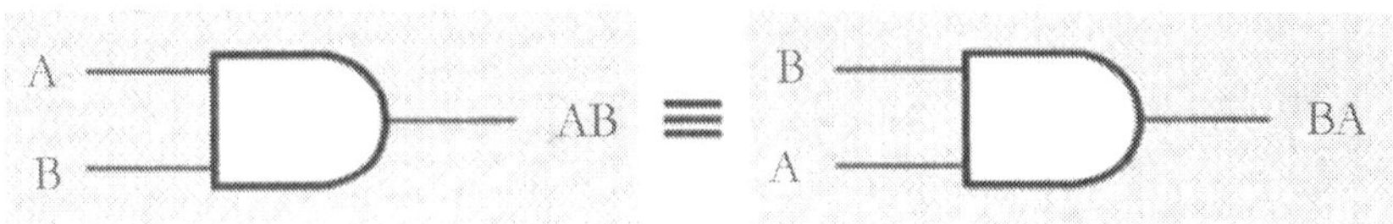

Associative Law

The "Associative Laws" say that it does not matter how we group the numbers (i.e. which we calculate first).

This law allows the removal of brackets from an expression and regrouping of the variables.

$A + (B + C) = (A + B) + C$ (OR Associate Law)

$A.(B.C) = (A.B)C = A.B.C$ (AND Associate Law)

Distributive Law

This law enables a reduction in a complicated expression to a simpler one by absorbing like terms.

$A.(B + C) = (A.B) + A.C$ (OR Distributive Law)

$A + (B.C) = (A.B) + A.C$ (AND Distributive Law)

Boolean Algebraic Theorems

The Boolean algebraic theorems are follows:

AND- Operation Theorem

A.A=A

A.0=0

A.1=A

$A.\overline{A} = 0$

OR-operation Theorem

A+A=A

A+0=0

A+1=A

$A + \overline{A} = 0$

De Morgan's Theorem

$\overline{A + B} = \overline{A}\,\overline{B}$

And

$\overline{AB} = \overline{A} + \overline{B}$

Transposition Theorem

(A+B) (A+C)= A + BC

Distribution Theorem

A + BC = (A+B) (A+C)

Boolean Algebraic Theorems

Theorm No	Theorm
1	$(A+B).(A+\overline{B})=A$
2	$AB+\overline{A}C=(A+C)(\overline{A}+B)$
3	$(A+B)(\overline{A}+C)=AC+\overline{A}B$
4	$AB+\overline{A}C+BC=AB+\overline{A}C$
5	$(A+B)(\overline{A}+C)(B+C)=(A+B)(\overline{A}+C)$
6	$\overline{A\cdot B\cdot C\ldots}=\overline{A}+\overline{B}+\overline{C}+\ldots$
7	$\overline{A+B+C\ldots}=\overline{A}\cdot\overline{B}\cdot\overline{C}+\ldots$

Duality Theorem

The duality property of Boolean algebra state that all binary expressions remain valid when following two steps are performed:

Step 1: Interchange OR and AND operators.

Step 2: Replace all 1's by 0's and 0's by 1's.

Boolean Function Representation

The Boolean function representation are follows:

Canonical Form

A canonical form of a mathematical object is a standard way of presenting that object as a mathematical expression. The distinction between "canonical" and "normal" forms varies by subfield. In most fields, a canonical form specifies a unique representation for every object, while a normal form simply specifies its form, without the requirement of uniqueness.

The canonical form of a positive integer in decimal representation is a finite sequence of digits that does not begin with zero.

Standard Form

- Implementations of logic functions based on canonical forms are called two-level implementations (inverters of input variables are not counted)
- It is often possible to simplify canonical forms into standard forms:
- The Sum-of-Minterms (SoM) form can be simplified into a Sum-of-Products (SoP) form.
- The Product-of-Maxterm (PoM) form can be simplified into a Product-of-Sums (PoS) form

Minterms and Maxterms

A minterm is a Boolean expression resulting in 1 for the output of a single cell, and 0s for all other cells in a Karnaugh map, or truth table. If a minterm has a single 1 and the remaining cells as 0s, it would appear to cover a minimum area of 1s. The illustration above left shows the minterm ABC, a single product term, as a single 1 in a map that is otherwise 0s. We have not shown the 0s in our Karnaugh maps up to this point, as it is customary to omit them unless specifically needed.

A maxterm is a Boolean expression resulting in a 0 for the output of a single cell expression, and 1s for all other cells in the Karnaugh map, or truth table. The illustration above left shows the maxterm (A+B+C), a single sum term, as a single 0 in a map that is otherwise 1s. If a maxterm has a single 0 and the remaining cells as 1s, it would appear to cover a maximum area of 1s.

Complete Simplification Rules

Boolean algebra finds its most practical use in the simplification of logic circuits. If we translate a logic circuit's function into symbolic (Boolean) form, and apply certain algebraic rules to the resulting equation to reduce the number of terms and/or arithmetic operations, the simplified equation may be translated back into circuit form for a logic circuit performing the same function with fewer components. If equivalent function may be achieved with fewer components, the result will be increased reliability and decreased cost of manufacture.

- Construct the K-map and place 1's in those cells corresponding to the 1's in the truth table. Place 0's in the other cells.
- Examine the map for adjacent 1's and loop those 1's which are not adjacent to any other 1's. these are called isolated 1's.
- Next, look for those 1's which are adjacent to only one other 1. Loop any pair containing such a 1.
- Loop any octet even it contains some 1's that have already been looped.
- Loop any quad that contains one or more 1's which habe not already been looped, making sure to use the minimum number of loops.
- Loop any pairs necessarily to inculde any 1's that have not yet been looped, making sure to use the minimum number of loops.
- Form the OR sum of all the terms generated by each loop.

4 Arithmetic Operation and Circuit

The arithmetic operation and circuit have two types of operation that are performed on binary data include arithmetic and logic operations. Basic arithmetic operations include addition, subtraction, multiplication and division. AND, OR and NOT are the basic logic functions.

The Half Adder

Binary arithmetic is carried out by combinational logic circuits, the simplest of which is the half adder, shown in following figure. This circuit consists, in its most basic form of two gates, an XOR gate that produces a logic 1 output whenever A is 1 and B is 0 or when B is 1 and A is 0. The AND gate produces a logic 1 at the carry output when both A and B are 1. The half adder truth table is shown in following table and describes the result of binary addition.

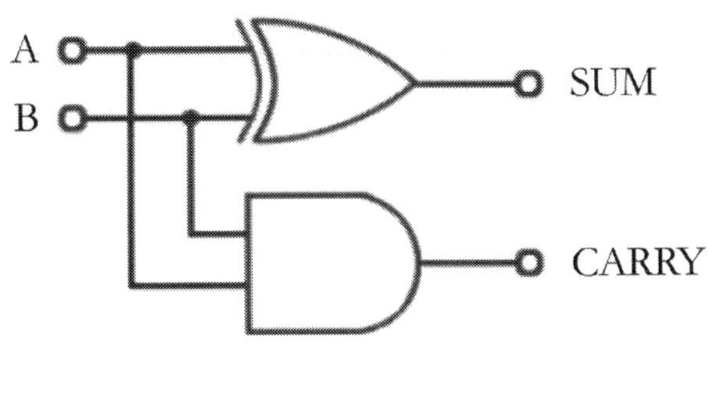

A	B	Sum	Carry
0	0	1	0
0	1	0	0
1	1	0	0
1	0	1	1

1 plus $0=1_2(1_{10})$

And

1 plus $1=10_2(2_{10})$

The half adder is fine for adding two 1-bit numbers together, but for binary numbers containing several bits, a carry may be produced at some time (as a result of adding 1 and 1) that must be added to the next column. As the half adder has only two inputs it cannot add in a carry bit from a previous column, so it is not practical for anything other than 1-bit additions.

The Full Adder

When 2 or more bits are to be added, the circuit used is the Full Adder, shown in following figure, (blue background) together with its simplified block diagram symbol. This circuit simply comprises two half adders, the sum of A and B from the first half adder is used as input S_1 on the second half adder, which now produces a sum of the first half adder sum (S_1) plus any 'carry in' from the C_{IN} terminal. Any carries produced by the two half adders are then 'ORed' together to produce a single C_{OUT} output. The truth table for the circuit is given in following table.

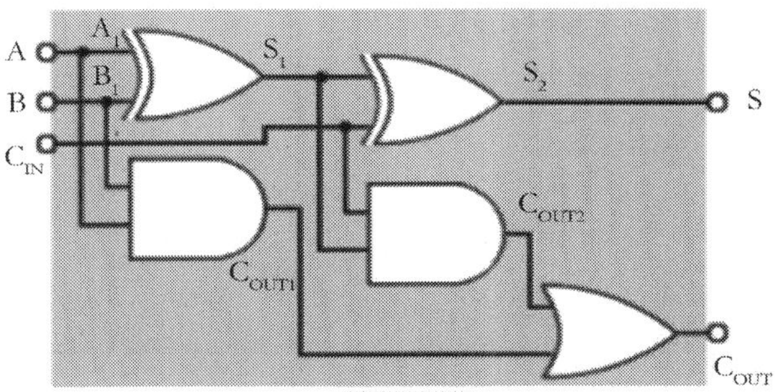

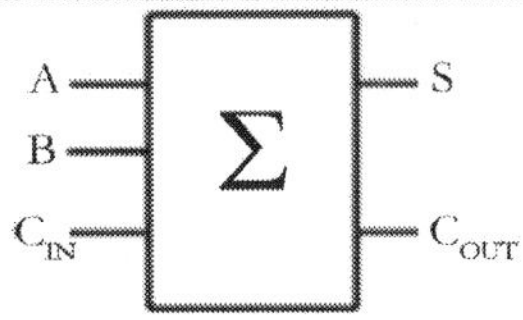

Half Adder$_1$					Half Adder$_2$		
A_1	B_1	C_{IN}	C_{OUT1}	S_1	C_{OUT2}	S_2	C_{OUT}
			$A \cdot B$	$A \oplus B$	$A_2 \cdot B_2$	$A_2 \oplus C_{IN}$	$C_{OUT1} + C_{OUT2}$
0	0	0	0	0	0	0	0
0	0	1	0	0	0	1	0
0	1	0	0	1	0	1	0
0	1	1	0	1	1	0	1
1	0	0	0	1	0	1	0
1	0	1	0	1	1	0	1
1	1	0	1	0	0	0	1
1	1	1	1	0	0	1	1

Parallel Adders

Even the full adder is only adding two single bit binary numbers, but full adders may be combined to form parallel adders, which will add two multi bit numbers. Parallel adders can be built in several forms to add multi-bit binary numbers, each bit of the parallel adder using a single full adder circuit. As parallel adder circuits would look quite complex if drawn showing all the individual gates, it is common to replace the full adder schematic diagram with a simplified block diagram version.

4 Bit Parallel Adder

Following figure illustrates how a number of full adders can be combined to make a parallel adder, also called a 'Ripple Carry Adder' because of the way that any carry appearing at the carry in input (C_{IN}) or produced when adding any of the 4-bit inputs, 'ripples' along the adder stages until a final carry out appears at the carry out output (C_{OUT}) of full adder for bit A_3+B_3.

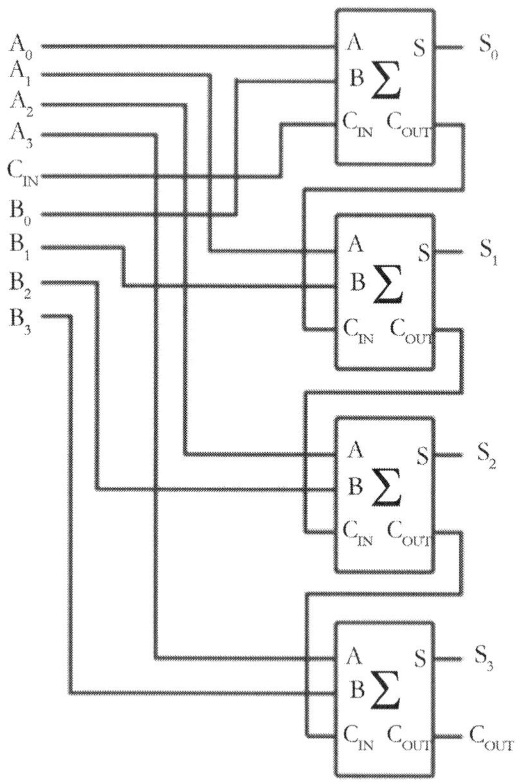

8 Bit Parallel Adder/Subtractor

To carry out arithmetic however, it is also necessary to be able to subtract. A further development of the parallel adder is shown in following figure. This is an 8-bit parallel adder/subtractor.

When subtraction is required, the control input is set to logic 1, which causes the bit at any particular B input to be complemented by an XOR gate before being fed to input B of the full adder circuit. Twos complement subtraction in an 8-bit adder/subtractor requires that the 8-bit number at input B is complemented (inverted) and has 1 added to it, before being added to the 8-bit number at input A. The result of this will be an 8-bit number in twos complement format, i.e. with its value represented by the lower 7 bits (bit 0 to bit 6) and the sign represented by the most significant bit (bit 7). The logic 1 on the control input is therefore also fed to the first carry input of the adder to be included in the addition, which for subtraction is therefore:

$$A + \overline{B} + 1$$

Here+ signifies addition rather than OR

Alternatively, if addition of A and B is required, then the control input is at logic 0 and number B is fed to the adder without complementing.

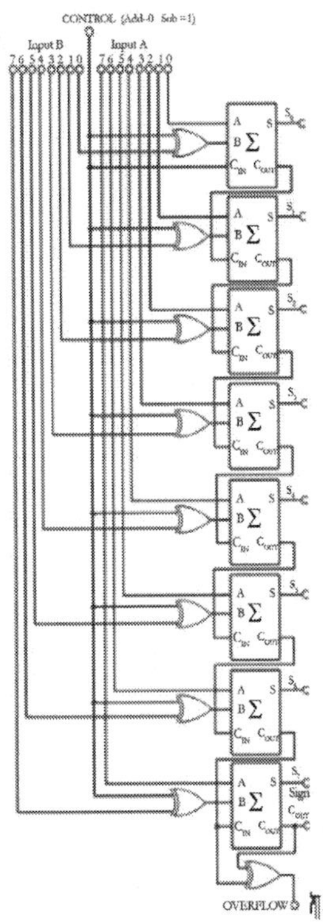

5 Code Converters, Multiplexers and Demultiplexers

The code converters, multiplexers and demultiplexers are follows:

Code Converters

- A code converter circuit will convert coded information in one form to a different coding form.
- One example of code conversion is to convert BCD to straight binary.
- The weighting of BCD bits is not the same as straight binary. The weight of each bit within a BCD digit is 1, 2, 4, 8 respectively; but, each digit must be multiplied by its decimal positional value (1, 10, 100, 1000, etc.).
- The 74184 IC will convert 1 $^{1}/_{2}$ BCD digits to straight binary. (The LSB is the same in BCD and in binary).
- The 74158 IC is used to convert 6 binary bits into a BCD code (again the LSB is feed around the chip).
- Converters are also available that convert BCD input to a seven segment code output such as the 7447 IC.

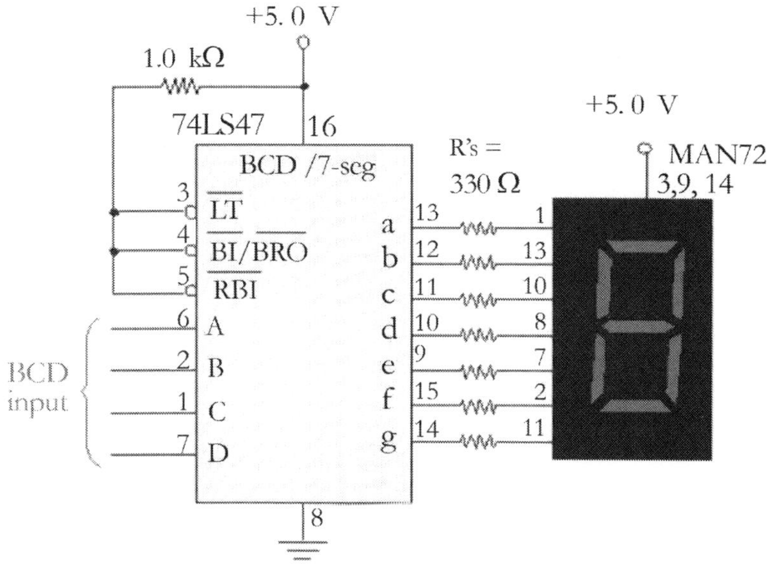

Multiplexers

- The multiplexer (also known as a data selector) will select data from several transmission lines to be gated to the output transmission line.
- The multiplexer will also have a number of control inputs that are used to select the appropriate data channel for input.
- The number of data inputs is equal to 2^N where N is the number of control select leads.
- The 74151 is a eight-line multiplexer with three select lines.
- An additional input on the 74151 is an enable input which can be used to expand the size of the multiplexer.
- A multiplexer can also be used to implement a SOP Boolean Expression in combinational logic.

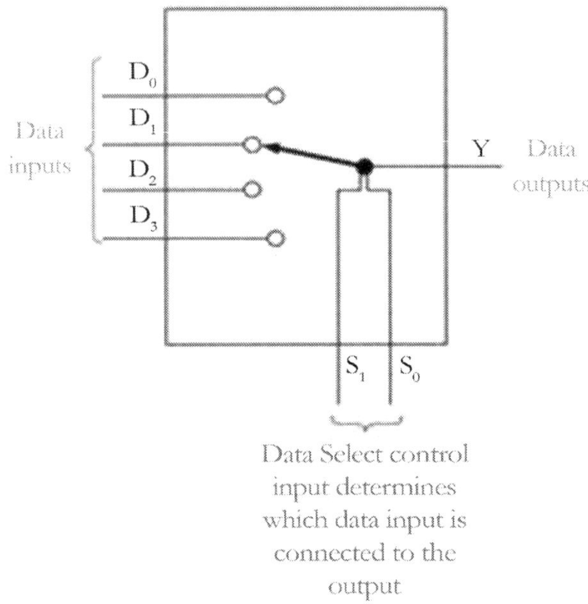

Demultiplexers

- A demultiplexer (data distributor) will select a specific output data transmission path from many for the input data to be sent to.
- The demultiplexer has several control select lines which are used to determine the selected output transmission path.
- The number of data outputs is equal to 2^N where N is the number of control select leads.
- The 74139 is a 1-of-4 decoder that can also be used as a demultiplexer.
- To use the 74139 as a demultiplexer, connect the incoming data to the enable input and use the A0 and A1 inputs as the control select lines.
- The 74154 1-of-16 decoder can also be used as a 16 line demultiplexer.
- Several CMOS chips use a transmission gate as part of the multiplexer/demultiplexer. This means that the gate is bidirectional and can also handle AC signals as well as binary 1's and 0's.

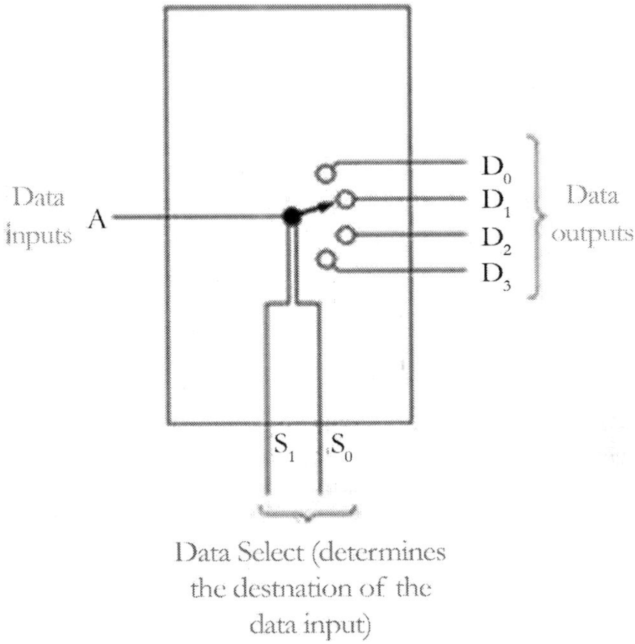

1:2 Demux

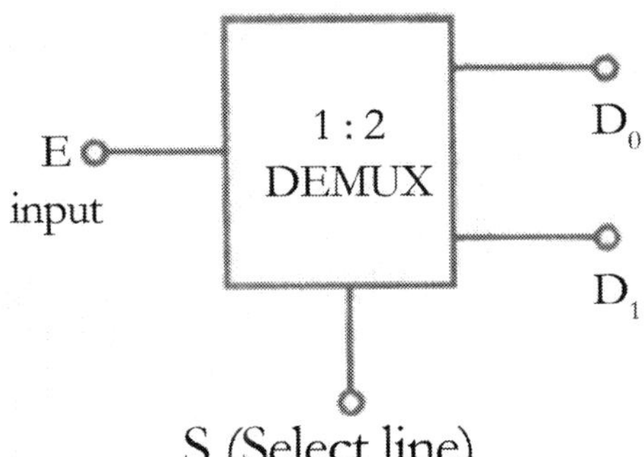

$D_0 = \bar{S}E$ and $D_1 = SE$

1:4 Demux

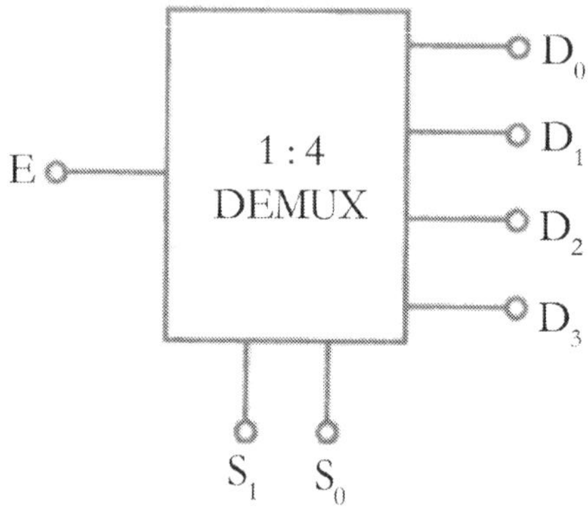

Select lines		Outputs			
S1	S0	D_3	D_2	D_1	D_0
0	0	0	0	0	E
0	1	0	0	E	0
1	0	0	E	0	0
1	1	E	0	0	0

$D_0 = \bar{S}_1\bar{S}_0\ E,\ D_1 = \bar{S}_1\bar{S}_0\ E,\ D_2 = \bar{S}_1\bar{S}_0\ E,\ D_4 = \bar{S}_1\bar{S}_0\ E$

Given DE-MUX	To be Implemented DEMUX	Required No. of DEMUX
1:2	1:4	1+2=3
1:2	1:8	1+2+4=7
1:2	1:16	1+2+4+8=15
1:2	1:64	1+2+4+8+16+32=63
1:4	1:16	1+4=5

Decoders

Decoders are simply a collection of logic gates which are arranged in a specific way so as to breakdown any combination of inputs to a set of terms that are all set to '0' apart from one term. Therefore when one input changes, two output terms will change.

If we have N inputs to a decoder, the number of outputs will be equal to 2^N.

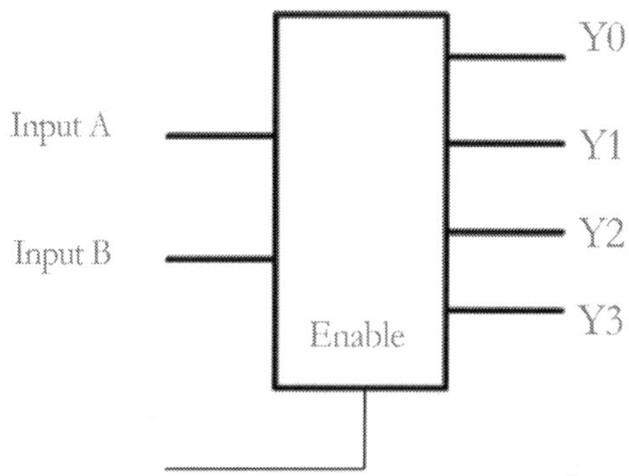

A	B	Y0	Y1	Y2	Y3
0	0	0	1	1	1
0	1	1	0	1	1
1	0	1	1	0	1
1	1	1	1	1	0

Decoders contains NAND gates or AND gates.

Encoder

An encoder is a device, circuit, transducer, software program, algorithm or person that converts information from one format or code to another, for the purposes of standardization, speed, secrecy, security or compressions.

CODE CONVERTERS

Code converter is a circuit that makes the two systems compatible even though each uses the different code.

BCD to Excess 3 Code

To convert from binary code A to binary code B, the input lines must supply the bit combination of elements as specified by code A and the output lines must generate the corresponding bit combination of code B. A combinational circuit performs this transformation by means of logic gates.

Input (BCD)				Output (Excess-3)			
A	B	C	D	W	X	Y	Z
0	0	0	0	0	0	1	1
0	0	0	1	0	1	0	0
0	0	1	0	0	1	0	1
0	0	1	1	0	1	1	0
0	1	0	0	0	1	1	1
0	1	0	1	1	0	0	0
0	1	1	0	1	0	0	1
0	1	1	1	1	0	1	0
1	0	0	0	1	0	1	1
1	0	0	1	1	1	0	0

Minimized Boolean function

z = D

y = CD+C'D'=CD(C+D)'

x = B'C+B'D+BC'D'= B'(C+D) +BC'D'

x = B'(C+D) +B(C+D)'

w = A+BC+BD=A+B(C+D)

Binary to Gray Converter

Most significant bit (B_1) is same as the most significant bit in Gray Code ($B_1 = G_1$). To find next bit perform Ex-OR (Exclusive OR) between the Current binary bit and previous bit.

Follow the below Steps to convert Binary number to gray code. Let's Consider

The Binary number B_1 B_2 B_3 B_4 ... B_n and the Gray code is G_1 G_2 G_3 G_4 ... G_n

1. Most significant bit (B_1) is same as the most significant bit in Gray Code ($B_1 = G_1$)
2. To find next bit perform Ex-OR (Exclusive OR) between the Current binary bit and previous bit. $G_n = B_n$ (Ex-OR) B_{n-1}

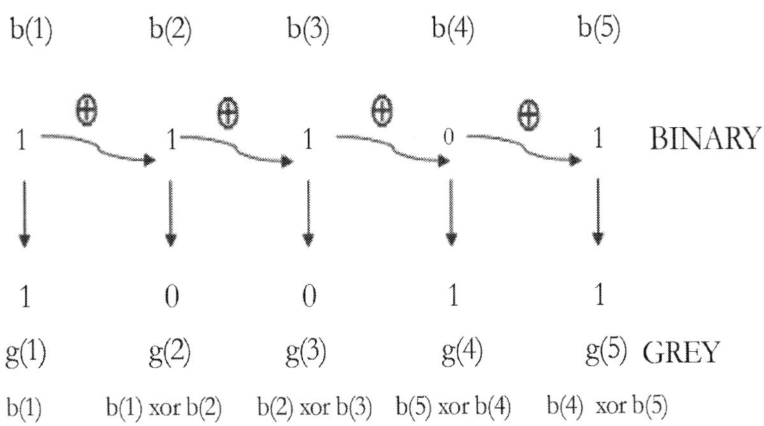

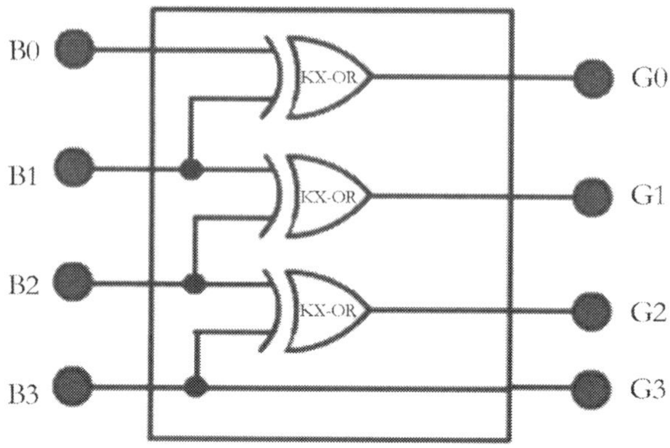

$G_3 = B3$

$G_2 = B_3 \oplus B_2$

$G_1 = B_2 \oplus B_1$

$G_0 = B_1 \oplus B_0$

Gray to Binary Converter

Most significant bit (G_1) is same as the most significant bit in Binary Code ($G_1 = B_1$). The next number can be obtain by taking Exclusive OR operation between the previous binary bit, and the current gray code bit and write down the value.

Follow the steps to convert Gray Code to Binary

1. Most significant bit (G_1) is same as the most significant bit in Binary Code ($G_1 = B_1$)
2. The next number can be obtain by taking Exclusive OR operation between the previous binary bit, and the current gray code bit and write down the value.

Repeat the Step until you find B_n

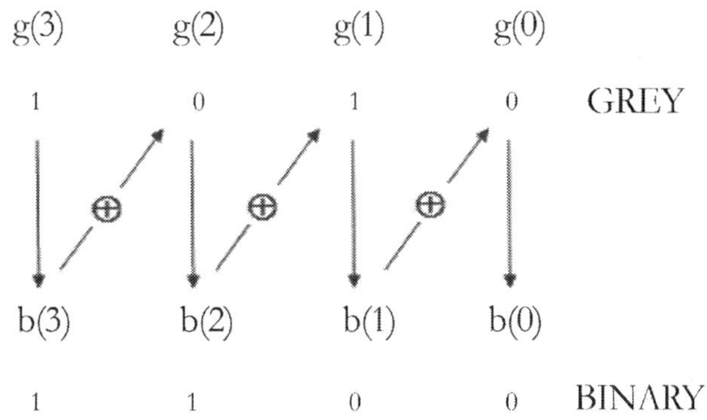

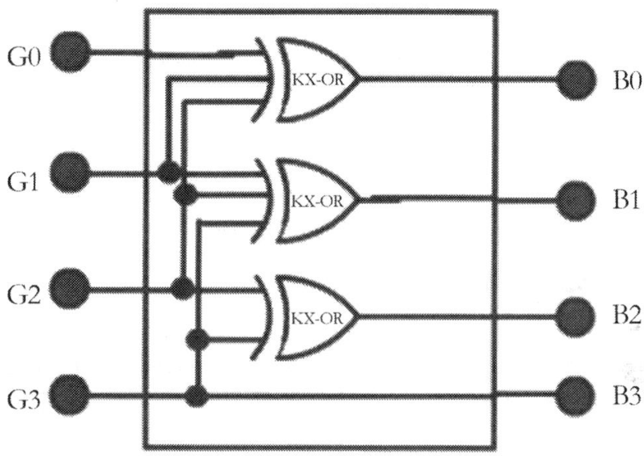

$B_3 = G_3$

$B_2 = B_3 \oplus G_2$

$B_1 = B_2 \oplus G_1$

$B_0 = B_1 \oplus G_0$

6 Sequential Circuits (FFs and Latches)

A sequential circuit is a combinational circuit with some feedback from the outputs. The output state depends on both the inputs and the outputs. The term "sequential" comes from the fact that the output depends not only on the current states, but on the states immediately preceding.

Flip-flop

Flip flop is a sequential circuit which generally samples its inputs and changes its outputs only at particular instants of time and not continuously. Flip flop is said to be edge sensitive or edge triggered rather than being level triggered like latches.

SR Flip-Flop

The SR flip-flop, also known as a SR Latch. The inputs of an SR flip-flop are labeled R (reset) and S (set) and the outputs are labeled as Q and Q'.

Block diagram of SR flip flop

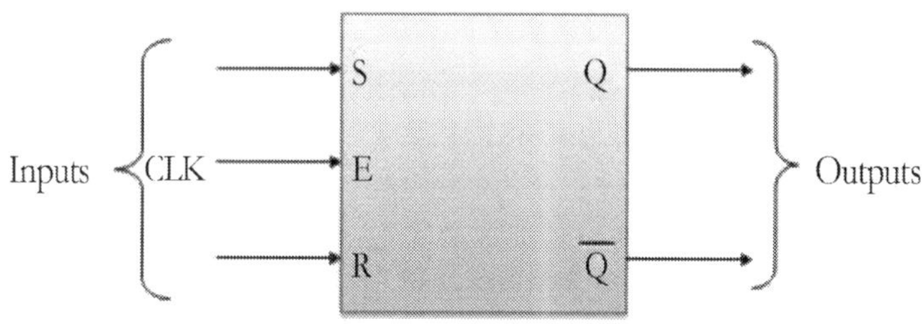

Circuit diagram of SR flip flop

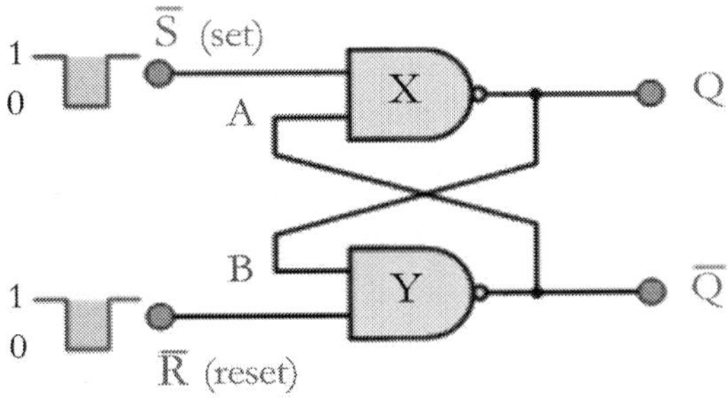

Truth table of SR flip flop

Clock	S_n	R_n	Q_{n+1}
0	X	X	Q_n
1	0	0	Q_n(HoldState)
1	0	1	0(Reset)
1	1	0	1(Set)
1	1	1	Invailid

where

S_n and R_n denotes the input

Q_n the output during the bit time n

Q_{n+1} denotes the output Q after CLK passes

Characteristics equation of SR FF

$$Q_{n+1} = S + \overline{R}Q_n$$

J-K Flip-flop

The J-K flip-flop is the most versatile of the basic flip-flops. It has two inputs, traditionally labeled J and K. If J and K are different then the output Q takes the value of J at the next clock edge.

Logic diagram of SR flip flop

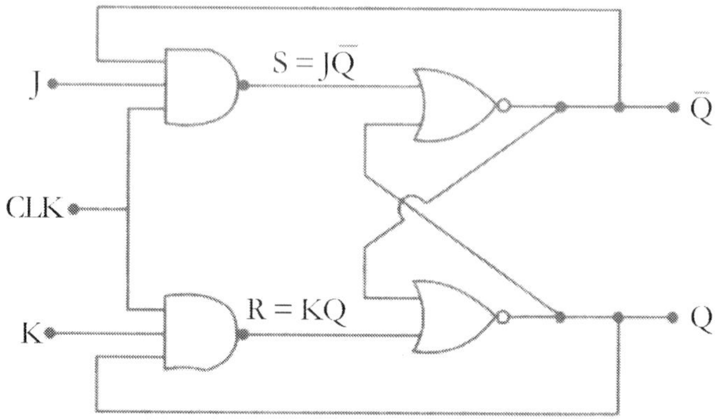

The truth table for the JK flip flop

Initial Conditions	Inputs (Pulsed)		Final Output
Q	S	R	Q (t + 1)
0	0	0	0
0	0	1	0
0	1	0	1
0	1	1	1
1	0	0	1
1	0	1	0
1	1	0	1
1	1	1	0

Excitation table for JK Flip Flop

S	R	Q
0	0	No Change
0	1	0
1	0	0
1	1	Toggle

T flip-flop

If the T input is high, the T flip-flop changes state ("toggles") whenever the clock input is strobe. If the T input is low, the flip-flop holds the previous value.

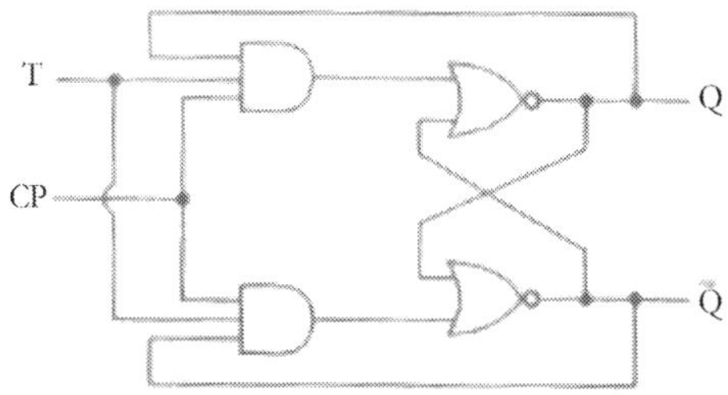

Truth table for T Flip Flop

CLK	T	Qn+1
0	X	Qn
1	0	Qn(Hold)
1	1	$\overline{Qn}$ (Reset)

Excitation table for T Flip Flop

Q_n	T	$Q_n + 1$
0	0	0
0	1	1
1	0	1
1	1	0

7 Shift Registers

Shift registers are a form of sequential logic. Shift registers produce a discrete delay of a digital signal or waveform. A shift register basically consists of several single bit "D-Type Data Latches", one for each data bit, either a logic "0" or a "1", connected together in a serial type daisy-chain arrangement so that the output from one data latch becomes the input of the next latch and so on.

Shift Registers are used for data storage or for the movement of data and are therefore commonly used inside calculators or computers to store data such as two binary numbers before they are added together, or to convert the data from either a serial to parallel or parallel to serial format.

Serial-in/Serial-out (SISO)

The data is shifted serially "IN" and "OUT" of the register, one bit at a time in either a left or right direction under clock control.

4-bit Serial-in to Serial-out Shift Register

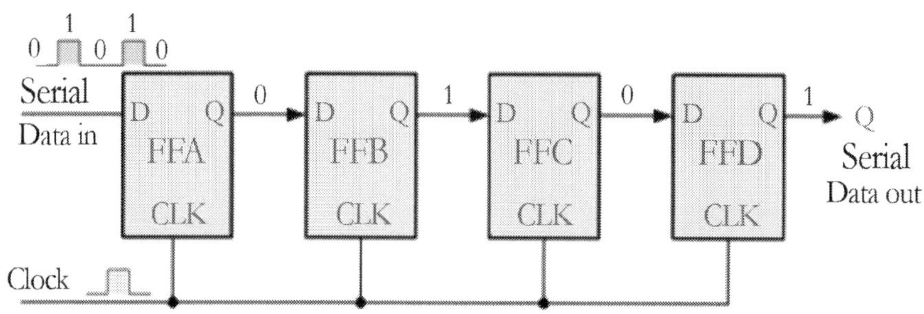

Serial-in/ parallel-out (SIPO)

The register is loaded with serial data, one bit at a time, with the stored data being available at the output in parallel form.

4-bit Serial-in to Parallel-out Shift Register

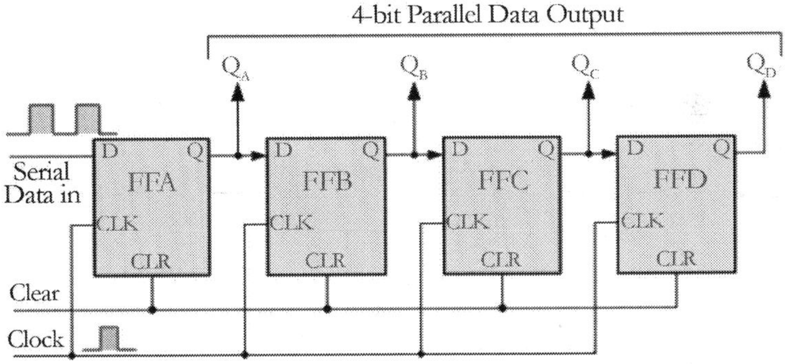

Parallel-in to Serial-out (PISO)

The parallel data is loaded into the register simultaneously and is shifted out of the register serially one bit at a time under clock control.

4-bit Parallel-in to Serial-out Shift Register

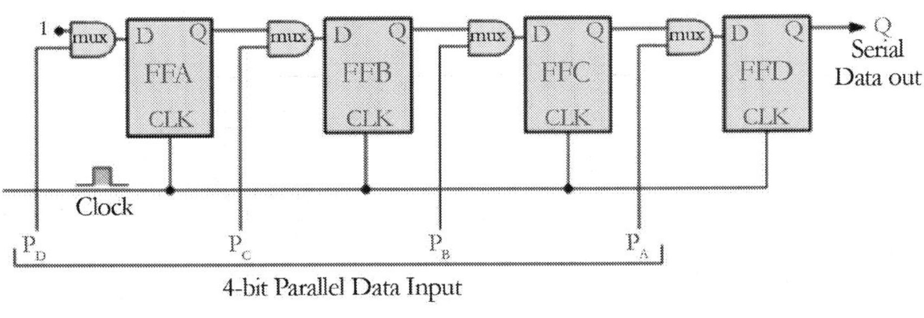

Parallel-in to Parallel-out (PIPO)

The parallel data is loaded simultaneously into the register, and transferred together to their respective outputs by the same clock pulse.

4-bit Parallel-in to Parallel-out Shift Register

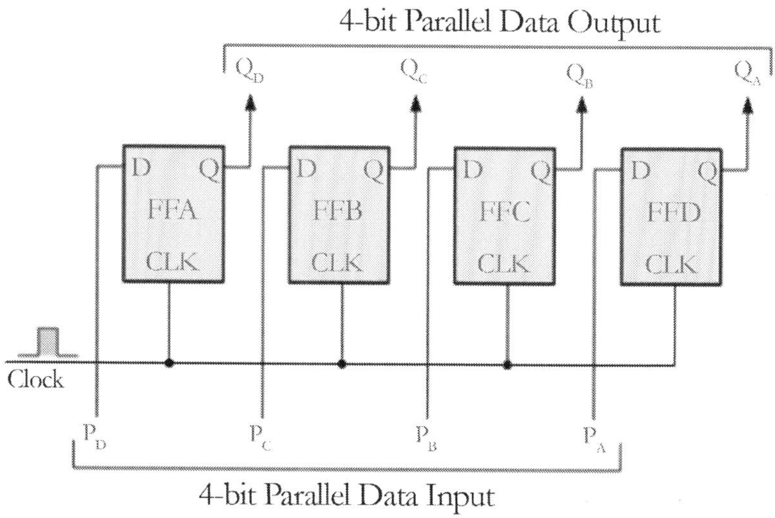

8 Counters

Counter is a device which stores (and sometimes displays) the number of times a particular event or process has occurred, often in relationship to a clock signal. The most common type is a sequential digital logic circuit with an input line called the "clock" and multiple output lines. The values on the output lines represent a number in the binary or BCD number system.

Synchronous Counters

The synchronous counter provides a more reliable circuit for counting purposes, and for high-speed operation, as the clock pulses in this circuit are fed to every flip-flop in the chain at exactly the same time. Synchronous counters use JK flip-flops, as the programmable J and K inputs allow the toggling of individual flip-flops to be enabled or disabled at various stages of the count.

Up-Down Counter

An up down counter is a bi-directional counter and it can be made to count upwards as well as downwards. In other words an up down counter is one which can provide n^{th} count up and down counts operations in a single unit.

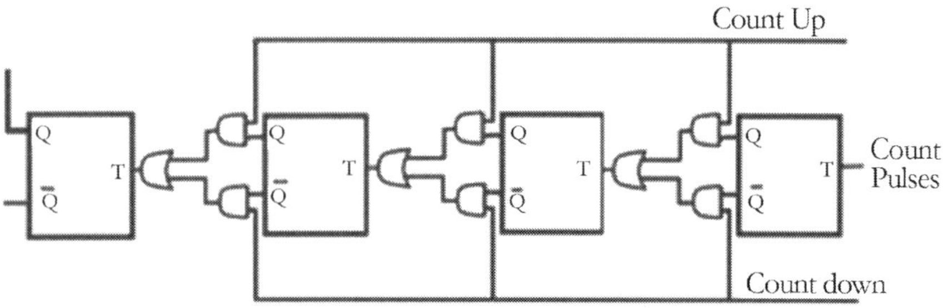

Ripple Counter

A ripple counter is an asynchronous counter where only the first flip-flop is clocked by an external clock. All subsequent flip-flops are clocked by the output of the preceding flip-flop. Asynchronous counters are also called ripple-counters because of the way the clock pulse ripples it way through the flip-flops.

The MOD of the ripple counter or asynchronous counter is 2^n if n flip-flops are used. For a 4-bit counter, the range of the count is 0000 to 1111 (2^4-1).

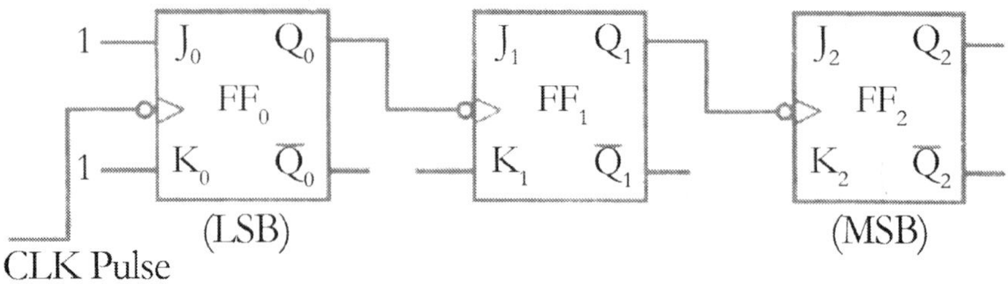

Decade Counter

A decade counter is the one which goes through 10 unique combinations of outputs and then resets as the clock proceeds. We may use some sort of a feedback in a 4-bit binary counter to skip any six of the sixteen possible output states from 0000 to 1111 to get to a decade counter.

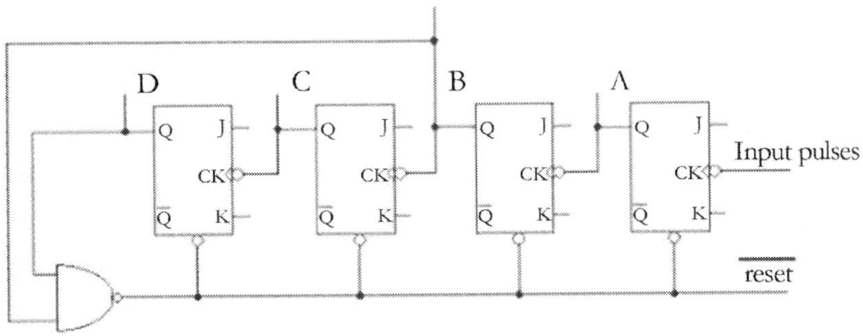

Ring Counter

A ring counter is a circular shift register which is initiated such that only one of its flip-flops is the state one while others are in their zero states. The ring counter contains only one logical 1 or 0 which it circulates. The total cycle length is equal to the number of stages.

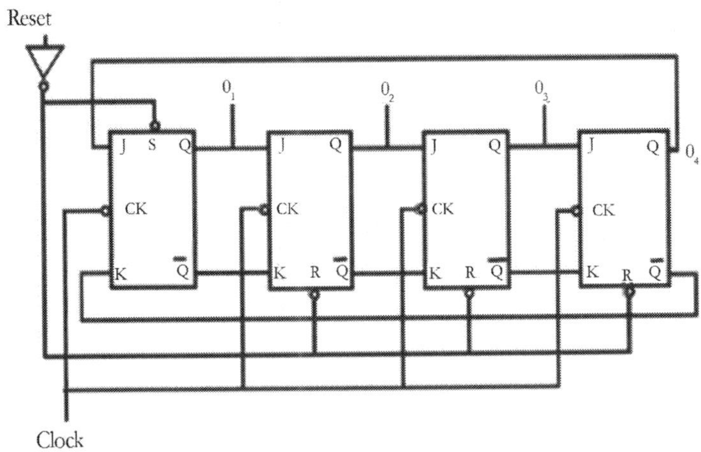

Johnson Counter

A Johnson counter is a modified ring counter, where the inverted output from the last flip flop is connected to the input to the first. The register cycles through a sequence of bit-patterns. The MOD of the Johnson counter is 2^n if n flip-flops are used. The main advantage of the Johnson counter is that it only needs half the number of flip-flops compared to the standard ring counter for the same MOD.

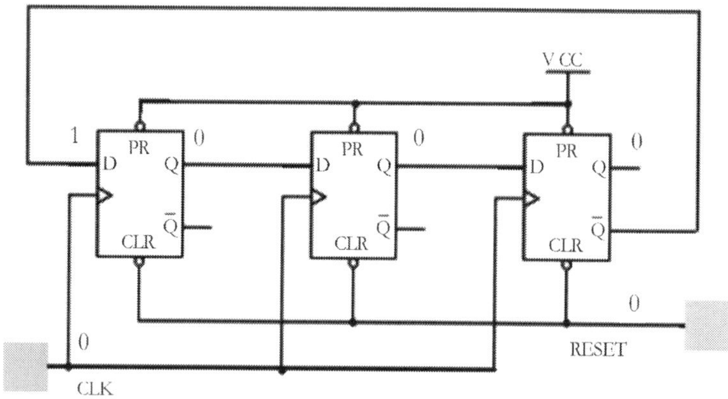

9 Digital ICs Family

A logic family refers to digital integrated circuit devices which are constructed with a combination of electronic gates. A family has its own power supply voltage and distinct logic levels. Some families were also introduced as individual components which were used to design integrated circuits and also to interconnect them.

Characteristics Digital ICs

Characteristics digital ICs are follows:

Fan-out

Fan out specifies the number of standard loads that the output of the gate can drive without impairment of its normal operation

Power Dissipation

Power dissipation is measure of power consumed by the gate when fully driven by all its inputs.

Propagation Delay

Propagation delay is the average transition delay time for the signal to propagate from input to output when the signals change in value. It is expressed in ns.

Noise Margin

It is the maximum noise voltage added to an input signal of a digital circuit that does not cause an undesirable change in the circuit output. It is expressed in volts.

High state noise margin is:

$$V_{NH} = V_{OH(min)} - V_{IH(min)}$$

Low state noise margin is:

$$V_{NL} = V_{IL(max)} - V_{OL(max)}$$

Fan in

Fan in is the number of inputs connected to the gate without any degradation in the voltage level.

Operating temperature

All the gates or semiconductor devices are temperature sensitive in nature. The temperature in which the performance of the IC is effective is called as operating temperature. Operating temperature of the IC vary from 00 C to 700 c.

MOS Logic Family

MOS logic uses NMOS and PMOS to implement logic gates. One needs to know the operation of the FET and MOS transistors to understand the operation of MOS logic circuits.

NMOS Logic

NMOS logic uses n-channel MOSFETs exclusively.

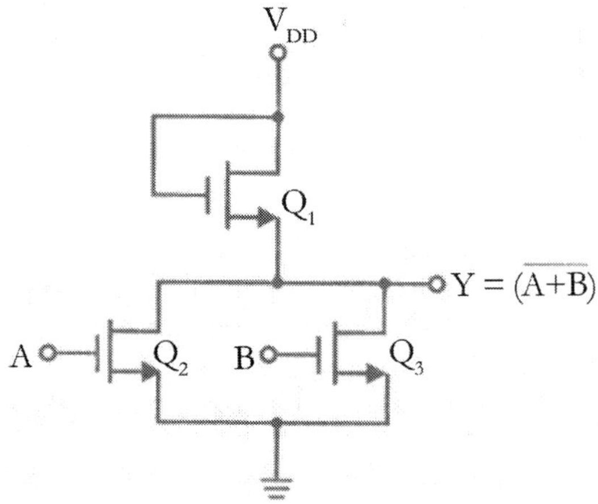

Symbol

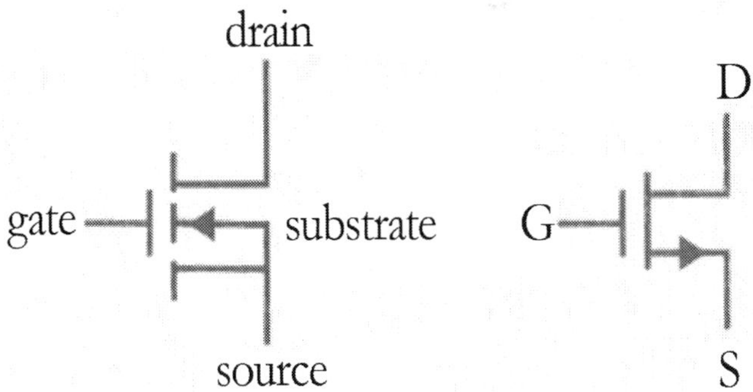

N-MOS inverter

When input is LOW, NMOS transistor does not conduct, and thus output is HIGH but when the input is HIGH, NMOS transistor conducts and thus output is LOW.

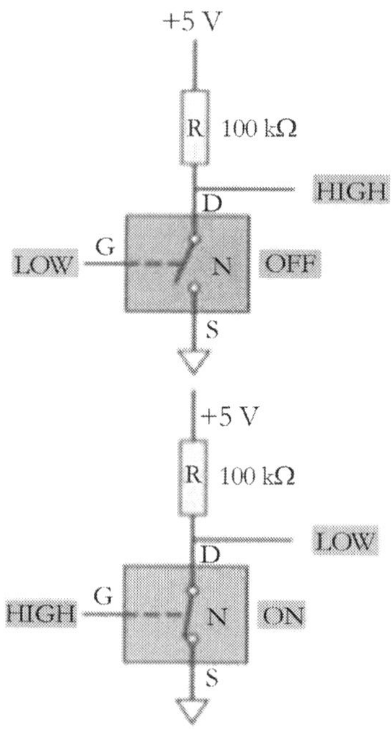

NMOS NAND Gate

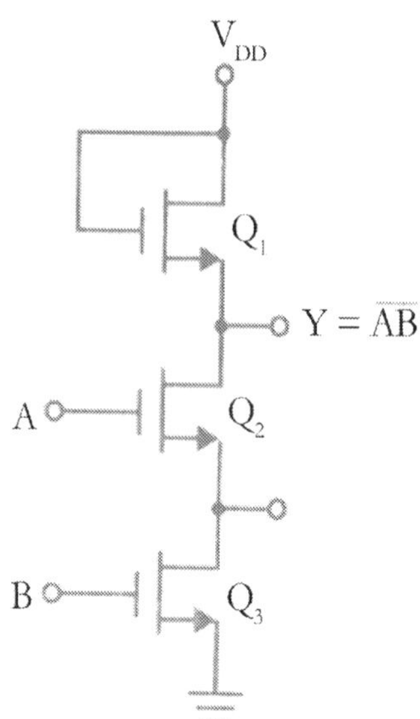

NMOS NOR Gate

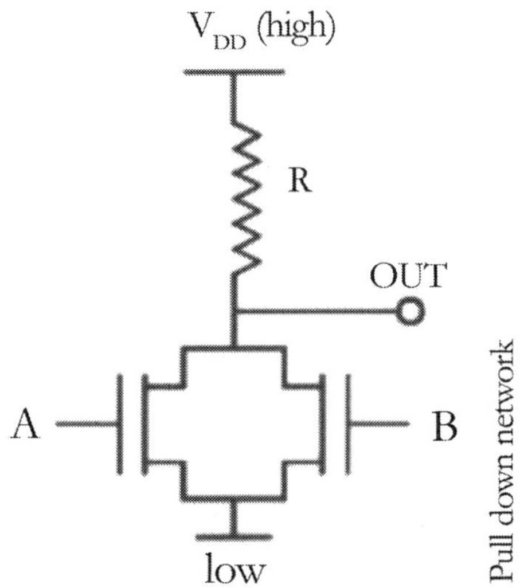

PMOS Logic

The P-channel MOSFET used as a switch.

Symbol

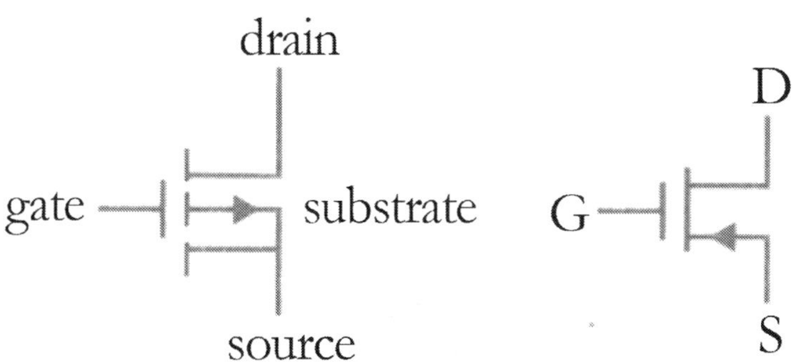

P-MOS Inverter

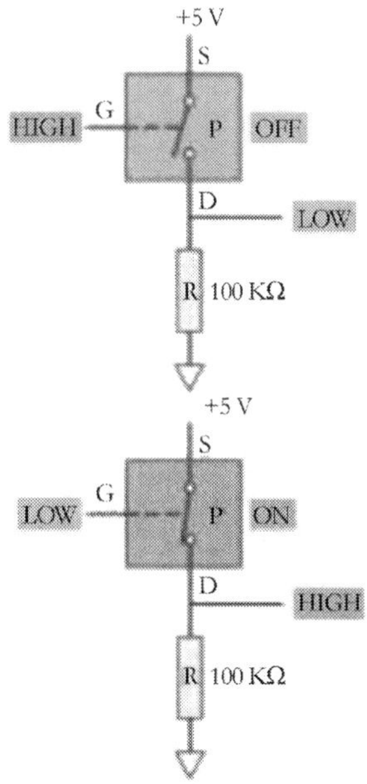

NAND by PMOS

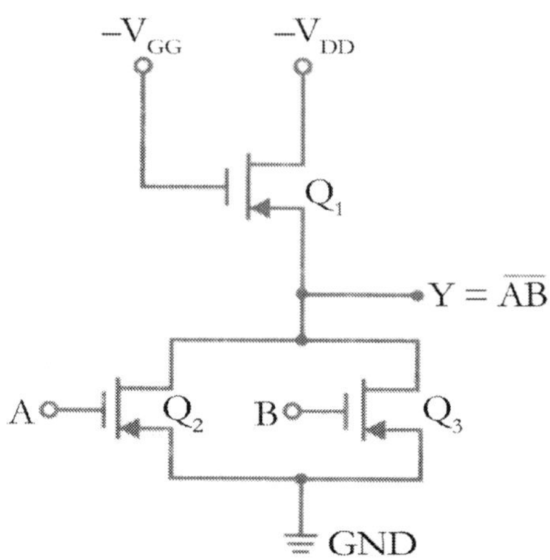

CMOS Logic

CMOS or Complementary Metal Oxide Semiconductor logic is built using both NMOS and PMOS. CMOS logic is faster and consumes less power.

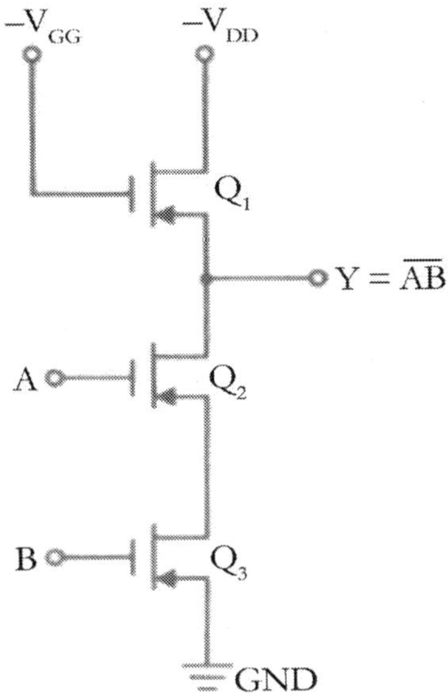

CMOS Inverter

The CMOS inverter is designed to use one NMOS and one PMOS, both having an equal threshold voltage (in magnitude).

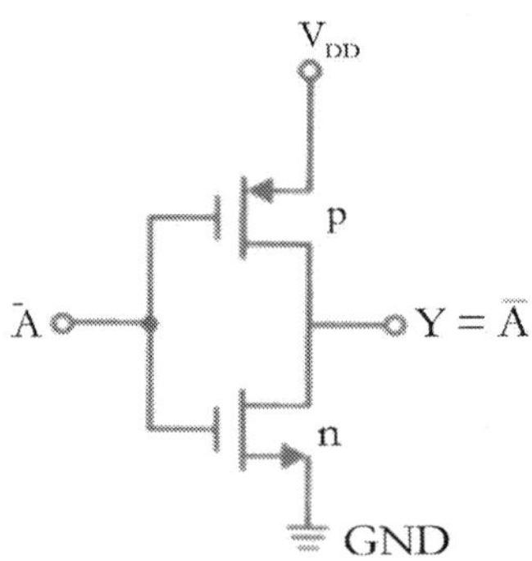

CMOS NAND Gate

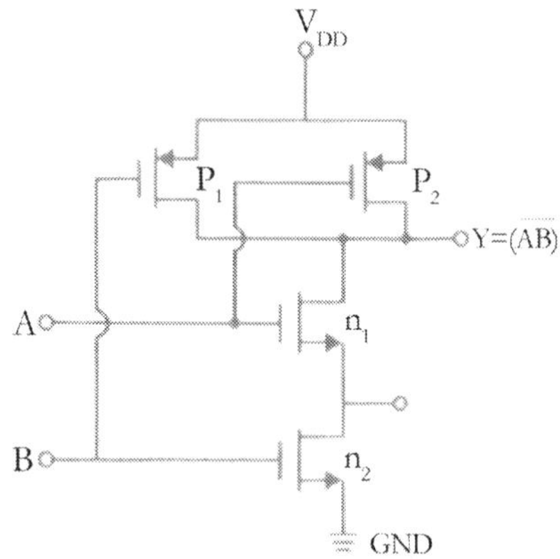

CMOS NOR Gate

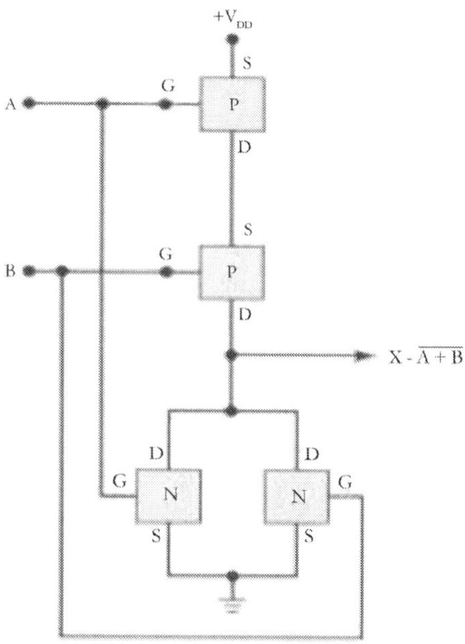

10 ADCs and DACs

Digital to Analog Converter

Digital-to-analog converter (DAC, D/A, D2A or D-to-A) is a function that converts digital data (usually binary) into an analog signal (current, voltage, or electric charge). Most real world signals are analog, these two converting interfaces are necessary to allow digital electronic equipment to process the analog signals.

A DAC can be constructed by using a Summing Amplifier and a set of resistors R, 2R, 4R and 8R as its inputs.

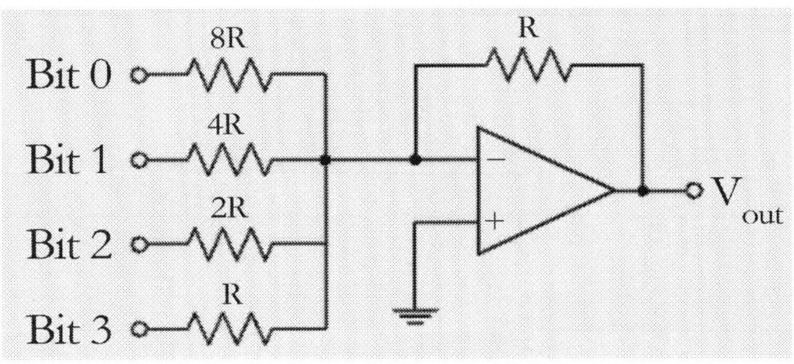

The resistor with the lowest value R corresponds to the highest weighted binary input Bit 3 (MSB) [23 = 8], and 2R, 4R, 8R correspond to the binary weights of Bit 2 ($2^2 = 4$), Bit 1 ($2^1 = 2$), and Bit 0 (LSB) [$2^0 = 1$]

respectively. The relationship between the digital inputs (Bit 0 to Bit 3) and the analog output V_{OUT} is as follow:

$$V_{OUT} = -V_{ref}\left(\frac{1}{1}Bit3 + \frac{1}{2}Bit2 + \frac{1}{4}Bit1 + \frac{1}{8}Bit0\right)$$

where V_{ref} is the Reference Voltage of the circuit.

Assuming the value of V_{ref} as 5 V, the Analog Output Voltages corresponding to the Digital Input Codes as follows:

Digital Input Code				Analog Output Voltage
Bit 3	Bit 2	Bit 1	Bit 0	(V)
0	0	0	0	0.000
0	0	0	1	-0.625
0	0	1	0	-1.250
0	0	1	1	-1.875
0	1	0	0	-2.500
0	1	0	1	-3.125
0	1	1	0	-3.750
0	1	1	1	-4.375
1	0	0	0	-5.000
1	0	0	1	-5.625
1	0	1	0	-6.250
1	0	1	1	-6.875
1	1	0	0	-7.500
1	1	0	1	-8.125
1	1	1	0	-8.750
1	1	1	1	-9.375

R-2R Ladder DAC by Using Inverting OP-AMP

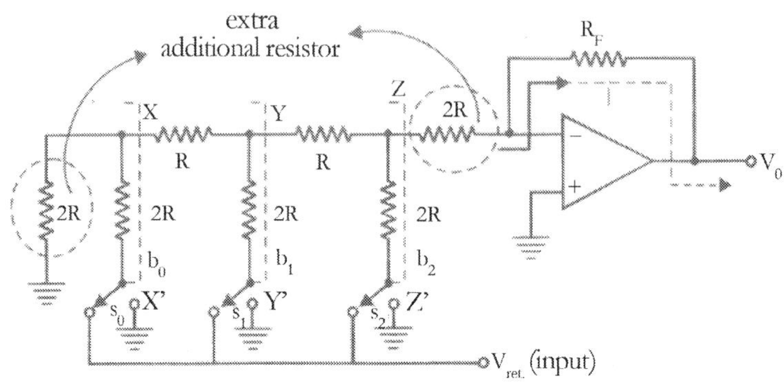

Output Voltage

For 3 bit DAC

$$V_0 = -\left(\frac{R_F}{3R}\right)\left(\frac{V_{ref}}{2^3}\right)\left[4b_2 + 2b_1 + b_0\right]$$

For n bit DAC

$$V_0 = \frac{V_{ref}}{2^N} \times \left[\sum_{i=0}^{N-1} 2^i b_i\right] \times \left[\frac{-R_F}{3R}\right]$$

Output Current

$$I = \frac{V_{ref}}{2^N} \times \left[\sum_{i=0}^{N-1} 2^i b_i\right] \times \left[\frac{1}{3R}\right]$$

Inverted Ladder (R-2R) type DAC

Output Voltage

For 4 bit DAC

$$V_0 = \frac{V_{ref}}{2^4}\left[b_0 + 2b_1 + 4b_2 + 8b_3\right]$$

For n bit DAC

$$V_0 = \frac{V_{ref}}{2^N} \times \left[\sum_{i=0}^{N-1} 2^i b_i\right] \times \left[\frac{-R_F}{R}\right]$$

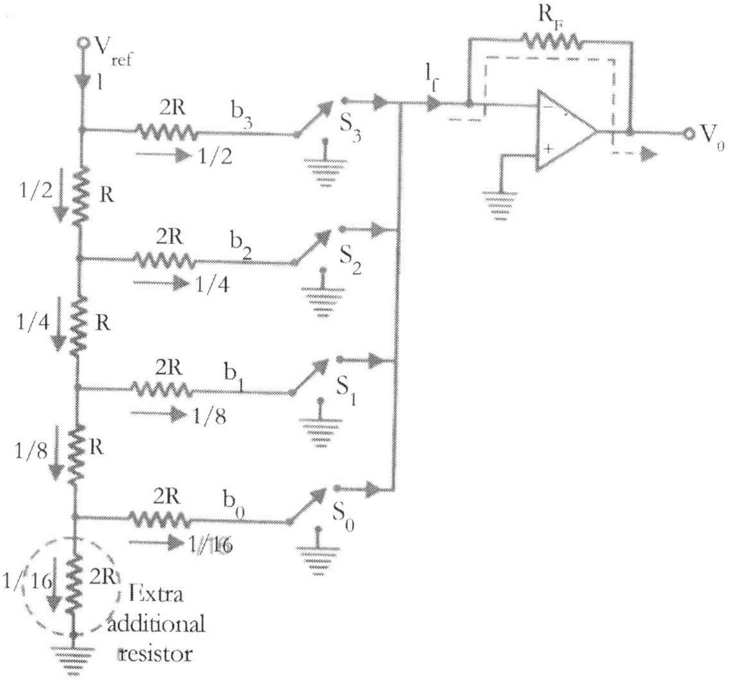

Forward Current

$$I_f = \frac{V_{ref}}{2^N} \times \left[\sum_{i=0}^{N-1} 2^i b_i \right] \times \left[\frac{1}{R} \right]$$

Analog to Digital Converter

An analog-to-digital converter (ADC, A/D, or A to D) is a device that converts a continuous physical quantity (usually voltage) to a digital number that represents the quantity's amplitude. An ADC is defined by its bandwidth (the range of frequencies it can measure) and its signal to noise ratio.

The performance of ADCs and DACs mainly depends on their Resolution and Speed. An n-bit ADC has the resolution of 1 / 2n.

Counter type ADC

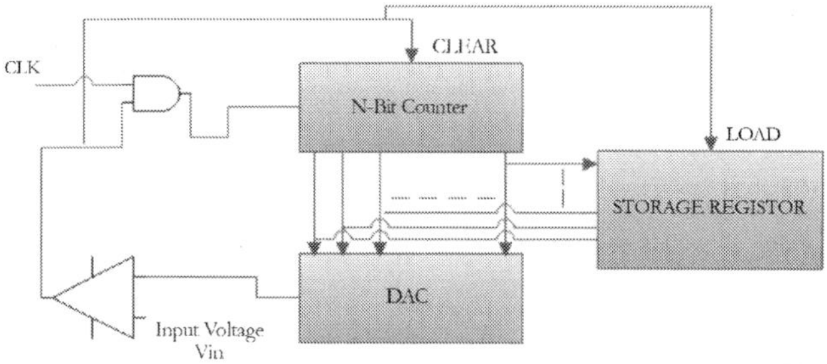

This circuit consists of N bit counter, DAC and Op-amp comparator.

The maximum conversion of Counter type ADC is = (2^N-1) T.

Where, T is the time period of clock pulse.

Parallel Comparator Type ADC

It is formed of a series of comparators, each one comparing the input signal to a unique reference voltage. The comparator outputs connect to the inputs of a priority encoder circuit, which then produces a binary output.

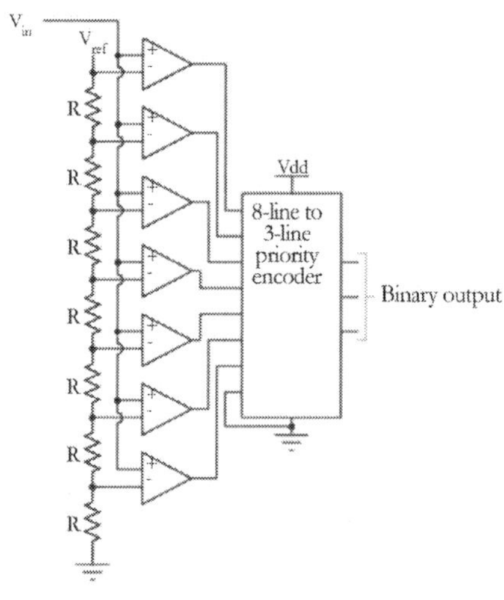

SAR Type ADC

Successive approximation ADC is the advanced version of Digital ramp type ADC which is designed to reduce the conversion and to increase speed of operation. In successive approximation ADC the normal counter is replaced with successive approximation register

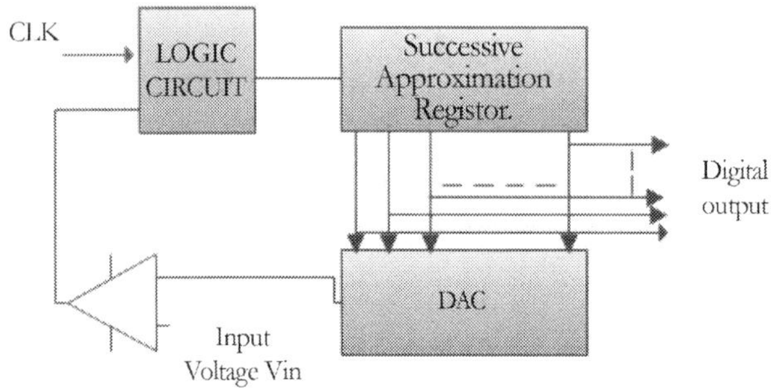

N bit Successive Approximation ADC conversion time = 3T (T- clock pulse).

Dual Slope Integrating Type ADC

A dual-slope ADC (DS-ADC) integrates an unknown input voltage (V_{IN}) for a fixed amount of time (T_{INT}), then "de-integrates" (T_{DEINT}) using a known reference voltage (V_{REF}) for a variable amount of time.

Dual-slope integration:

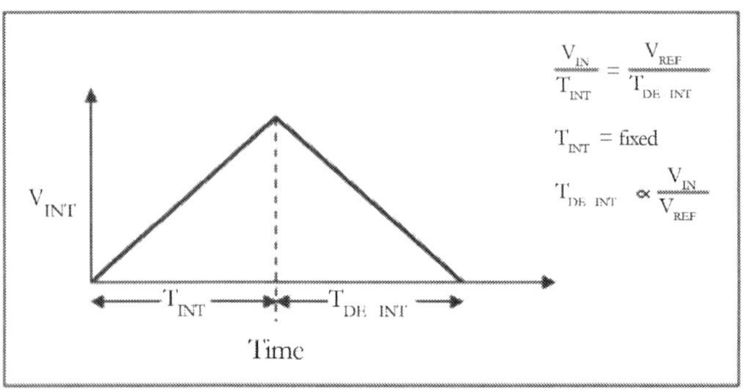

Dual-slope converter:

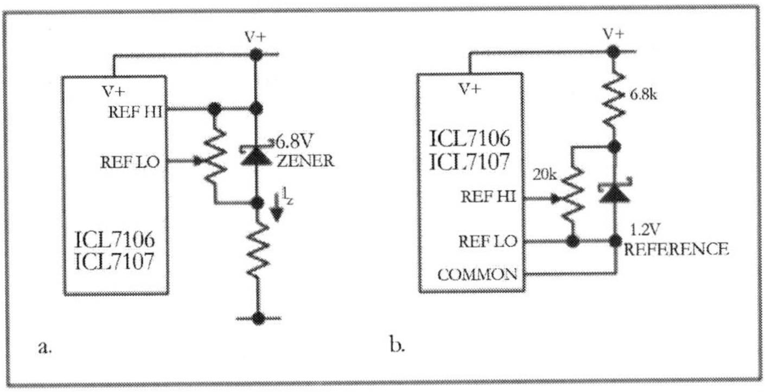

11 Miscellaneous

Monostable Multivibrators Using NOR and NAND Gates

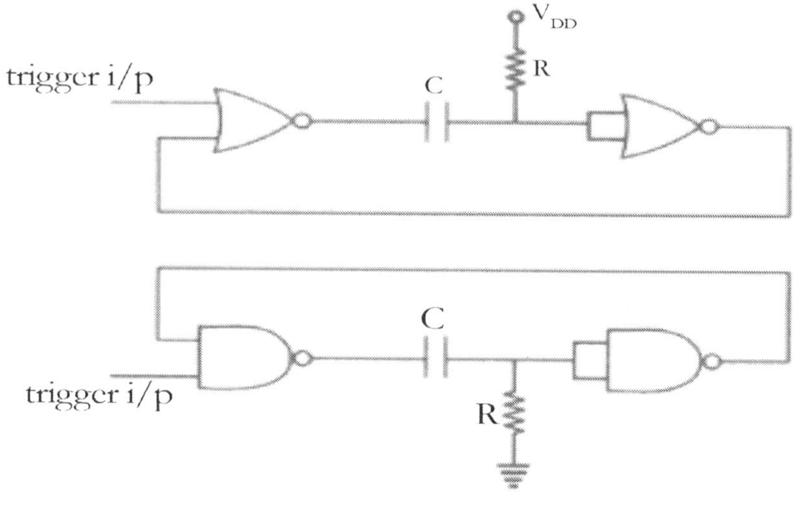

Memories

They are of two types

1. Primary memories

   RAM

   ROM

2. Secondary memories

   Semi random

   Serial access

A Handbook on Electronics and Communications Engineering

9

Microprocessor and Computer Organization

CONTENTS

1 Basics Microprocessors

The microprocessor contains all, or most of, the central processing unit (CPU) functions and is the "engine" that goes into motion when you turn your computer on. A microprocessor is designed to perform arithmetic and logic operations that make use of small number-holding areas called registers. Typical microprocessor operations include adding, subtracting, comparing two numbers, and fetching numbers from one area to another. These operations are the result of a set of instructions that are part of the microprocessor design.

8085 Microprocessors

Microprocessor is an electronic chip that functions as the central processing unit of a computer.

For example: Washing machines, microwave ovens, mobile phones etc.

Its advance applications are Radar, Satellites, and flights.

All processors are using the basic concept of stored program execution. Program or instructions are stored sequentially in the memory. Every microprocessor has its own associated set of instructions. Instruction set for microprocessor is in two forms one in mnemonic, which is comparatively easy to understand and the other is binary machine code.

8 bit microprocessor (8085 microprocessor can read or write or perform arithmetic and logical operations on 8-bit data at time).

- It has 8 data lines and 16 address lines hence capacity is 216 = 64 kB of memory
- Cock frequency is 3 MHz
- It requires +5V power supply.
- It is a single chip NMOS device implemented with 6200 transistors.
- It provides 74 instructions with five addressing modes.
- It provides 5 hardware interrupt and 8 software interrupts.

2 8085 Architecture

The internal architecture of 8085 includes the ALU, timing and control unit, instruction register and decoder, register array, interrupt control and serial I/O control. The main heart of microprocessor is CPU. CPU comprises of ALU, timing and control unit, instruction register and decoder, register array, interrupt control and serial I/O control.

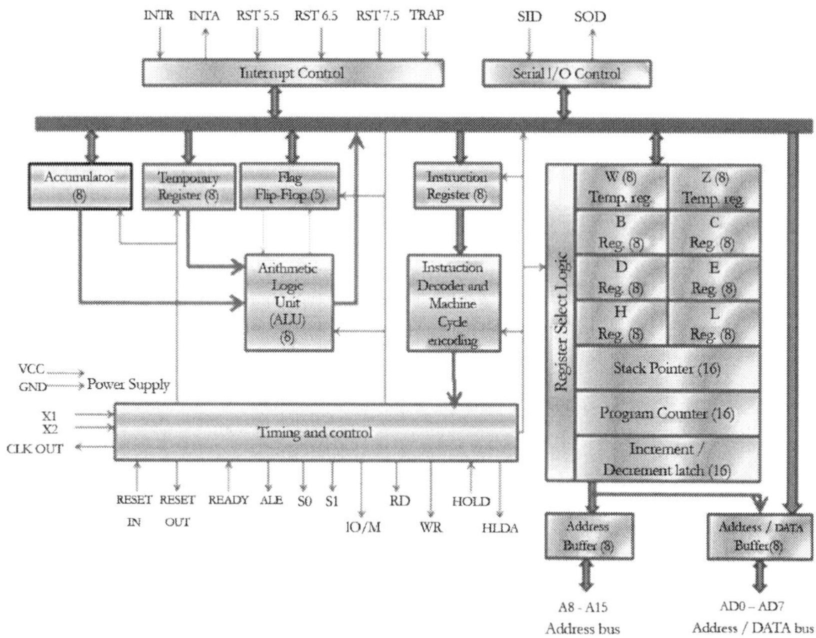

Flag Registers

It consists of 5 flip flop which changes its status according to the result stored in an accumulator. It is also known as status registers. It is connected to the ALU. There are five flip-flops in the flag register are as follows:

1. Sign(S)
2. Zero(z)
3. Auxiliary carry(AC)
4. Parity(P)
5. Carry(C)

The bit position of the flip flop in flag register is:

D_7	D_6	D_5	D_4	D_3	D_2	D_1	D_0
S	Z		AC		P		CY

All of the three flip flop set and reset according to the stored result in the accumulator.

1. Sign- If D7 of the result is 1 then sign flag is set otherwise reset. As we know that a number on the D7 always desides the sign of the number.

   If D7 is 1: the number is negative.

   If D7 is 0: the number is positive.
2. Zeros (Z)-If the result stored in an accumulator is zero then this flip flop is set otherwise it is reset.
3. Auxiliary carry (AC)-If any carry goes from D3 to D4 in the output then it is set otherwise it is reset.
4. Parity (P)-If the no of 1's is even in the output stored in the accumulator then it is set otherwise it is reset for the odd.
5. Carry(C)-If the result stored in an accumulator generates a carry in its final output then it is set otherwise it is reset.

Program Counter

It is a 16 bit register used as memory pointer. It stores the memory address of the next instruction to be executed. So we can say that this register is used to sequencing the program. Generally the memories have 16 bit addresses so that it has 16 bit memory. The program counter is set to 0000H.

Stack Pointer

It is also a 16 bit register used as memory pointer. It points to the memory location called stack. Generally stack is a reserved portion of memory where information can be stores or taken back together.

8085 Machine Cycle Status and Control Signals

		Status			
	Machine cycle	$IO/\overline{M}$	S_1	S_0	Control signals
1.	Opcode fetch	0	1	1	$\overline{RD} = 0$
2.	Memory read	0	1	0	$\overline{RD} = 0$
3.	Memory write	0	0	1	$\overline{WR} = 0$
4.	I/O read	1	1	0	$\overline{RD} = 0$
5.	I/O write	1	0	1	$\overline{WR} = 0$
6.	Interrupt Acknowledge	1	1	1	$\overline{INTA} = 0$
7.	Halt	Z	0	0	$\overline{RD} = 0$
8.	Hold	Z	X	X	$\overline{WR} = 0$
9.	Reset	Z	X	X	$\overline{INTA} = 1$

where Z = Tri state (high impedance)
X = Unspecified

Interfacing I/O Devices

- There are memory mapped I/O devices and I/O mapped I/O.
- For I/O mapped I/O only IN and OUT instruction are used for data transfer.
- For memory mapped I/O, all memory related instructions can be us for data transfer.
- Maximum 256 I/O mapped I/O devices can be connected to 80 microprocessor.
- If control devices exceed the number 256 then memory mapped I/O is used.
- Arithmetical or logical operation can be performed directly in case of memory-mapped I/O while it is not possible in case of I/O mapped I/O.
- An input and output port can have same address. They will be differentiated by control signals. $\overline{RD}$ is used to enable the input port and the $\overline{WR}$ is used to enable the output port.
- A tri-state buffer is commonly used to interface input devices.

3 Introduction to 8085 Programming

The processor will start execution this program starting from memory location 0100H and it keeps on doing the execution job till it encounters a halt instruction.

If we program the microprocessor in this way, i.e. writing the machine code directly, it is known as machine language programming.

Assembly Language programming:

The assembly language programming is simpler as compared to the machine language programming.

The instruction mnemonics are directly used in the assembly language programming. There is an one-to –one correspondence between instruction mnemonics and machine instruction.

An assembler is used to translate the assembly language programming to machine code.

The program written in assembly language programming is more readable than machine language programming.

The main improvement in assembly language over machine language is that the address value and the contents can be identified by labels.

The assembly language instruction sequence for 8085 microprocessor of the previously discussed program will look like:

LDA 1000*H* // Load accumulator from memory address 1000*H*

MOV B, *A* // Move the content from accumulator to *B*

LDA 1001*H* // Load accumulator from memory address 1001*H*

ADD B // add the content of B to accumulator and store the result in accumulator

STA 1002*H* // Store the content of accumulator in memory location 1002*H*

HLT // Halt

The 8085 Addressing Modes

The instructions MOV B, A or MVI A, 82H are to copy data from a source into a destination. In these instructions the source can be a register, an input port, or an 8-bit number (00H to FFH). Similarly, a destination can be a register or an output port. The sources and destination are operands. The various formats for specifying operands are called the ADDRESSING MODES. For 8085, they are:

1. Immediate addressing.
2. Register addressing.
3. Direct addressing.
4. Indirect addressing.

Immediate Addressing

Data is present in the instruction. Load the immediate data to the destination provided.

Example: MVI R,data

Register Addressing

Data is provided through the registers.

Example: MOV Rd, Rs

Direct Addressing

Used to accept data from outside devices to store in the accumulator or send the data stored in the accumulator to the outside device. Accept the data from the port 00H and store them into the accumulator or Send the data from the accumulator to the port 01H.

Example: IN 00H or OUT 01H

Indirect Addressing

This means that the Effective Address is calculated by the processor. And the content of the address is used to form a second address. The second address is where the data is stored. Note that this requires several memory accesses; two accesses to retrieve the 16-bit address and a further access (or accesses) to retrieve the data which is to be loaded into the register.

Instruction Set

An instruction is a binary pattern designed inside a microprocessor to perform a specific function. The entire group of instructions, called the instruction set, determines what functions the microprocessor can perform.

The instruction set, also called instruction set architecture (ISA), is a part of the computer that pertains to programming, which is basically machine language. The instruction set provides commands to the processor, to tell it what it needs to do. The instruction set consists of multiple pieces, including addressing modes, instructions, native data types, registers, memory architecture, interrupt, and exception handling, and external I/O.

Types of Instruction Set

These instructions can be classified into the following five functional categories: data transfer operations, arithmetic operations, logical operations, branching operations, and machine-control operations.

Data Transfer Operations

This group of instructions copy data from a location called a source to another location called a destination, without modifying the contents of the source. In technical manuals, the term data transfer is used for this copying function. However, the term transfer is misleading; it creates the impression that the contents of the source are destroyed when, in fact, the contents are retained without any modification. The various types of data transfer (copy) are listed below together with examples of each type:

Type	Examples
1. Between Registers.	1.Copy the contents of the registers B into register D
2. Specific data byte to a registers or a memory location.	2. Load register B with the data byte 32H.
3. Between a Memory location and a registers.	3.From a memory Location 2000H to register B
4. Between an I/O device and the accumulator.	From an input keyboard to the accumulator.

Arithmetic Operations

These instructions perform arithmetic operations such as addition, subtraction, increment, and decrement.

Addition

Any 8-bit number, or the contents of a register or the contents of a memory location can be added to the contents of the accumulator and the sum is stored in the accumulator. No two other 8-bit registers can be added directly (e.g., the contents of register B cannot be added directly to the contents of the register C). The instruction DAD is an exception; it adds 16-bit data directly in register pairs.

Subtraction

Any 8-bit number, or the contents of a register, or the contents of a memory location can be subtracted from the contents of the accumulator and the results stored in the accumulator. The subtraction is performed in 2's compliment, and the results if negative, are expressed in 2's complement. No two other registers can be subtracted directly.

Increment/Decrement

The 8-bit contents of a register or a memory location can be incremented or decrement by 1. Similarly, the 16-bit contents of a register pair (such as BC) can be incremented or decrement by 1. These increment and decrement operations differ from addition and subtraction in an important way; i.e., they can be performed in any one of the registers or in a memory location.

Logical Operations

These instructions perform various logical operations with the contents of the accumulator.

AND, OR Exclusive-OR

Any 8-bit number, or the contents of a register, or of a memory location can be logically ANDed, Ored, or Exclusive-ORed with the contents of the accumulator. The results are stored in the accumulator.

Rotate

Each bit in the accumulator can be shifted either left or right to the next position.

Compare

Any 8-bit number, or the contents of a register, or a memory location can be compared for equality, greater than, or less than, with the contents of the accumulator.

Complement

The contents of the accumulator can be complemented. All 0s are replaced by 1s and all 1s are replaced by 0s.

Branching Operations

This group of instructions alters the sequence of program execution either conditionally or unconditionally.

Jump

Conditional jumps are an important aspect of the decision-making process in the programming. These instructions test for a certain conditions (e.g., Zero or Carry flag) and alter the program sequence when the condition is met. In addition, the instruction set includes an instruction called unconditional jump.

Call, Return, and Restart

These instructions change the sequence of a program either by calling a subroutine or returning from a subroutine. The conditional Call and Return instructions also can test condition flags.

Machine Control Operations

These instructions control machine functions such as Halt, Interrupt, or do nothing.

The microprocessor operations related to data manipulation can be summarized in four functions:

- copying data
- performing arithmetic operations
- performing logical operations
- testing for a given condition and alerting the program sequence

Some important aspects of the instruction set are noted below:

1. In data transfer, the contents of the source are not destroyed; only the contents of the destination are changed. The data copy instructions do not affect the flags.
2. Arithmetic and Logical operations are performed with the contents of the accumulator, and the results are stored in the accumulator (with some expectations). The flags are affected according to the results.

3. Any register including the memory can be used for increment and decrement.
4. A program sequence can be changed either conditionally or by testing for a given data condition

4 Interrupts

When a Process is executed by the CPU and when a user Request for another Process then this will create disturbance for the Running Process. This is also called as the Interrupt.

Interrupts can be generated by User, Some Error Conditions and also by Software's and the hardware's. But CPU will handle all the Interrupts very carefully because when Interrupts are generated then the CPU must handle all the Interrupts Very carefully means the CPU will also Provide Response to the Various Interrupts those are generated. When an interrupt has Occurred then the CPU will handle by using the Fetch, decode and Execute Operations.

Types of Interrupts

Generally there are three types of Interrupts those are occurred:

1. External Interrupt.
2. Internal Interrupt
3. Software Interrupt.

External Interrupts

These types of interrupts generally come from external input/output devices which are connected externally to the processor. They are generally independent and oblivious of any programming that is currently running on the processor.

Internal Interrupts

They are also known as traps and their causes could be due to some illegal operation or the erroneous use of data. Instead of being triggered by an external event they are usually triggered due to any exception that has been caused by the program itself. Some of the causes of these types of interrupts can be due to attempting a division by zero or an invalid opcode etc.

Software Interrupts

These types if interrupts can occur only during the execution of an instruction. They can be used by a programmer to cause interrupts if need be. The primary purpose of such interrupts is to switch from user mode to supervisor mode.

5 8155 I/O & Timer and 8279 Keyboard Display

Control word definition in the 8155

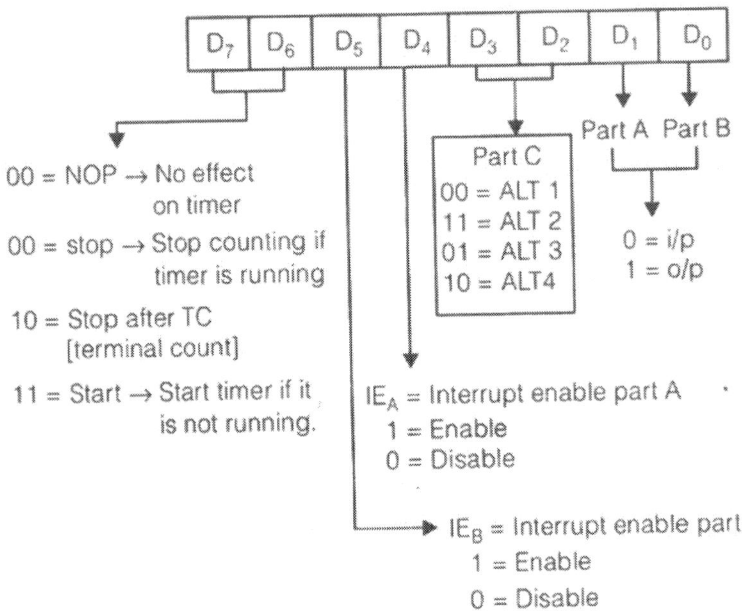

$$\text{Timer count} = \frac{\text{Pulse period}}{\text{Clock period}}$$

- The timer section of 8155 has two 8-bit registers; 14 bits are used for the counter, two bits for the timer mode.

M_2 M_1

0 0 → Mode 0 → Remains high for half of the count and goes low for rest of count, providing single square wave.

0 1 → Mode 1 → For generating continuous square wave.

1 0 → Mode 2 → In this mode a single clock pulse is provided at the end of the count like monostable multivibrator.

1 1 → Mode 3 → This is similar to mode 2 except the initial count is reloaded to provide a continuous wave form.

COMPUTER ORGANIZATION

6 Basics

A computer system has three main components: a central processing unit (CPU) or processor, a memory unit, and input/output (I/O) devices. These three components are interconnected by a system bus. The term "bus" is used to represent a group of electrical signals or the wires that carry these signals.

Machine Instructions and Addressing Modes

Computer Architecture and Organization is the study of internal working, structuring and implementation of a computer system. Architecture in computer system, same as anywhere else, refers to the externally visual attributes of the system. Externally visual attributes, here in computer science, mean the way a system is visible to the logic of programs. Organization of computer system is the way of practical implementation which results in realization of architectural specifications of a computer system.

Machine Instructions

Operation commands (program) written in the native code (machine code) of a machine (such as a computer or robot) that it can recognize and execute.

- Zero address instructions: NOP
- One address instructions: PUSH, POP
- Two address instruction: ADD, MULT, SUB

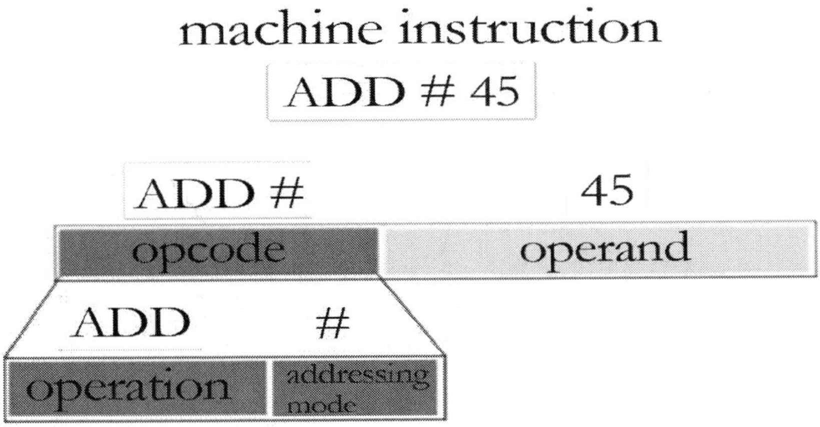

Load and Store Instructions

Load/store Instructions only allows memory to be accessed by load and store operations, and all values for an operation need to be loaded from memory and be present in registers. Following the operation, the result needs to be stored back to memory.

Addressing Modes

The term addressing modes refers to the way in which the operand of an instruction is specified. Information contained in the instruction code is the value of the operand or the address of the result/operand. Following are the main addressing modes that are used on various platforms and architectures.

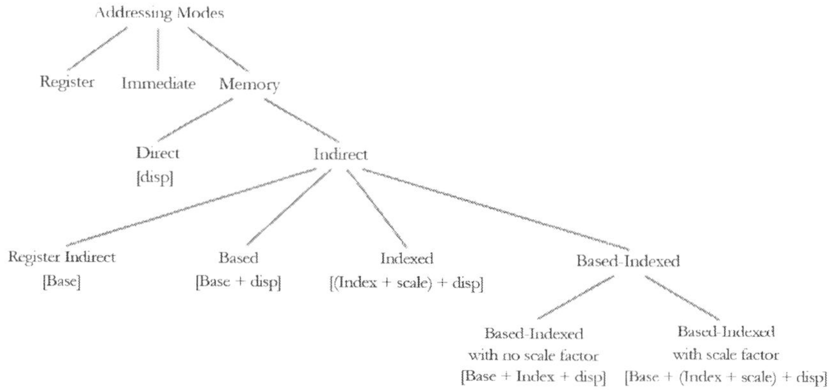

Sequential Control Flow Addressing Mode

Sequential control flow addressing mode is describe as follows:

Register addressing mode

Register mode is similar to direct mode. The key difference between the two modes is that the address field of the instruction refers to a register rather than a memory location.

- Operands are located in registers
- It is the most efficient addressing mode

Immediate Addressing Mode

The immediate mode is the simplest form of addressing. The operand is part of the instruction and therefore no memory reference, other than the instruction, is required to retrieve the operand. This fast mode is used to define constants or set initial variable values. Immediate mode has a limited range because it is restricted to the size of the address field, which for most instruction sets is small compared to word length.

- Operand is stored as part of the instruction
- This mode is used mostly for constants

- It imposes several restrictions
- Efficient as the data comes with the instructions

Auto Indexed Addressing Mode

- This mode is used to access the linear array element.
- The base address is maintained in the base register.
- Step size is dependency on the amount of the data to be accessed from the memory.

Data= [EA]=[[base register]+[step size]]

Memory Addressing Modes

- Pentium offers several addressing modes to access operands located in memory
- Primary reason: To efficiently support high-level language constructs and data structures
- Available addressing modes depend on the address size used
- 16-bit modes (shown before)
- same as those supported
- 32-bit modes (shown before)
- supported by Pentium
- more flexible set

Direct Memory Addressing

When operands are specified in memory addressing mode, direct access to main memory, usually to the data segment, is required. This way of addressing results in slower processing of data. To locate the exact location of data in memory, we need the segment start address, which is typically found in the DS register and an offset value. This offset value is also called effective address.

In direct addressing mode, the offset value is specified directly as part of the instruction, usually indicated by the variable name. The assembler calculates the offset value and maintains a symbol table, which stores the offset values of all the variables used in the program.

Indirect Memory Addressing

This addressing mode utilizes the computer's ability of Segment: Offset addressing. Generally, the base registers EBX, EBP (or BX, BP) and the index registers (DI, SI), coded within square brackets for memory references, and is used for this purpose.

Indirect addressing is generally used for variables containing several elements like, arrays. Starting address of the array is stored in, say, the EBX register.

ALU and Data Path, CPU Control Design

ALU and data path, CPU control design are follows:

Arithmetic logic unit (ALU)

Arithmetic logic unit is a digital circuit used to perform arithmetic and logic operations. It represents the fundamental building block of the central processing unit of a computer. Modern CPUs contain very powerful and complex ALUs. In addition to ALUs, modern CPUs contain a control unit. Most of the operations of a CPU are performed by one or more ALUs, which load data from input registers. A register is a small amount of storage available as part of a CPU. The control unit tells the ALU what operation to perform on that data and the ALU stores the result in an output register. The control unit moves the data between these registers, the ALU and memory.

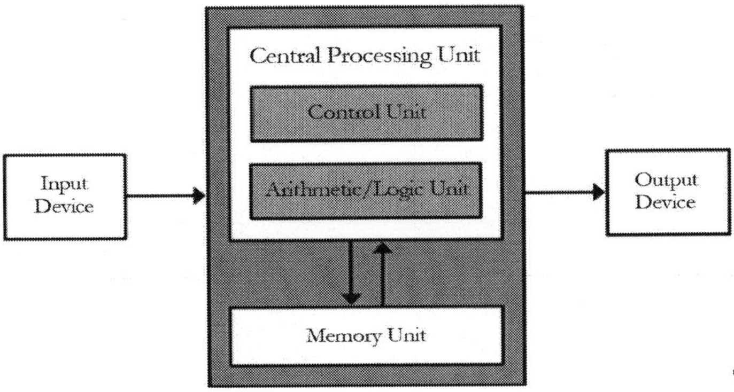

Data path

A data path is a collection of functional units, as arithmetic logic units or multipliers that perform data processing operations, registers, and buses. Along with the control unit it composes the central processing unit (CPU).

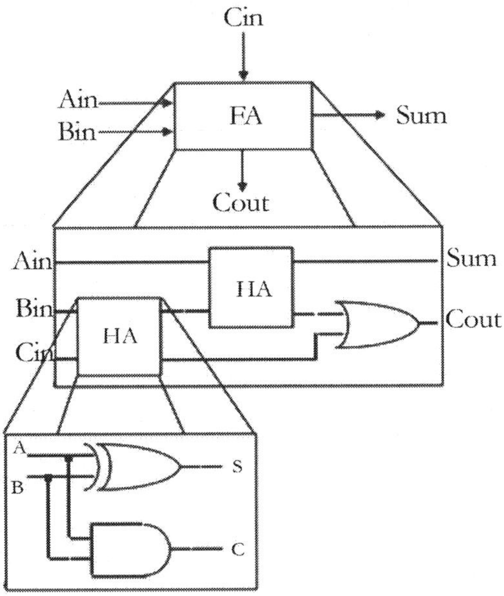

- Elements that process data and addresses in the CPU
- We will build a MIPS data path incrementally
- considering only a subset of instructions

CPU Registers

In computer architecture, a processor register is a very fast computer memory used to speed the execution of computer programs by providing quick access to commonly use values-typically, the values being in the midst of a calculation at a given point in time.

These registers are the top of the memory hierarchy, and are the fastest way for the system to manipulate data. In a very simple microprocessor, it consists of a single memory location, usually called an accumulator. Registers are built from fast multi-ported memory cell. They must be able to drive its data onto an internal bus in a single clock cycle. The result of ALU operation is stored here and could be re-used in a subsequent operation or saved into memory.

- **Memory Address Register (MAR)**: It stores address of data or instructions to be fetched from memory.
- **Memory Buffer Register (MBR)**: It stores instruction and data received from the memory and sent from the memory.
- **Instruction Register (IR):** Instructions are stored in instruction register. When one instruction is completed, next instruction is fetched in memory for processing.
- **Program Counter (PC):** It counts instructions.

CPU Control Design

CPU-Control is a small yet powerful software solution designed to boost performance of multi-core systems, setting up the processes to run on separate processors. The CPU has two major components: the Control Unit and the ALU. If the design did include an instruction pre-fetch unit, that unit would independently fetch instructions and place them in an instruction queue for use by the execute unit, which might then fetch and execute an instruction in a single step.

Design of Control Unit

To execute an instruction, the control unit of the CPU must generate the required control signal in the proper sequence. As for example, during the fetch phase, CPU has to generate PCout signal along with other required signal in the first clock pulse.

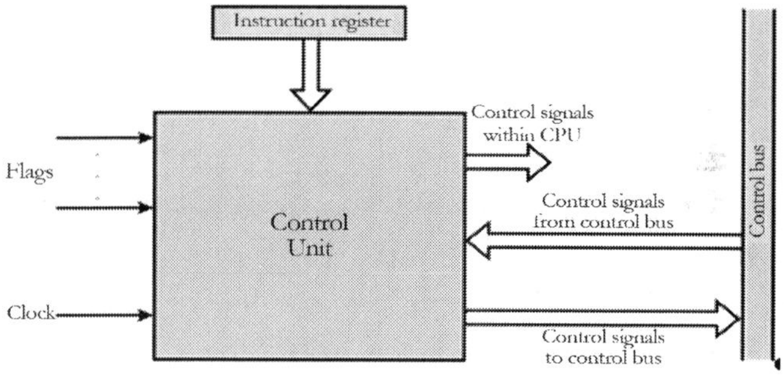

In the second clock pulse CPU has to generate PCin signal along with other required signals. So, during fetch phase, the proper sequence for generating the signal to retrieve from and store to PC is PCout and PCin. To generate the control signal in proper signal, a write variety of techniques exists. Most of these techniques, however, fall into one of the two categories.

- Hardwired control
- Micro programmed Control

Hardwired Control Unit

In this hardwired control techniques the control signals are generated by means of hardwired circuit. The main object of control unit is to generate the control signal in proper sequence.

Consider the sequence of control signal required to execute the add instruction that is explained in previous lecture. It is obvious that eight non-overlapping time slots are required for proper execution of the instruction represented by this sequence.

Hence the required control signals are uniquely determined by the following

- Contents of the control counter
- Contents of instruction register
- Contents of the condition code and other status flags.

Similarly, the Boolean logic function for add signal is

Add = T1 + T6.ADD_MD + T5.BR + ...

These logic functions can be implemented by a two level combinational circuit of AND and OR gates.

Similarly, the End control signal is generated by the logic function:

End = T8.ADD_MD + M7.BR + (T7.N + T4.Nbar).BRN + ...

Micro-program Control Unit

Control unit (CU) design focused on hardwired control units - i.e., ones where the control functions for the various components were synthesized using combinational logic. In this document, we will consider an alternate design technique: the micro programmed CU.

Memory and I/O Interface

Memory and I/O interface are discuss in follows:

Memory

In computing, memory refers to the devices used to store information for use in a computer. The term primary memory is used for storage systems which function at high-speed (i.e. RAM), as a distinction from secondary memory, which provides program and data storage that is slow to access but offer higher memory capacity. If needed, primary memory can be stored in secondary memory, through a memory management technique called "virtual memory". An archaic synonym for memory is store.

The term "memory", meaning primary memory is often associated with addressable semiconductor memory, i.e. integrated circuits consisting of silicon-based transistors, used for example as primary memory but also other purposes in computers and other digital electronic devices. There are two main types of semiconductor memory: volatile and non-volatile. Examples of non-volatile memory are flash memory (sometimes used as secondary, sometimes primary computer memory) and ROM/PROM/EPROM/EEPROM memory (used for firmware such as boot programs). Examples of volatile memory are primary memory (typically dynamic RAM, DRAM), and fast CPU cache memory (typically static RAM, SRAM, which is fast but energy-consuming and offer lower memory capacity per area unit than DRAM).

Most semiconductor memory is organized into memory cells or bistable flip-flops, each storing one bit (0 or 1). Flash memory organization includes both one bit per memory cell and multiple bits per cell (called MLC, Multiple Level Cell). The memory cells are grouped into words of fixed word length, for example 1, 2, 4, 8, 16, 32, 64 or 128 bit. Each word can be accessed by a binary address of N bit, making it possible to store 2 raised by N words in the memory. This implies that processor registers normally are not considered as memory, since they only store one word and do not include an addressing mechanism.

The term storage is often used to describe secondary memory such as tape, magnetic disks and optical discs (CD-ROM and DVD-ROM).

Memory Interface

Memory interface is a bus protocol for communication from an integrated circuit, such as a microprocessor, to an external memory device located on a circuit board. The memory is referred to as external because it is not contained within the internal circuitry of the integrated circuit and thus is externally located on the circuit board.

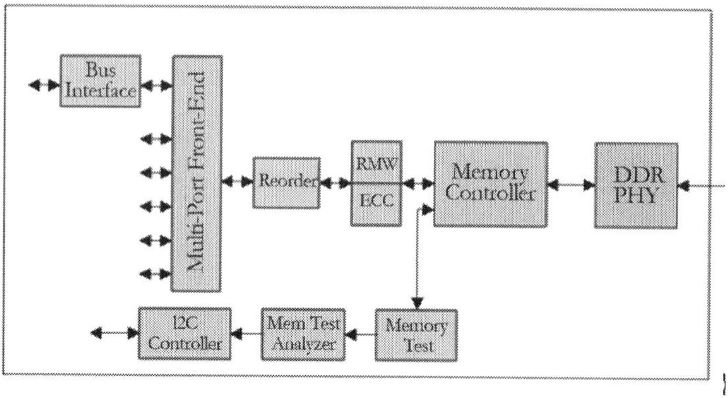

Input and Output Interface

Provides a method for transferring information between internal storage (such as memory and CPU registers) and external I/O devices

Resolves the differences between the computer and peripheral devices

- Peripherals - Electromechanical Devices
- CPU or Memory - Electronic Device
- Data Transfer Rate
- Peripherals - Usually slower
- CPU or Memory - Usually faster than peripherals

Some kinds of Synchronization mechanism may be needed

Unit of Information

- Peripherals – Byte, Block, …
- CPU or Memory – Word

Input Device

- Keyboard
- Optical input devices

  Card Reader

  Paper Tape Reader

Bar code reader

Digitizer

Optical Mark Reader

- Magnetic Input Devices

Magnetic Stripe Reader

- Screen Input Devices

Touch Screen

Light Pen

Mouse

- Analog Input Devices

Output Devices

- Card Puncher, Paper Tape Puncher
- CRT
- Printer (Impact, Ink Jet, Laser, Dot Matrix)
- Plotter
- Analog
- Voice

I/O Interface Module

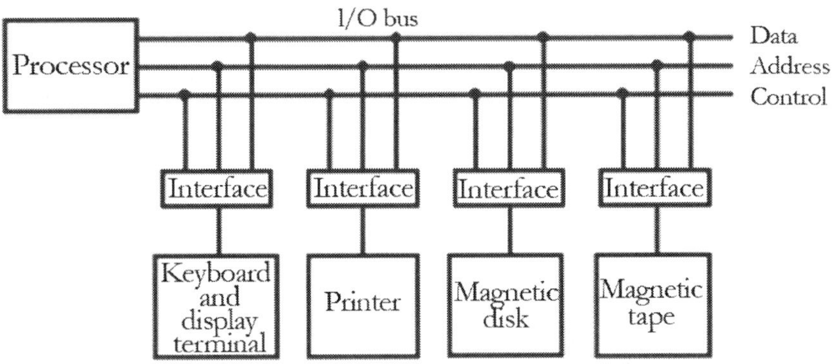

Each peripheral has an interface module associated with it Interface

- Decodes the device address (device code)
- Decodes the commands (operation)
- Provides signals for the peripheral controller
- Synchronizes the data flow and supervises the transfer rate between peripheral and CPU or Memory

Interrupted

The CPU issues commands to the I/O module then proceeds with its normal work until interrupted by I/O device on completion of its work.

For input, the device interrupts the CPU when new data has arrived and is ready to be retrieved by the system processor. The actual actions to perform depend on whether the device uses I/O ports, memory mapping.

Direct Memory Access (DMA)

Direct Memory Access (DMA) means CPU grants I/O module authority to read from or write to memory without involvement. DMA module controls exchange of data between main memory and the I/O device. Because of DMA device can transfer data directly to and from memory, rather than using the CPU as an intermediary, and can thus relieve congestion on the bus. CPU is only involved at the beginning and end of the transfer and interrupted only after entire block has been transferred.

Instruction Pipelining

The instruction pipeline represents the stages in which an instruction is moved through the processor, including its being fetched, perhaps buffered, and then executed.

- Instruction fetch
- Instruction decode and register fetch
- Execute

- Memory access
- Register write back

Instruction Processing

The first step in applying pipelining techniques to instruction processing is to divide the task into steps that may be performed with independent hardware. The most obvious division is between the FETCH cycle (fetch and interpret instructions) and the EXECUTE cycle (access operands and perform operation). If these two activities are to run simultaneously, they must use independent registers and processing circuits, including independent access to memory

Non-linear Pipelines

More sophisticated instruction pipelines can sometimes be nonlinear or no sequential. One *example* is branch processing in which the pipeline has two forks to process two possible paths at once. Sequential processing can be relaxed by a pipeline which allows a later instruction to enter when a previous one is stalled by a data conflict. This, of course, introduces much more difficult timing and consistency problems.

Cache Memory and Main Memory, Secondary Storage

Cache Memory and main memory, secondary storage are discuss in follows:

Cache Memory

A small, high speed memory close to the CPU

- Each line of cache memory can be "occupied" or "empty"
- "Occupied" lines map to a memory location
- Hardware cache management mechanism (potentially with software hints)

- Typically used for both programs and data

Also, by extension, other specialized hardware or software buffers

- TLB for mapping physical to virtual memory addresses
- Disk sector storage
- Network machine name to IP address translation
- Read and write buffers

Works because of locality

- (Does not work when there is not locality)

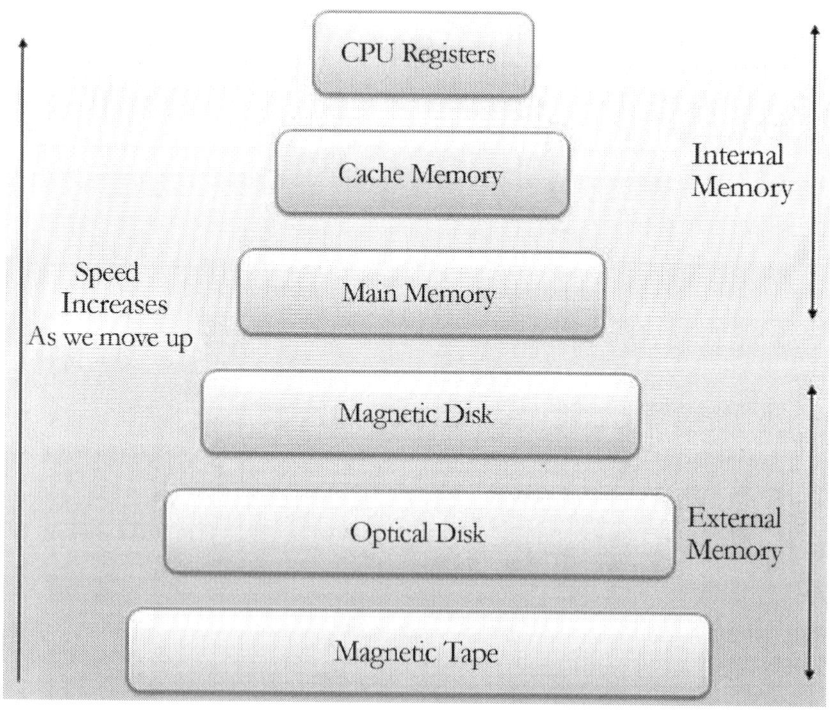

Main Memory

The main memory of the computer is also known as RAM, standing for Random Access Memory. It is constructed from integrated circuits and needs to have electrical power in order to maintain its information. When power is lost, the information is lost too! It can be directly accessed by the CPU. Main memory is where programs and data are kept when the processor is actively using them. When programs and data become active, they are copied from secondary memory into main memory where the processor can interact with them. A copy remains in secondary memory.

Secondary Memory

Secondary memory is where programs and data are kept on a long-term basis. Common secondary storage devices are the hard disk and optical disks.

- The hard disk has enormous storage capacity compared to main memory.
- The hard disk is usually contained inside the case of a computer.
- The hard disk is used for long-term storage of programs and data.
- Data and programs on the hard disk are organized into files.
- A **file** is a collection of data on the disk that has a name.

Secondary Storage

Secondary storage does not lose the data when the device is powered down—it is non-volatile. Per unit, it is typically also an order of magnitude less expensive than primary storage. Consequently, modern computer systems typically have an order of magnitude more secondary storage than primary storage and data is kept for a longer time there.

7 Graph

Graph theory is the study of graphs, which are mathematical structures used to model pairwise relations between objects. A "graph" in this context is made up of "vertices" or "nodes" and lines called edges that connect them.

In the most common sense of the term, a graph is an ordered pair $G = (V, E)$ A comprising a set V V of vertices or nodes together with a set E of edges or lines, which are 2-element subsets of V (i.e., an edge is related with two vertices, and the relation is represented as an unordered pair of the vertices with respect to the particular edge). To avoid ambiguity, this type of graph may be described precisely as undirected and simple.

Other senses of *graph* stem from different conceptions of the edge set. In one more generalized notion, V is a set together with a relation of incidence that associates with each edge two vertices. In another generalized notion, E is a multiset of unordered pairs of (not necessarily distinct) vertices? Many authors call this type of object a multigraph or pseudo graph.

All of these variants and others are described more fully below.

The vertices belonging to an edge are called the ends, endpoints, or end vertices of the edge. A vertex may exist in a graph and not belong to an edge.

V and E are usually taken to be finite, and many of the well-known results are not true (or are rather different) for infinite graphs because many of the arguments fail in the infinite case. The order of a graph is $|V|$ (the number of vertices). A graph's size is $|E|$, the number of edges. The degree or valence of a vertex is the number of edges that connect to it, where an edge that connects to the vertex at both ends (a loop) is counted twice.

For an edge $\{u, v\}$ s, graph theorists usually use the somewhat shorter notation uv.

Problems in Graph Theory

Problems in graph theory are follows:

Enumeration

There is a large literature on graphical enumeration: the problem of counting graphs meeting specified conditions. Some of this work is found in Harary and Palmer (1973).

Subgraphs, induced subgraphs, and minors

A common problem, called the subgraph isomorphism problem, is finding a fixed graph as a subgraph in a given graph. One reason to be interested in such a question is that many graph properties are hereditary for subgraphs, which means that a graph has the property if and only if all subgraphs have it too. Unfortunately, finding maximal subgraphs of a certain kind is often an NP-complete problem.

- Finding the largest complete subgraph is called the clique problem (NP-complete).

A similar problem is finding induced subgraphs in a given graph. Again, some important graph properties are hereditary with respect to induced subgraphs, which means that a graph has a property if and only if all induced subgraphs also have it. Finding maximal induced subgraphs of a certain kind is also often NP-complete. For example,

- Finding the largest edgeless induced subgraph, or independent set, called the independent set problem (NP-complete).

Still another such problem, the minor containment problem, is to find a fixed graph as a minor of a given graph. A minor or sub-contraction of a graph is any graph obtained by taking a subgraph and contracting some (or no) edges. Many graph properties are hereditary for minors, which mean that a graph has a property if and only if all minors have it too. A famous example:

- A graph is planar if it contains as a minor neither the complete bipartite graph $K_{3,3}$ (See the Three-cottage problem) nor the complete graph K_5.

Another class of problems has to do with the extent to which various species and generalizations of graphs are determined by their point-deleted subgraphs, for example:

- The reconstruction conjecture.

Graph coloring

Many problems have to do with various ways of coloring graphs, for example:

- The four-color theorem
- The strong perfect graph theorem
- The Erdős–Faber–Lovász conjecture(unsolved)
- The total coloring conjecture, also called Behzad's conjecture) (unsolved)
- The list coloring conjecture (unsolved)
- The Hadwiger conjecture (graph theory) (unsolved)

Subsumption and unification

Constraint modeling theories concern families of directed graphs related by a partial order. In these applications, graphs are ordered by specificity, meaning that more constrained graphs—which are more specific and thus contain a greater amount of information—are subsumed by those that are more general. Operations between graphs include evaluating the direction of a subsumption relationship between two graphs, if any, and computing graph unification. The unification of two argument graphs is defined as the most general graph (or the computation thereof) that is consistent with (i.e. contains all of the information in) the inputs, if such a graph exists; efficient unification algorithms are known.

For constraint frameworks which are strictly compositional, graph unification is the sufficient satisfiability and combination function. Well-known applications include automatic theorem proving and modeling the elaboration of linguistic structure.

Route problems

- Hamiltonian path and cycle problems
- Minimum spanning tree
- Route inspection problem (also called the "Chinese Postman Problem")
- Seven Bridges of Königsberg
- Shortest path problem
- Steiner tree
- Three-cottage problem
- Traveling salesman problem (NP-hard)

Network flow

There are numerous problems arising especially from applications that have to do with various notions of flows in networks, for example:

- Max flow min cut theorem

Visibility problem

- Museum guard problem

Covering problems

Covering problems in graphs are specific instances of subgraph-finding problems, and they tend to be closely related to the clique problem or the independent set problem.

- Set cover problem
- Vertex cover problem

Decomposition problems

Decomposition, defined as partitioning the edge set of a graph (with as many vertices as necessary accompanying the edges of each part of the partition), has a wide variety of question. Often, it is required to decompose a graph into sub graphs isomorphic to a fixed graph; for instance, decomposing a complete graph into Hamiltonian cycles. Other problems specify a family of graphs into which a given graph should be decomposed, for instance, a family of cycles, or decomposing a complete graph K_n into $n - 1$ specified trees having, respectively, 1, 2, 3, ..., $n - 1$ edges.

Some specific decomposition problems that have been studied include:

- Arbor city, a decomposition into as few forests as possible
- Cycle double cover, a decomposition into a collection of cycles covering each edge exactly twice
- Edge coloring, a decomposition into as few matching's as possible
- Graph factorization, a decomposition of a regular graph into regular subgraphs of given degrees

Graph classes

Many problems involve characterizing the members of various classes of graphs. Some examples of such questions are below:

- Enumerating the members of a class
- Characterizing a class in terms of forbidden substructures

- Ascertaining relationships among classes (e.g., does one property of graphs imply another)
- Finding efficient algorithms to decide membership in a class
- Finding representations for members of a class.

8 Tree

A tree is a widely used abstract data type (ADT) or data structure implementing this ADT that simulates a hierarchical tree structure, with a root value and sub trees of children with a parent node, represented as a set of linked nodes.

Directed Acyclic Graph

A directed acyclic graph is a directed graph with no directed cycles. That is, it is formed by a collection of vertices and directed edges, each edge connecting one vertex to another, such that there is no way to start at some vertex v and follow a sequence of edges that eventually loops back to v again.

Binary Trees

A binary tree is made of nodes, where each node contains a "left" reference, a "right" reference, and a data element. The topmost node in the tree is called the root. Every node (excluding a root) in a tree is connected by a directed edge from exactly one other node. This node is called a parent. On the other hand, each node can be connected to arbitrary number of nodes, called children. Nodes with no children are called leaves, or external nodes. Nodes which are not leaves are called internal nodes. Nodes with the same parent are called siblings.

More tree terminology:

- The depth of a node is the number of edges from the root to the node.
- The height of a node is the number of edges from the node to the deepest leaf.
- The height of a tree is a height of the root.
- A full binary tree.is a binary tree in which each node has exactly zero or two children.
- A complete binary tree is a binary tree, which is completely filled, with the possible exception of the bottom level, which is filled from left to right.

Traversals

A traversal is a process that visits all the nodes in the tree. Since a tree is a nonlinear data structure, there is no unique traversal. We will consider several traversal algorithms with we group in the following two kinds

- depth-first traversal
- breadth-first traversal

There are three different types of depth-first traversals:

- Pre Order traversal - visit the parent first and then left and right children;
- In Order traversal - visit the left child, then the parent and the right child;
- Post Order traversal - visit left child, then the right child and then the parent;

Binary Search Trees

A BST is a binary tree where nodes are ordered in the following way:

- each node contains one key (also known as data)
- the keys in the left sub tree are less than the key in its parent node, in short L < P;

- the keys in the right sub tree are greater the key in its parent node, in short P < R;
- duplicate keys are not allowed

Spanning Tree

In the mathematical field of graph theory, a spanning tree T of an undirected graph G is a subgraph that includes all of the vertices of G that is a tree. In general, a graph may have several spanning trees, but a graph that is not connected will not contain a spanning tree.

Spanning Tree Protocol

The Spanning Tree Protocol (STP) is a network protocol that ensures a loop-free topology for any bridged Ethernet local area network. Spanning-Tree Protocol (STP) prevents loops from being formed when switches or bridges are interconnected via multiple paths. Spanning-Tree Protocol implements the 802.1D IEEE algorithm by exchanging BPDU messages with other switches to detect loops, and then removes the loop by shutting down selected bridge interfaces. This algorithm guarantees that there is one and only one active path between two network devices.

Spanning Tree in a Nutshell

- STP provides a means to prevent loops by blocking links in an Ethernet network. Blocked links can be brought in to service if active links fail.
- The root bridge in a spanning tree is the logical center and sees all traffic on a network.
- Spanning tree recalculations are performed automatically when the network changes but cause a temporary network outage.
- Newer protocols, such as TRILL, prevent loops while keeping links that would be blocked by STP in service.

9 Stacks and Queues

An array is a random access data structure, where each element can be accessed directly and in constant time. A typical illustration of random access is a book - each page of the book can be open independently of others. Random access is critical to many algorithms, for example binary search.

A linked list is a sequential access data structure, where each element can be accessed only in particular order. A typical illustration of sequential access is a roll of paper or tape - all prior material must be unrolled in order to get to data you want.

In this note we consider a subcase of sequential data structures, so-called limited access data structures.

Stacks

A stack is a container of objects that are inserted and removed according to the last-in first-out (LIFO) principle. In the pushdown stacks only two operations are allowed: push the item into the stack, and pop the item out of the stack. A stack is a limited access data structure - elements can be added and removed from the stack only at the top. Push adds an item to the top of the stack, pop removes the item from the top. A helpful analogy is to think of a stack of books; you can remove only the top book, also you can add a new book on the top.

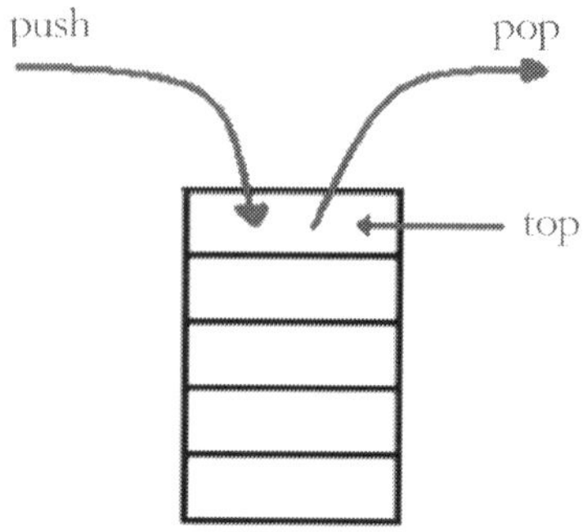

A stack is a recursive data structure. Here is a structural definition of a Stack:

A stack is either empty or it consists of a top and the rest which is a stack;

Applications

The simplest application of a stack is to reverse a word. You push a given word to stack - letter by letter - and then pop letters from the stack.

Another application is an "undo" mechanism in text editors; this operation is accomplished by keeping all text changes in a stack.

Backtracking: This is a process when you need to access the most recent data element in a series of elements. Think of a labyrinth or maze - how do you find a way from an entrance to an exit?

Once you reach a dead end, you must backtrack. But backtrack to where? To the previous choice point. Therefore, at each choice point you store on a stack all possible choices. Then backtracking simply means popping a next choice from the stack.

Language processing:

- Space for parameters and local variables is created internally using a stack.
- Compiler's syntax check for matching braces is implemented by using stack.
- support for recursion

Implementation

In the standard library of classes, the data type stack is an adapter class, meaning that a stack is built on top of other data structures. The underlying structure for a stack could be an array, a vector, an Array List, a linked list, or any other collection. Regardless of the type of the underlying data structure, a Stack must implement the same functionality. This is achieved by providing a unique interface:

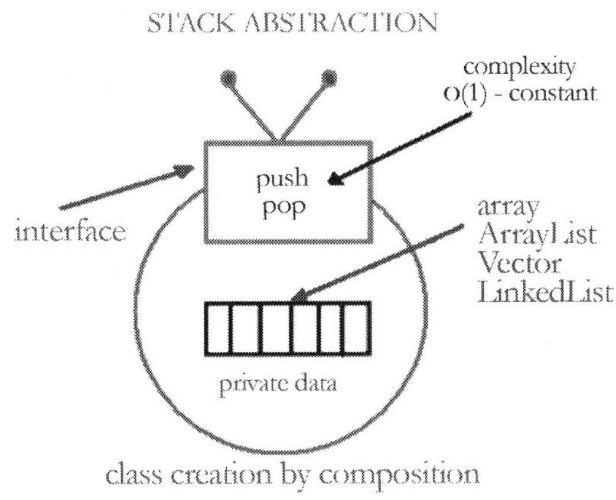

Queue

A queue is a container of objects (a linear collection) that are inserted and removed according to the first-in first-out (FIFO) principle. An excellent example of a queue is a line of students in the food court of the UC. New additions to a line made to the back of the queue, while removal (or serving) happens in the front. In the queue only two operations are allowed enqueue and dequeue. Enqueue means to insert an item into the back of the queue, dequeue means removing the front item. The picture demonstrates the FIFO access.

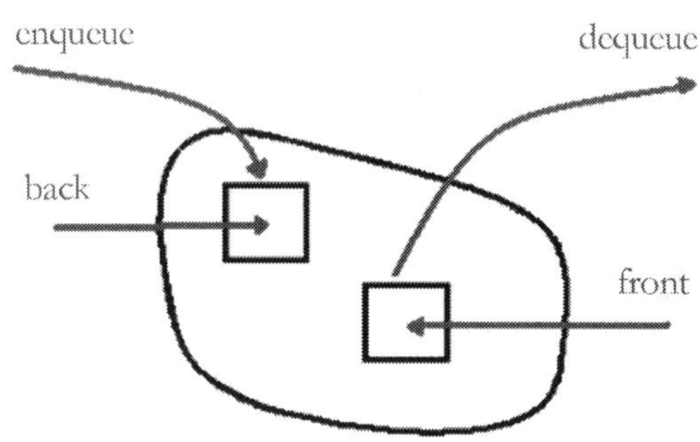

Implementation

In the standard library of classes, the data type queue is an adapter class, meaning that a queue is built on top of other data structures. The underlying structure for a queue could be an array, a Vector, an Array List, a Linked List, or any other collection. Regardless of the type of the underlying data structure, a queue must implement the same functionality. This is achieved by providing a unique interface.

10 Memory Management

Memory management is the act of managing computer memory at the system level. The essential requirement of memory management is to provide ways to dynamically allocate portions of memory to programs at their request, and free it for reuse when no longer needed. This is critical to any advanced computer system where more than a single process might be underway at any time.

Virtual memory does more than just make your computer's memory go further. The memory management subsystem provides:

Large Address Spaces

The operating system makes the system appear as if it has a larger amount of memory than it actually has. The virtual memory can be many times larger than the physical memory in the system,

Protection

Each process in the system has its own virtual address space. These virtual address spaces are completely separate from each other and so a process running one application cannot affect another. Also, the hardware virtual memory mechanisms allow areas of memory to be protected against writing. This protects code and data from being overwritten by rogue applications.

Memory Mapping

Memory mapping is used to map image and data files into a processes address space. In memory mapping, the contents of a file are linked directly into the virtual address space of a process.

Fair Physical Memory Allocation

The memory management subsystem allows each running process in the system a fair share of the physical memory of the system,

Shared Virtual Memory

Although virtual memory allows processes to have separate (virtual) address spaces, there are times when you need processes to share memory. For example there could be several processes in the system running the bash command shell. Rather than have several copies of bash, one in each processes virtual address space, it is better to have only one copy in physical memory and all of the processes running bash share it. Dynamic libraries are another common example of executing code shared between several processes.

Shared memory can also be used as an Inter Process Communication (IPC) mechanism, with two or more processes exchanging information via memory common to all of them. Linux supports the Unix TM System V shared memory IPC.

An Abstract Model of Virtual Memory

As the processor executes a program it reads an instruction from memory and decodes it. In decoding the instruction it may need to fetch or store the contents of a location in memory. The processor then executes the instruction and moves onto the next instruction in the program. In this way the processor is always accessing memory either to fetch instructions or to fetch and store data.

In a virtual memory system all of these addresses are virtual addresses and not physical addresses. These virtual addresses are converted into physical addresses by the processor based on information held in a set of tables maintained by the operating system.

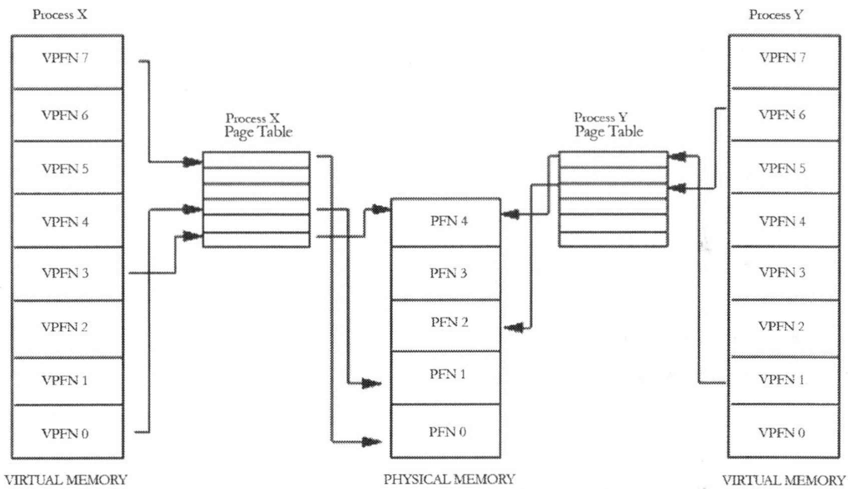

To make this translation easier, virtual and physical memory are divided into handy sized chunks called pages. These pages are all the same size, they need not be but if they were not, the system would be very hard to administer. Linux on Alpha AXP systems uses 8 Kbyte pages and on Intel x86 systems it uses 4 Kbyte pages. Each of these pages is given a unique number; the page frame number (PFN).

In this paged model, a virtual address is composed of two parts; an offset and a virtual page frame number. If the page size is 4 Kbytes, bits 11:0 of the virtual address contain the offset and bits 12 and above are the virtual page frame number. Each time the processor encounters a virtual address it must extract the offset and the virtual page frame number. The processor must translate the virtual page frame number into a physical one and then access the location at the correct offset into that physical page. To do this the processor uses page tables.

Demand Paging

As there is much less physical memory than virtual memory the operating system must be careful that it does not use the physical memory inefficiently. One way to save physical memory is to only load virtual pages that are currently being used by the executing program. For example, a database program may be run to query a database. In this case not all of the database needs to be loaded into memory, just those data records that are being examined. If the database query is a search query then it does not make sense to load the code from the database program that deals with adding new records. This technique of only loading virtual pages into memory as they are accessed is known as demand paging.

When a process attempts to access a virtual address that is not currently in memory the processor cannot find a page table entry for the virtual page referenced. For example, in Figure 3.1 there is no entry in process X's page table for virtual page frame number 2 and so if process X attempts to read from an address within virtual page frame number 2 the processor cannot translate the address into a physical one. At this point the processor notifies the operating system that a page fault has occurred.

Swapping

If a process needs to bring a virtual page into physical memory and there are no free physical pages available, the operating system must make room for this page by discarding another page from physical memory.

If the page to be discarded from physical memory came from an image or data file and has not been written to then the page does not need to be saved. Instead it can be discarded and if the process needs that page again it can be brought back into memory from the image or data file.

However, if the page has been modified, the operating system must preserve the contents of that page so that it can be accessed at a later time. This type of page is known as a dirty page and when it is removed from memory it is saved in a special sort of file called the swap file. Accesses to the swap file are very long relative to the speed of the processor and physical memory and the operating system must juggle the need to write pages to disk with the need to retain them in memory to be used again.

Shared Virtual Memory

Virtual memory makes it easy for several processes to share memory. All memory access are made via page tables and each process has its own separate page table. For two processes sharing a physical page of memory, its physical page frame number must appear in a page table entry in both of their page tables.

Physical and Virtual Addressing Modes

It does not make much sense for the operating system itself to run in virtual memory. This would be a nightmare situation where the operating system must maintain page tables for itself. Most multi-purpose processors support the notion of a physical address mode as well as a virtual address mode. Physical addressing mode requires no page tables and the processor does not attempt to perform any address translations in this mode. The Linux kernel is linked to run in physical address space.

Access Control

The page table entries also contain access control information. As the processor is already using the page table entry to map a processes virtual address to a physical one, it can easily use the access control information to check that the process is not accessing memory in a way that it should not.

There are many reasons why you would want to restrict access to areas of memory. Some memory, such as that containing executable code, is naturally read only memory; the operating system should not allow a process to write data over its executable code. By contrast, pages containing data can be written to but attempts to execute that memory as instructions should fail. Most processors have at least two modes of execution: kernel and user. You would not want kernel code executing by a user or kernel data structures to be accessible except when the processor is running in kernel mode.

Caches

If you were to implement a system using the above theoretical model then it would work, but not particularly efficiently. Both operating system and processor designers try hard to extract more performance from the system. Apart from making the processors, memory and so on faster the best approach is to maintain caches of useful information and data that make some operations faster. Linux uses a number of memory management related caches:

Buffer Cache

The buffer cache contains data buffers that are used by the block device drivers.

These buffers are of fixed sizes (for example 512 bytes) and contain blocks of information that have either been read from a block device or are being written to it. A block device is one that can only be accessed by reading and writing fixed sized blocks of data. All hard disks are block devices.

The buffer cache is indexed via the device identifier and the desired block number and is used to quickly find a block of data. Block devices are only ever accessed via the buffer cache. If data can be found in the buffer cache then it does not need to be read from the physical block device, for example a hard disk, and access to it is much faster.

Page Cache

This is used to speed up access to images and data on disk.

It is used to cache the logical contents of a file a page at a time and is accessed via the file and offset within the file. As pages are read into memory from disk, they are cached in the page cache.

Swap Cache

Only modified (or dirty) pages are saved in the swap file.

So long as these pages are not modified after they have been written to the swap file then the next time the page is swapped out there is no need to write it to the swap file as the page is already in the swap file. Instead the page can simply be discarded. In a heavily swapping system this saves many unnecessary and costly disk operations.

Hardware Caches

One commonly implemented hardware cache is in the processor; a cache of Page Table Entries. In this case, the processor does not always read the page table directly but instead caches translations for pages as it needs them. These are the Translation Look-aside Buffers and contain cached copies of the page table entries from one or more processes in the system.

11 Miscellaneous

Structures and Linked Lists

You may be wondering why malloc has been introduced right after the structure. The answer is that the dynamic allocation of memory and the struct go together a bit like the array and the for loop. The best way to explain how this all fits together is via a simple example. You can use malloc to create as many variables as you want as the program runs, but how do you keep track of them? For every new variable you create you also need an extra pointer to keep track of it. The solution to this otherwise tricky problem is to define a struct which has a pointer as one of its components.

```
struct list
{
int data;
struct list *ptr;
};
```

This defines a structure which contains a single int and - something that looks almost paradoxical - a pointer to the structure that is being defined. All you really need to know is that this is reasonable and it works. Now if you use malloc to create a new struct you also automatically get a new pointer to the struct. The final part of the solution is how to make use of the pointers. If you start off with a single 'starter' pointer to the struct you can create the first new struct using malloc as:

```
struct list *start;
start = (*struct list) malloc(sizeof(struct list))
```

After this start points to the first and only example of the struct. You can store data in the struct using statements like:

```
start-££££data=value;
```

The next step is to create a second example of the struct:

```
start = (*struct list) malloc(sizeof(list));
```

This does indeed give us a new struct but we have now lost the original because the pointer to it has been overwritten by the pointer to the new struct. To avoid losing the original the simplest solution is to use:

```
struct list *start,newitem;
newitem = (*struct list) malloc(sizeof(struct list));
start-££££prt=start;
start=newitem;
```

This stores the location of the new struct in newitem. Then it stores the pointer to the existing struct into the newitem›s pointer and sets the start of the list to be the newitem. Finally the start of the list is set to point at the new struct. This procedure is repeated each time a new structure is created with the result that a linked list of structures is created. The pointer start always points to the first struct in the list and the prt component of this struct points to the next and so on. You should be able to see how to write a program that examines or prints the data in each of the structures. For example:

```
thisptr=start;
while (1==1)
{
printf(«%d»,thisprt-££££ data);
thisprt=thisprt-££££prt;
}
```

This first sets thisptr to the start of the list, prints the data in the first element and then gets the pointer to the next struct in the list and so on. How does the program know it has reached the end of the list? At the moment it just keeps going into the deep and uncharted regions of your machine»s memory! To stop it we have to mark the end of the list using a null pointer. Usually a pointer value of 0 is special in that it never occurs in a pointer pointing at a valid area of memory. You can use 0 to initialise a pointer so that you know it is not pointing at anything real. So all we have to do is set the last pointer in the list to 0 and then test for it that is:

```
Thisptr = start;
while (thisptr!=0)
{
printf(«%d»,thisprt-££££data);
thisprt=thisprt-££££ prt;
}
```

To be completely correct you should TYPE cast 0 to be a pointer to the struct in question. That is:

```
while (thisptr!=(struct list*)0)
```

By generally mucking about with pointers stored in the list you can rearrange it, access it, sort it, delete items and do anything you want to. Notice that the structures in the list can be as complicated as you like and, subject to there being enough memory, you can create as many structures as you like.

You can use the same sort of technique to create even more complicated list structures. For example you can introduce another pointer into each structure and a pointer to the end of the list so that you can work your way along it in the other direction - a doubly linked list. You can create stacks, queues, trees and so on.

UNION

A union is a collection of variables of different types, just like a structure. However, with unions, you can only store information in one field at any one time.

You can picture a union as like a chunk of memory that is used to store variables of different types. Once a new value is assigned to a field, the existing data is wiped over with the new data.

A union can also be viewed as a variable type that can contain many different variables (like a structure), but only actually holds one of them at a time (not like a structure). This can save memory if you have a group of data where only one of the types is used at a time. The size of a union is equal to the size of its largest data member. In other words, the C compiler allocates just enough space for the largest member. This is because only one member can be used at a time, so the size of the largest, is the most you will need.

Accessing Union Fields

To access the fields of a union, use the dot operator (.) just as you would for a structure. When a value is assigned to one member, the other member(s) get whipped out since they share the same memory. Using the example above, the precise time can be accessed like this:

```
printTime( mytime.perciseDate );
```

A Handbook on

Electronics and Communications Engineering

10

Electrical Matrial

CONTENTS

1 Dielectric Properties of Insulators Static Field

Dielectric materials are non-conducting materials. When dielectric materials are placed in an electric field, they store electrical charges. If the main function of non-conducting materials is to provide electrical insulation then, they are called insulators

Dielectric Parameters

A dielectric material is an electrical insulator that can be polarized by an applied electric field. When a dielectric is placed in an electric field, electric charges do not flow through the material as they do in a conductor, but only slightly shift from their average equilibrium positions causing dielectric polarization. Because of dielectric polarization, positive charges are displaced toward the field and negative charges shift in the opposite direction.

Dielectric Constant

Dielectric constant, property of an electrical insulating material a dielectric equal to the ratio of the capacitance of a capacitor filled with the given material to the capacitance of an identical capacitor in a vacuum without the dielec tricmaterial.

$$\varepsilon = \varepsilon_r \varepsilon_0$$

ε_r = Re lative permittivity . ε_0 = Permittivity of vacume . $\varepsilon_0 = 8.854 \times 10^{-12}$ farad – metre^{-1}

Relative permittivity is typically denoted as $\varepsilon_r(\omega)$ (sometimes $\varkappa$ or K) and is defined as χ_e, as $\varepsilon_r(\omega) = 1 + \chi_e$.

$$\varepsilon_r(\omega) = \frac{\varepsilon(\omega)}{\varepsilon_0}$$

$$\varepsilon_r(\omega) = \varepsilon'_r(\omega) + i\varepsilon''_r(\omega).$$

Dipole Moment

Electric dipole moment is a measure of the separation of positive and negative electrical charges in a system of electric charges. That is, a measure of the charge system's overall polarity n the simple case of two point charges, one with charge $+q$ and the other one with charge $-q$, the electric dipole moment p is:

$$p = qd$$

$\bar{p} = N\bar{d}$ debay and $1\text{Deye} = 3.33 \times 10^{-30}$ coulomb-meter.

Polarization

Polarization occurs when an electric field distorts the negative cloud of electrons around positive atomic nuclei in a direction opposite the field. This slight separation of charge makes one side of the atom somewhat positive and the opposite side somewhat negative. P=Np

$$P = \frac{p}{V}$$

Polarizability

Distortion of the electron cloud of a molecular entity by an electric field (such as that due to the proximity of a charged reagent). It is experimentally measured as the ratio of induced dipole moment (uind) to the field E which induces the electronic polarizability □ is defined as the ratio of the induced dipole moment *P* of an atom to the electric field *E* that produces this dipole moment. $\alpha = \mu_{ind}/E$

$$p = \alpha E$$

Electric Flux Density

Dielectric material the presence of an electric field E causes the bound charges in the material atomic nuclei and their electrons to slightly separate, inducing a local electric dipole moment. The electric displacement field D is defined as

$$D \equiv \varepsilon_0 E + P$$

Electric Flux Density when electric fields applied

Electric flux density A measure of the strength of an electric field generated by a free electric charge, corresponding to the number of electric lines of force passing through a given area. It is equal to the electric field strength multiplied by the permittivity of the material through which the electric field extends. It is measured in coulombs per square meter.

$$D = \varepsilon E$$

$$E = \frac{q_1}{4\pi\varepsilon R^2}$$

$$D = \varepsilon E = \frac{q_1}{4\pi R^2}$$

Polarization

Polarization, slight relative shift of positive and negative electric charge in opposite directions within an insulator, or dielectric, induced by an external electric field. Polarization occurs when an electric field distorts the negative cloud of electrons around positive atomic nuclei in a direction opposite the field

$$\chi_e = \frac{\text{Bound charge density}}{\text{Free charge density}}$$

$$\alpha = \frac{\epsilon_0 \chi_e}{N} = \frac{\epsilon_0(\epsilon_r - 1)}{N}$$

The Stored Energy per Unit Volume in a Dielectric Medium Due to the Polarization

Capacitors are often considered to be charge storage devices. We will consider them to be energy storage devices which use charge as a storage mechanism. For a capacitor with charge +Q on one plate, -Q on the other, and voltage V between them, the stored energy.

$$W = \frac{1}{2} PE$$

POLARIZATION

The amount of charge stored in a capacitor is the product of the voltage and the capacity. What limits the amount of charge that can be stored on a capacitor the voltage can be increased, but electric breakdown will occur if the electric field inside the capacitor becomes too large. The capacity can be increased by expanding the electrode areas and by reducing the gap between the electrodes.

Electronics or induced Polarization

In the electric field the positive charge in the nucleus and the center of the negative charges from the electron "cloud" will experience forces in different direction and will become separated. The separating force of the external field is balanced by the attractive force between the centers of charge

Observed in all types of Materials

All types of materials are follows:

Electronics Polarization

In the electric field the positive charge in the nucleus and the center of the negative charges from the electron "cloud" will experience forces in different direction and will become separated. The separating force of the external field is balanced by the attractive force between the centers of charge

$$P_e = N\alpha_e E \quad \text{where, } \alpha_e = \text{electronic polarizability}$$

$$\alpha_e = 4\pi \in_0 R^3 \qquad \text{where, } R = \text{radius of atom}$$

$$\alpha_e = \frac{\in_0(\in_r - 1)}{N}; \qquad \in_r = 1 + 4\pi N R^3 \ ...(\text{for a rare gas})$$

Ionic Polarization

In this case a material must have ionic character of bonds. It then by definition has internal dipoles, but the sebuilt in dipoles exactly cancel each other (and are unable to rotate). The external field then induces net dipoles by slightly displacing the ions from their rest position. The model materials are simple ionic crystals like NaCl. $\alpha_i \rightarrow$ Ionic Polarization

$$P_i = N\alpha_i E$$

Orientation Polarization

There the material must have molecules with their own dipolar momentum (which can rotate freely the external field aligns these dipoles to some extent and thus induces a polarization of the material. The paradigmatic material is water.

$$P = \frac{Np_p^{2} E}{3kT}$$

$$P = \frac{\sum_i p_0 \cos \alpha_i}{V}$$

$$P = n_0 p_0 \overline{\cos \alpha}$$

$$\alpha_0 = \frac{p^2_p}{3kT}$$

$$P = P_e + P_i + p_o$$

$$P = N\left(\alpha e + \alpha \iota + \frac{p^2_p}{3kT} \right) E$$

$$\overline{E}_i = \overline{E}_{ext} + \frac{\gamma}{\epsilon_0} \overline{P}$$

$$\overline{E}_{i(Lorentz)} = \overline{E}_{ext} + \frac{\overline{P}}{3 \epsilon_0}$$

Orianted Polarizability

Orientational polarization arises when there is a permanent dipole moment in the material. Materials such as HCl and H_2O will have a net permanent dipole moment because the charge distributions of these molecules are skewed. For example, in a molecule of HCl, the chlorine atom will be negatively charged and the hydrogen atoms will be positively charged causing the molecule to be dipolar.

$$\alpha_0 = \frac{p^2_p}{3kT}$$

Total Polarization of a polyatomic gas

$$P = P_e + P_i + p_o$$

$$P = N\left(\alpha e + \alpha \iota + \frac{p^2{}_p}{3kT}\right)E$$

INTERNAL FIELD IN SOLID AND LIQUIDS

Consider a dipole with charges '+q 'and '–q' separated by a small distance 'dx' as shown in fig. The dipole moments given by μ= q dx. Consider a point 'P' at a distance 'r' from the center of dipole

The Internal Field

When dielectric material is placed in the external electric field, it is polarized creating electric dipoles. Each dipole sets electric field in the vicinity. Hence the electric field at any point within the dielectric material is given by The sum of external field and the field due to all dipoles surrounding that point"the field due to all dipoles surrounding that point "the field due to all dipoles surrounding that point"the field due to all dipoles surrounding that point. $\overline{P}$ =Dipole moment per unit volume

γ =Proportionality constant which is reffered to as the internal constant

$$\overline{E}_i = \overline{E}_{ext} + \frac{\gamma}{\epsilon_0}\overline{P}$$

$$\overline{E}_{i(Lorentz)} = \overline{E}_{ext} + \frac{\overline{P}}{3\epsilon_0}$$

Note: Whene the atom in a solid are sounded cubically by other atom

$\gamma = \frac{1}{3}$ and internal field is called Lorentz intenal field

CLOUSSIUS MOSTTI RELATION

The Clausius–Mossotti relation is named after the Italian physicist Ottaviano-Fabrizio Mossotti, whose 1850 book analyzed the relationship between the dielectric constants of two different media, and the German physicist Rudolf Clausius, who gave the formula explicitly in his 1879 bookin the context not of dielectric constants but of indices of refraction.

The Clausius–Mossotti law applies to the dielectric constant of a dielectric that is perfect, homogeneous and isotropic. It is the second of the following three equalities:

$$\frac{\epsilon - \epsilon_0}{\epsilon + 2\epsilon_0} \cdot \frac{M}{d} = \frac{\epsilon_r - 1}{\epsilon_r + 2} \cdot \frac{M}{d} = \frac{4\pi N_A \alpha'}{3} = \frac{N_A \alpha}{3\epsilon_0};$$

$$N\alpha = 3\frac{n^2 - 1}{n^2 + 2}$$

Where

$\epsilon = \epsilon_r \epsilon_0$ is the dielectric constant of a substance

ϵ_0 is the permittivity of a vacuum

M is the molar mass of the substance

d is its density

N_A is Avogadro's number,

α is the molecular polarizability in SI-units (C m²/V) and

$\alpha' = \alpha / (4\pi\varepsilon_0)$ is the molecular polarizability *volume* (m³ in SI) or the polarizability in the CGS system of units

Maxwell's Theory of Materials

A Maxwell material is a viscoelastic material having the properties both of elasticity and viscosity It is named for James Clerk Maxwell who proposed the model in 1867. It is also known as a Maxwell fluid.

$$\in_{re} = n^2$$

In Cloussius Mostti Relation for Gasses at Low Pressure

The Clausius Mossotti equation is strictly realized for nonpolar gases at low pressures (less than 200–500 mm Hg, or 26.6–66.5 kilo newton's per sq. m[kN/m^2]) or medium pressures (from 500 mm Hg to 5 atm, or 66.5–500 kN/m^2). The equation is approximately realized for nonpolar gases at high pressures (above 5–10 atm, or 0.5–1 MN/m^2), for nonpolar liquids, and for many nonpolar crystals

$$\frac{N\alpha_e}{\in_0} = \in_r - 1 = \chi_e$$

DEBAY'S GENERALIZATION CLOUSSIUS MOSTTI RELATION

It basically relates the polarizability of a material to its relative permittivity and hence it is refractive index the induced dipole moment of an atom/molecule in the presence of an external electric field to be given by

$$\vec{p} = \alpha \vec{E}$$

$$\alpha = \frac{3\in_0}{N}\left(\frac{\in_r - 1}{\in_r + 2}\right)$$

Polarization per Kilogram Molecule

Polarization P in its quantitative meaning is the amount of dipole moment p per unit volume V of a polarized material,

$$P = p/V.$$

$$\pi = \frac{N_A \alpha}{3\in_0} = \frac{\in_r - 1}{\in_r + 2} \cdot \frac{M}{d}$$

Where α is the polarizability. From this definition, he derives the Clausius-Mossotti relation to be

π = Molar polarizability M = Molecular weight of matrials, kg

π = Molar polarizability N = Avogadro's Number

$= 6.023 \times 10^{26}$ α = Polarizability which indudes includes effect of orientational polarization

Maxwell's Equetions of Optical Frequecies or Lorentz-Lorentz Equation

The lorentz–lorenz equation, also known as the clausius–mossotti relation and maxwell's formula, relates the refractive index of a substance to its polarizability.

The most general form of the lorentz–lorenz equation is

$$\frac{n_2 - 1}{n_2 + 2} = \frac{4\pi}{3} N\alpha,$$

$$\pi = \frac{N_A \alpha}{3\in_0} = \frac{n^2 - 1}{n^2 + 2} \cdot \frac{M}{d}$$

N=Refractive Index $\alpha = \alpha_e$

Note: Lorentz-Lorentz equation will only for the case

Condition for Spontaneos Polarization

Spontaneous polarization of Triglycine Selenate and Triglycine Fluoberyllate was investigated with crystals produced under various growing conditions. Spontaneous polarization of TGSe crystals grown at the Curie temperature is of the greatest value. The higher the supersaturation of the solution from which TGSe crystals were grown, the greater the spontaneous polarization. TGFB crystals grown at high temperatures are of greater value of spontaneous polarization than those grown at low temperatures $N = \text{Number of molecules per unit volume, m}^{-3}$ $\alpha = \text{Polarizability, Farad} - m^2$ $\gamma = \text{Internal field constant}$

$$\frac{N\alpha\gamma}{\epsilon_0} = 1$$

CURIE WEISS LAW

A relation between magnetic or electric susceptibilities and the absolute temperature which is followed by fer=romagnets, antiferromagnets, non-polar-ferroelectrics and anti-ferroelectrics, and some paramagnets $C = \text{Curie constant}$ $T = \text{Temprature}$

θ = Characteristics temprature which usually a few degree smaller than the ferrcelectric curie temprature θ_T

$$\chi_e = \epsilon_r - 1 = \frac{C}{T - \theta}$$

PIEZOELECTRICITY

The piezoelectric effect describes the relation between a mechanical stress and an electrical voltage in solids. An applied mechanical stress will generate a voltage and an applied voltage will change the shape of the solid by a small amount.

Direct Effect

The direct piezoelectric effect is that these materials, when subjected to mechanical stress, generate an electric charge proportional to that stress. An example of the use of the direct effect is found in gas lighters.

Piezoelectric sensors, like acceleration sensors and pressure sensors, also exploit the direct piezoelectric effect.

Inverse Effects

The inverse piezoelectric effect is that the same materials become strained when an electric field is applied, the strain again being proportional to the applied field. An example of the use of the inverse effect is found in buzzers. Piezoelectric actuators, which can be used for micro-positioning, also rely on the inverse piezoelectric effect.

$$\left.\begin{array}{l} T = cS \\ S = sT \end{array}\right]$$

$$P = dT$$

Stress

It is derived from the greek piezo or piezein which means to squeeze or press, and electric or electron which means amber, an ancient source of electric charge piezoelectricity was discovered in 1880 by french physicists Jacques and pierre curie

$$D = \in E + dT$$

Dielectric Displacement in the presence of stress

Piezoelectric nanocomposites, surface/interface stress and electric displacement have a great effect on the strength of structures. In this work, predicated on nanoscale size-effects, the dynamic stress along the interface of a cylindrical piezoelectric nano-fiber in piezoelectric nanocomposites under electro-elastic waves is studied.

$$S = sT + dE$$

2 Dielectric Breakdown of Gases

At normal temperature and pressure, the gases are excellent insulators. The current conduction is of the order of 10–10 A/cm2. This current conduction results from the ionisation of air by the cosmic radiation and the radioactive substances present in the atmosphere and the earth. At higher fields, charged particles may gain sufficient energy between collisions to cause ionization on impact with neutral molecules. It is known that during an elastic collision, an electron loses little energy and rapidly builds up its kinetic energy which is supplied by an external electric field. On the other hand, during elastic collision, a large part of the kinetic energy is transformed into potential energy by ionizing the molecule struck by the electron. Ionization by electron impact under strong electric field is the most important process leading to breakdown of gases.

Average Velocity Of Charge Career In A Gas

Electrons do not exactly travel in straight lines, sorry to ruin the illusion. They also move pretty slowly too. The drift velocity is basically the average velocity, taking into consideration all the silly paths the electrons take. It is therefore slower than the intrinsic electrons' speed.semiconductors these sound more exciting have fewer charge carriers, so to achieve the same current as your bog standard metal, you will need a bigger drift velocity (v). If you want no current, like in a perfect insulator, you would require no charge carriers (n=0), but this does not happen in real life, as you would expect, and so instead you will just have a very small number of charge carriers. $A+h\nu \rightarrow A+e$

$$V = \mu E$$

where, μ = Mobility of the charge carrier

E=Applied electric field

Condition of Ionization

A townsend discharge can be sustained only over a limited range of gas pressure and electric field intensity. The accompanying plot shows the variation of voltage drop and the different operating regions for a gas-filled tube with a constant pressure, but a varying current between its electrodes. At higher pressures, discharges occur more rapidly than the calculated time for ions to traverse the gap between electrodes, and the streamer theory of spark discharge of raether, meek and loeb is applicable. In highly non-uniform electric fields, the corona dischargeprocess is applicable. See electron avalanche for further description of these mechanisms. Discharges in vacuum require vaporization and ionization of electrode atoms. An arc can be initiated without a preliminary townsend discharge; for example when electrodes touch and are then separated.

$$E.\lambda = v_i$$

Where v_i = Ionization potential λ = Mean Free path

Electron Ionization Coefficient

$$\alpha \frac{1}{\lambda} e^{-\left(v_i / E.\lambda\right)}$$

Where

α = First townsend coefficient or ionization coefficient of elctron

$$\lambda \propto \frac{1}{\text{Pressure}} \quad \text{at constant temperature}$$

Secondary Ionization Coefficient

The number of electrons released from the cathode of an ionization chamber per initial ionizing collision in the gas. Also known as secondary ionizationcoefficient.

$$n = \frac{n^0 e^{\alpha d}}{1-\gamma(e^{\alpha d}-1)}$$

Where

n_0 = Number of primery photoelectrons per second emitted from cothode

n = Total number of electron arriving athteanode

γ = Townsend's second coefficient

d = distance between anode and cathode

Townsend criterion for spark breakdown

Townsend's coefficients are determined in an ionisation chamber which is first evacuated to a very high vacuum of the order of 10-4 and 10-6 torr before filling with the desired gas at a pressure of a few torr. The applied direct voltage is about 2 to 10 kV, and the electrode system consists of a plane high voltage electrode and a low voltage electrode surrounded by a guard electrode to maintain a uniform field. The low voltage electrode is earthed through an electrometer amplifier capable of measuring currents in the range 0.01 pA to 10 nA. The cathode is irradiated using an ultra-violet lamp from the outside to produce the initiation electron.

$$\gamma(e^{\alpha d}-1)=1$$

Dielectric Breakdown of Liquids

Liquid dielectrics are used for filling transformers, circuit breakers and as impregnants in high voltage cables and capacitors. For transformer, the liquid dielectric is used both for providing insulation between the live parts of the transformer and the grounded parts besides carrying out the heat from the transformer to the atmosphere thus providing cooling effect. For circuit breaker, again besides providing insulation between the live parts and the grounded parts, the liquid dielectric is used to quench the arc developed between the breaker contacts. The liquid dielectrics mostly used are petroleum oils. Other oils used are synthetic hydrocarbons and halogenated hydrocarbons and for very high temperature applications silicone oils and fluorinated hydrocarbons are also used.

Colloidal Theory

The breakdown in 'contaminated dielectrics occur due to the formation of conducting bridges between the electrodes by droplets of emulsified water and suspended particles especially by fibrous particles. The time taken to form the bridge depends on: (i) extent of contamination, (ii) shape of electrodes and (iii) gap between the electrodes. For instance if an impulse voltage is applied for a short duration, no continuous bridge will be formed in the liquid. At the place of breakdown, heat is evolved and the bridge breaks up due to thermal convection flow in the liquid. After sometime another conducting bridge may be formed and so on, thus making the breakdown process a rather indeterminate quantity in a highly contaminated liquid.

$$\frac{\varepsilon-\varepsilon'}{\varepsilon+2\varepsilon'}.r^3E \geq \frac{kT}{4}$$

Where, r=Radius of insulating particle

$\in$ = Permittivity of insulating material

$\in'$ = Permittivity of oil

k=Boltzmann's constant

E=Field strength

T=Absolute temperature

Bubble Theory

In technically pure di-electric the breakdown is initiated by ionization of gas contained in the liquid. All liquids dissolve a certain quantity of gas especially air. Bubbles of gases are formed by one or more of the following mechanisms:

- Gas pockets in the electrode surface;
- Electrostatic repulsion between the space charges may be sufficient to overcome the surface tension of the liquid to form bubbles;
- Vaporization of liquid by corona type discharge and (iv) Gaseous products of ionization of the molecule.

$$E_i = \frac{2 \in E_0}{2 \in +1}$$

Where, E_i=Electric field in a gas bubble which is immersed in a liquid

$\in$=Permittivity of liquid

E_0=Electric field in the liquid in absence of the bubble

Break Down Due to Liquid Globules

$$E = 487.7\sqrt{\frac{\sigma}{R \in_1}} \quad V/cm$$

Where, E=Critical field at which the globule loose its stability

$\in_1$ =Permittivity of the liquid medium

σ =Pressure due to surface tension, dyne/cm

R=Radius of globule

Dielectric Breakdown of Solids

Solid dielectrics have higher breakdown strength compared to liquids and gases. When breakdown occurs, solids get permanently damaged while gases fully and liquids partly recover their dielectric strength.

Theory of Von Hippel

$$E_c = \frac{2\pi^2 vem}{h}\left[\frac{1}{n^2_\infty} - \frac{1}{\in}\right] \quad V/m$$

Where, E_c=Critical field at which the breakdown occurs

V=Optical frequency of lattice

n_∞ = The index of refraction for infinite wave length

$\in$=Dielectric constant

e=Electron charge

m=Effective mass of electron

h=Plank's constant

Thermal Breakdown

- Heat generated per unit volume in unit time

$$W = E^2 f . \frac{\in_r \tan\delta}{1.8\times10^{12}} \quad \text{watts/cm}^3$$

Where,

E=Uniform electric field

f=Frequency of applied voltage

δ=Loss angle

$\in_r$=Relative permittivity

For direct voltage

$$W = \frac{E^2}{\rho}$$

Where, ρ =Resistivity of the insulation, ohm-cm

- Power loss

$$\in \tan\delta = \in' \tan\delta' \cdot e^{c(T-T')}$$

Where, $\in$ =$\tan\delta$=Loss factor of the dielectric at a temperature

$\in'$ tan δ'=Loss factor of the dielectric at the initial temperature T'

C=Coefficient depending on the properties concerned

3 Magnetic Properties of Materials

A magnet produces a vector field, the magnetic field, at all points in the space around it. It can be defined by measuring the force the field exerts on a moving charged particle, such as an electron. The force (F) is equal to the charge (q) times the speed of the particle times the magnitude of the field (B), or F = q*v x B, where the direction of F is at right angles to both v and B as a result of the cross product. This defines the magnetic field's strength and direction at any point.

- Force on a current element

Current element

$$\overline{dF} = I\left(d\bar{l} \times \overline{B}\right)\quad s$$

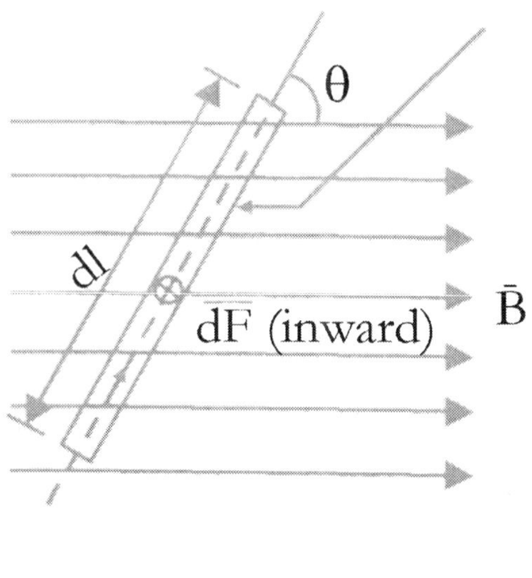

Where

B=Flux density, wb/m^2

dF=Force on conductor, Newtons

dl=Length of current carrying conductor, metres

I=Current in conductor

$$\overline{F} = 1\left(\overline{l} \times \overline{B}\right)$$

For liner conductor of length l in a uniform field B

- Biot savart law

$$\overline{dB} = \frac{\mu}{4\pi}\frac{1\left(\overline{dl} \times \overline{r}\right)}{|r|^3}$$

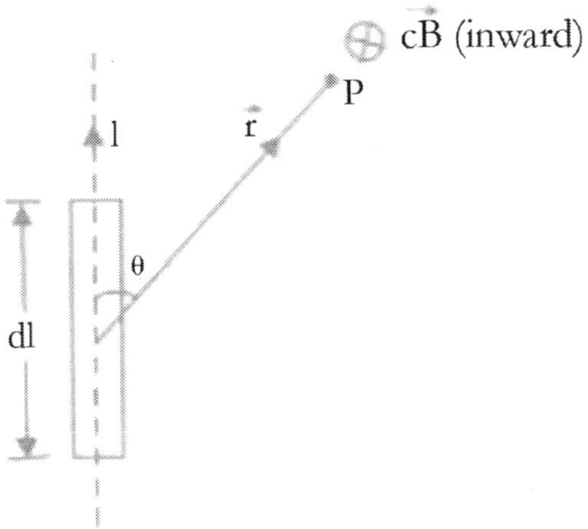

Where,

$\overrightarrow{dB}$ =Magnetic flux density produced by a current carrying element

$\vec{r}$ =Radius vector

Flux density produced by a infinitely long current carrying wire at a distance R $B_p = \frac{\mu l}{2\pi R}$

Torque in a current carrying loop

$T = BINA \sin\alpha$

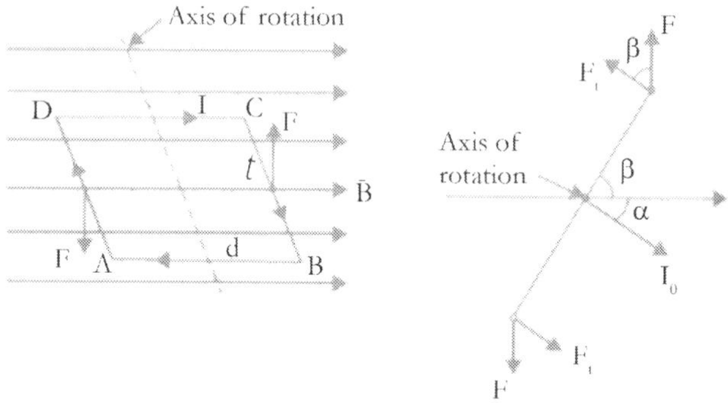

MAGNETIC PARAMETERS

Magnetic parameters are follows:

Permeability of the Medium

$\mu = \mu_r \mu_0$

Where,

μ_r =Relative permeability

μ_0 = Permeability of vacuum

$=4\pi \times 10^{-7}$ henry-metre^{-1}

Magnetic Moment

- Magnetic dipole moment

$\bar{p}_m = \text{current} \times \text{area of the coil} \quad \text{ampere} - \text{metre}^2$

Note: For a permanent bar magnet the dipole moment is defined as the product of pole strength and distance between them.

Magnetization

$$\overline{M} = \frac{\overline{p}_m}{volume}; \overline{M} = N\overline{p}_m$$

Where, N=Number of magnetic dipoles per unit volume

Magnetic Flux Density

$B = \mu H$ Where, H=Magnetic field intensity

- Magnetic flux density when a magnetic field is applied

$$B = \mu_0 H + \mu_0 M = \mu_0 \mu_r H$$

$$M = (\mu_r - 1) H = \chi_m H$$

Magnetization $\chi_m = \mu_r - 1$

Magnetic susceptibility

- For diamagnetic materials X_m is negative
- For paramagnetic materials X_m is very small and positive
- For ferromagnetic materials X_m is large and positive

MAGNETIC MOMENT

Magnetic Moment is a determination of its propensity to align through magnetic field. Equally the magnetic moment also might be measured towards vectors which contain a magnitude and the direction. The way of the magnetic moment point as of the south within the way of North Pole of a magnet. Magnetic field created through a magnet is relative toward its magnetic moment.

Orbital Dipole Moment

Currents in a wire loop create a magnetic dipole field perpendicular to the plane of the loop. Surely, the same is true for a charged particle in a circular orbit.

- Magnetic Dipole Moment of the Orbit

$$p_m = \frac{1}{2} e\omega_0 r^2$$

Where

ω_0 = Angular velocity of electron

r = Radius of the orbit

Angular Momentum

The angular momentum of a rigid object is defined as the product of the moment of inertia and the angular velocity. It is analogous to linear momentum and is subject to the fundamental constraints of the conservation of angular momentum principle if there is no external torque on the object. Angular momentum is a vector quantity. It is derivable from the expression for the angular momentum of a particle

- Orbital Angular Momentum

$$|M_a| = m\omega_0 r^2 ; \frac{p_m}{M_a} = \frac{e}{2m}$$

Where, M = Mass of the electron

Note

Ratio of P_m to M_a is independent of the angular velocity and the radius of the orbit.

Atomic unit of magnetic moment is called Bohr magneton.

1 Bohr magneton = 9.27×10^{-24} ampere $-m^2$.

- Larmor Angular Frequency

$$\omega_L = \frac{e}{2m} B$$

- Angular Velocity for any value of B

$$\omega = \omega_0 + \frac{e}{2m} B$$

- Magnetic Dipole Moment Include by the Field

$$p_{m,ind} = -\frac{e^2}{4m} \cdot r^2 B$$

- Bohr Magneton

$$\beta = \frac{eh}{4\pi m}$$

Ampere-metre^2

Electron Spin Dipole Moment

An electron spin s = 1/2 is an intrinsic property of electrons. Electrons have intrinsic angular momentum characterized by quantum number 1/2. In the pattern of other quantized angular momenta, this gives total angular momentum

$$S = \sqrt{\frac{1}{2}\left(\frac{1}{2}+1\right)}\,h = \frac{\sqrt{3}}{2}h$$

The resulting fine structure which is observed corresponds to two possibilities for the z-component of the angular momentum.

$$S_z = \pm\frac{1}{2}h$$

This causes an energy splitting because of the magnetic moment of the electron

$$\mu_s = -\frac{e}{2m}gS$$

- Generated due to Spin of Electron

$$p_{m,spn} = \left(\text{No. of unpaired } e^{-}s\right) \times \beta \; ; \; \left|\bar{p}_{m,spn}\right| > \left|\bar{p}_{m,orb}\right| ;$$

Dia Magnetic Materials

Diamagnetic materials have a weak, negative susceptibility to magnetic fields. Diamagnetic materials are slightly repelled by a magnetic field and the material does not retain the magnetic properties when the external field is removed. In diamagnetic materials all the electron are paired so there is no permanent net magnetic moment per atom. Diamagnetic properties arise from the realignment of the electron paths under the influence of an external magnetic field. Most elements in the periodic table, including copper, silver, and gold, are diamagnetic.

- Do not have Permanent Dipole Moment

$$\bar{p}_m \cong \bar{p}_{m,orb} + \bar{p}_{m,spin} = 0$$

Note:

For perfect diamagnetic materials, $\chi_m = -1$

In general, χ_m comes out to the order of -10^{-5} to -10^{-6}.

Magnetic Susceptibility

$$\chi_m = -\frac{Ne^2}{4m} r^2 \mu_0$$

Paramagnetic Materials

Paramagnetic materials have a small, positive susceptibility to magnetic fields. These materials are slightly attracted by a magnetic field and the material does not retain the magnetic properties when the external field is removed. Paramagnetic properties are due to the presence of some unpaired electrons, and from the realignment of the electron paths caused by the external magnetic field. Paramagnetic materials include magnesium, molybdenum, lithium, and tantalum.

- Spontaneous Magnetization for Paramagnetic Material is Zero
- Magnetic Susceptibility

$$\chi_m = -\frac{N\beta^2\mu_0}{kT}$$

- The magnetic field inside the material, when the effect of internal field is considered

$$H_i = H + \gamma M$$

Where,

$\gamma =$ Internal or molecular field constant

Curie Law

The curie law states that in a paramagnetic material, the material's magnetization is directly proportional to an applied magnetic field. But the case is not the same when the material is heated. When it is heated, the relation is reversed i.e. the magnetization becomes inversely proportional to temperature.

- Magnetic Susceptibility

$$\chi_m = \frac{C}{T}$$

Where,

C = Curie constant

T = Temperature

Curie-Weiss Law

A more accurate equation is obtained in many cases by modifying the above equation to $\chi = C/(T - \theta)$, where θ is a constant. This equation is called the Curie–Weiss law.

- Magnetic Susceptibility

$$\chi_m = \frac{C}{T - \theta}$$

Where,

$\theta =$ Paramagnetic Curie temperature

Ferromagnetism

- Ferromagnetism is characterized by the presence of parallel alignment of magnetic dipolc moment.
- Ferromagnetism occurs in Fe, Co, Ni, Gd and Dy.
- Common characteristic of these materials is partly filled inner electron shells.
- They follow Curie Weiss Law.
- Below Curie temperature they exhibit ferromagnetism but above curie temperature they behave as paramagnetic materials.

Total internal field which exists in these materials is

$H_i = H + \gamma M$ Where, γ = Molecular field constant

M = Magnetization

H = Applied field

- Spontaneous magnetization dies out as temperature increases.

Magnetic Anisotropy

- In some materials like iron magnetic properties depends on the thee in which they are measured, this phenomenon is called magnetic anisotropy

Magnetostriction

- When a ferromagnetic material is magnetized, its physical dimension changes, this phenomenon is called magnetostriction.

Villari Effect

- This is converse of magnetostriction, which is when strain is applied there is change in the magnetic property of the material.

Antiferromagnetic

These materials have a small susceptibility at all temperatures. As the temperature decreases x_m, increases but at a critical tempera T_N called Neel temperature goes maximum and if temperature is further lowered it further decreases. The variation of susceptibility with temperature of an antiferromagnetic material above critical temperature T_N is given by

$$\chi_m = \frac{C}{T-\theta}$$ Where, C = Curie constant

θ = paramagnetic Curie temperature

Classification of Magnetic Materials

	Type	Susceptibility χ_m	$\chi_m V_s T$ relation	Examples
1.	Diamagnetic	$\sim\sim 10^{-n}$ (negative)	independent	Atoms of solids having closed shells and some metals Au, Ge, etc.
2.	Paramagnetic	$\sim 10^{-5}$(positive)	$\chi_m = C/T$ curie law or $\chi_m = C/T\text{-}\theta$ curie weiss law	Atoms possessing odd number of electrons, ionic crystals etc.
3.	Ferromagnetic	Very large & positive	$(\chi_m \rightarrow \infty)$	Iron, cobalt, nickel, gadolinium.
4.	Antiferromagnetic	Small and positive	χ_m decreases with temperature	Salts and oxides of transition metals, e.g. NiO, MnF_2.
5.	Ferrimagnetic	Large and positive	$\chi_m \rightarrow \infty$	Ferrites, e.g. Fe_3O_4.

4 Conductive Materials

Conduction materials include metals, electrolytes, superconductors, semiconductors, plasmas and some nonmetallic conductors such as graphite and Conductive polymers.

Electrical Conductivity

Electrical conductivity is the measure of a material's ability to accommodate the transport of an electric charge. Its SI derived unit is the siemens per metre, ($A^2s^3m^{-3}kg^{-1}$) or, more simply, Sm^{-1}. It is the ratio of the current density to the electric field strength or, in more practical terms, is equivalent to the electrical conductance measured between opposite faces of a 1-metre cube of the material under test.

Ohm's Law

$$J = \sigma E = \frac{I}{A} \quad A/m^2$$

Where,

R=Resistance of conductor, Ω

ρ=Resistivity of the material, Ω-m

l=Length of conductor, m

A=Area of cross-section, m^2

J=Current density, A/m^2

σ = Conductivity of the material, $\Omega^{-1} m^{-1}$

E=Applied electric field, V/m

I= Current, A

- Conductivity

$$\sigma = \frac{ne^2\tau_c}{m} = ne\mu_e \quad \Omega^{-1}m^{-1}$$

Where,

n=No. of electrons per unit volume,

τ_c =Collision time, sec.

e=Charge of electron

m=mass of electron

μ_e =mobility of electron, m^2 $volt^{-1}$ sec^{-1}

$$v_d = \frac{e\tau_c}{m}E = \mu_e E$$

Drift velocity of electron , m/sec

Mean free path

$\lambda = v.\tau_c$ Where, v=Average electron velocity

- Velocity of an electron

$v_p = \sqrt{\frac{2W_F}{m}}$ Where, W_f= Fermi energy

Note:

- At absolute zero, all energy levels below a certain value W_F are filled, and all those above W_F being empty W_F is the Fermi level of electron.
- For isotropic materials, the mean time of collision is the same as the relaxation time.

Matthiessen's Rule

In pure metals and dilute alloys, the total resistivity is the sum of two terms. The thermal component ρ_{th}, which arises from lattice vibrations and residual resistivity ρ_r, caused by impurities i.e. $\rho = \rho_{th} + \rho_r$. This rule becomes less accurate at high temperature.

Joule's Law

Volume density of heat developed per second

$$W = \sigma E^2 = JE \quad \text{watts/m}^3$$

Thermal Conductivity

Thermal conductivity is the property of a material to conduct heat. It is evaluated primarily in terms of Fourier's Law for heat conduction. Heat transfer occurs at a lower rate across materials of low thermal conductivity than across materials of high thermal conductivity. Correspondingly, materials of high thermal conductivity are widely used in heat sink applications and materials of low thermal conductivity are used as thermal insulation. The thermal conductivity of a material may depend on temperature. The reciprocal of thermal conductivity is called thermal resistivity.

- Flux of thermal energy

$$Q = -K\left(\frac{dT}{dx}\right) \quad \text{watts/m}^2$$

Where, K=Thermal conductivity, watts m^{-1} degree

$\frac{dt}{dx}$=Temperature gradient, degree C per metra

Thermal conductivity

$$K = \frac{1}{3}\frac{n\pi^2 k^2 \tau T}{m}, \text{ watt/m.k}$$

Where,

n=Number of conduction electrons per m^3

T=Temperature

k=Boltzmann's constant

Weidemann-Frenz Law

- Lorentz number

$$L = \frac{K}{\sigma T} = \frac{1}{3}\frac{\pi^2 k^2}{e^2}$$

$\frac{k}{\sigma T}$ =universal constant for all metals

Lorentz number L=2.45 $\times$ 10^{-8} watt ohm degree^{-2}

Super Conductor

Superconductors are materials that conduct electricity with no resistance. This means that, unlike the more familiar conductors such as copper or steel, a superconductor can carry a current indefinitely without losing any energy. They also have several other very important properties, such as the fact that no magnetic field can exist within a superconductor.

- It is possible to destroy superconductivity by the application of sufficiently strong magnetic field.

$$H_{CT} = H_{CO}\left[1 - \left(\frac{T}{T_c}\right)^2\right]$$

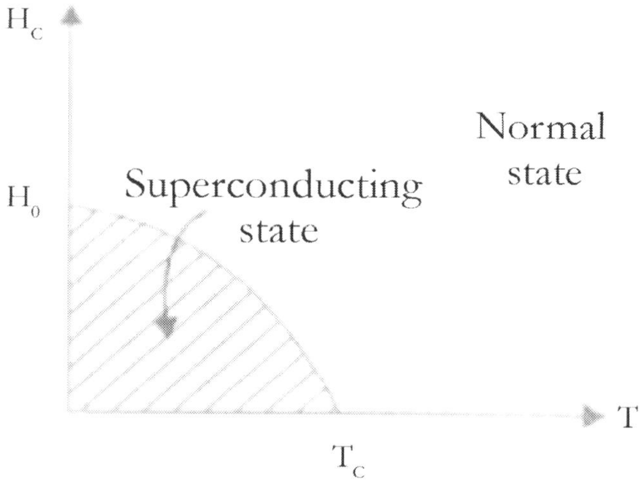

Where, H_{CO} =Critical at absolute zero

C_{CT}=Critical field at any temperature T

T_C=Transition temperature

Silsbee's Rule

- In a long superconductor wire of radius R, the super-conductivity may be destroyed when a current I exceeds the critical current value I_c

$I_c = 2\pi RH_{CT}$

- Magnetic susceptibility in a super conductor is negative. This is referred to as perfect diamagnetism. This phenomenon is called meissner effect.
- Transition temperature

$$T_c \propto \frac{1}{\sqrt{M}}$$

Where M is the mass of isotope

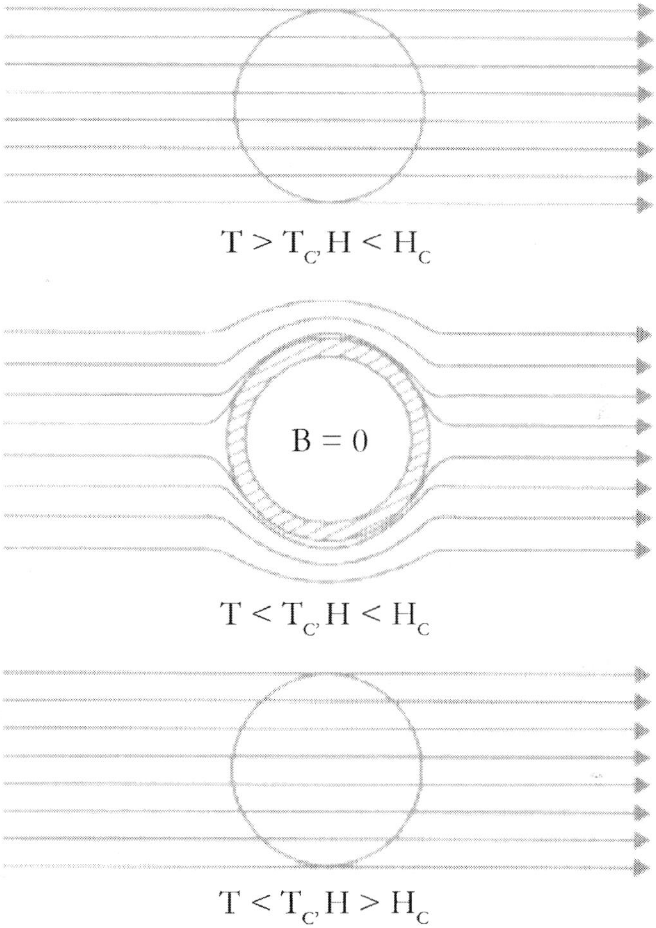
T > T_C, H < H_C
B = 0
T < T_C, H < H_C
T < T_C, H > H_C

5 Semi-conduction

A semiconductor material has an electrical conductivity value falling between that of a conductor, such as copper, and an insulator, such as glass. Semiconductors are the foundation of modern electronics. Semiconducting materials exist in two types' elemental materials and compound materials.

- The modern understanding of the properties of a semiconductor relies on quantum physics to explain the movement of electrons and holes in a crystal lattice.
- The unique arrangement of the crystal lattice makes silicon and germanium the most commonly used elements in the preparation of semiconducting materials. An increased knowledge of semiconductor materials and fabrication processes has made possible continuing increases in the complexity and speed of microprocessors and memory devices. Some of the information on this page may be outdated within a year, due to the fact that new discoveries are made in the field frequently

CONDUCTIVITY

The electrical conductivity of a metal or semiconductor is directly related to the mobility of the charge carriers

$$\sigma = \frac{1}{\rho} = e\mu N = \frac{\mu}{R_H}$$

The mobility of the charge carriers is the average drift velocity per unit electric field

$$\mu = \frac{}{kT}$$

The mobility of the carriers is related to the diffusion coefficient D by

$$\mu = \frac{eD}{kT}$$

Mobility

Mobility is a phenomenological parameter used in the context of transistor circuit design. It becomes relevant in different ways depending on the application. As Misha mentioned, you can get higher gain and speed for higher mobility.

$$\mu_e = \frac{e\tau_e}{m_e}$$

- Mobility of electron

Hall Effect

The Hall effect is the production of a voltage difference (the Hall voltage) across an electrical conductor, transverse to an electric current in the conductor and a magnetic field perpendicular to the current.

6 Insulating Materials

The materials which have very high resistivity i.e. offers a very high resistance to the flow of electric current. Insulating materials plays an important part in various electrical and electronic circuits. In domestic wiring insulating material protect us from shock and also prevent leakage current.

COMPLEX PERMITTIVITY

$\in_r * = \in'_r - j \in''_r$

Where,

$\in'_r$ =real part of dielectric constant

$\in''_r$ =imaginary part of dielectric constant

Debye Equations

$$\in'_r = \in_\infty + \frac{\in_s + \in_\infty}{1 + \omega^2 \tau^2}$$

$$\in''_r = \frac{(\in_s + \in_\infty)\omega\tau}{1 + \omega^2 \tau^2}$$

Where,

$\in_\infty$ = Dielectric constant at infinite frequency

$\in_s$ = Dielectric constant under static

Dielectric Losses

$$W(t) = \frac{1}{2}\omega \in_0 \in''_r E^2_0$$

- Energy absorbed per m^3
- Loss tangent

$$\tan\delta = \frac{\in''_r}{\in'_r}$$

Lossy Capacitor

Equivalent circuits

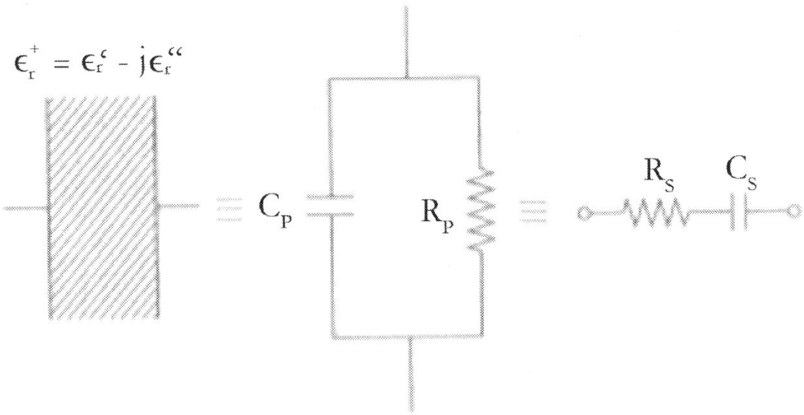

For parallel circuit

$$R_p = \frac{1}{\omega \in''_r C_0}\ ;\ C_p = \in'_r C_0$$

Where, C_0 =capacitance

$$C_0 = \frac{\epsilon_0 A}{d}$$

For series circuit

$$R_s = \frac{R_s}{1+\left(\omega C_p R_p\right)^2} ; C_s = \frac{1+\left(\omega C_p R_p\right)^2}{\omega^2 C_p R_1^2}$$

Loss tangent

For series circuit

$$\tan\delta = \omega C_s R_s$$

For parallel circuit

$$\tan\delta = \frac{1}{\omega C_p R_p}$$

Quartz

Quartz is a piezoelectric material, used for high frequency oscillation

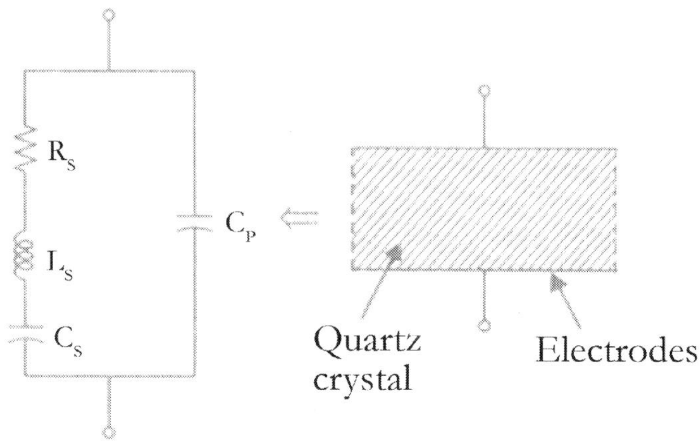

Q.factor

$$Q = \frac{\omega_s L_s}{R_s}$$

Series resonance frequency

$$\omega_s = \frac{1}{\sqrt{L_s C_s}}$$

Impendance at series resonance frequency

$$Z\big|_{\omega=\omega_s} = R_s$$

Parallel resonance frequency

$$\omega_p = \sqrt{\frac{1}{L_s}\left(\frac{1}{C_s} + \frac{1}{C_p}\right)}$$

Impedance at parallel resonance frequency

$$Z\big|_{\omega=\omega_p} = \frac{R^2_s + \left(\omega_p L_s - \dfrac{1}{\omega_p C_s}\right)}{R_s}; \; Z\big|_{\omega=\omega_p} > Z\big|_{\omega=\omega_s}$$

7 Structure of Materials

Structure of materials are follows:

Simple Cubic (SC)

- Distance between adjacent atom

$$d_{sc} = 2r = a$$

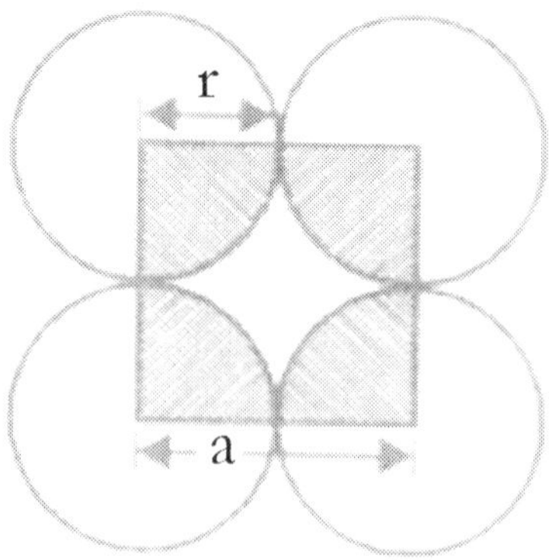

Where, r=Radius of an atom

a=Edge length of cube

Effective number of atom in a unit cell

$$Z_{sc} = 1$$

Packing fraction

$$PF_{sc} = 0.523$$

Body Centered Cubic (BCC)

The distance between adjacent atoms

$$d_{BCC} = 2r = \frac{\sqrt{3}}{2}a$$

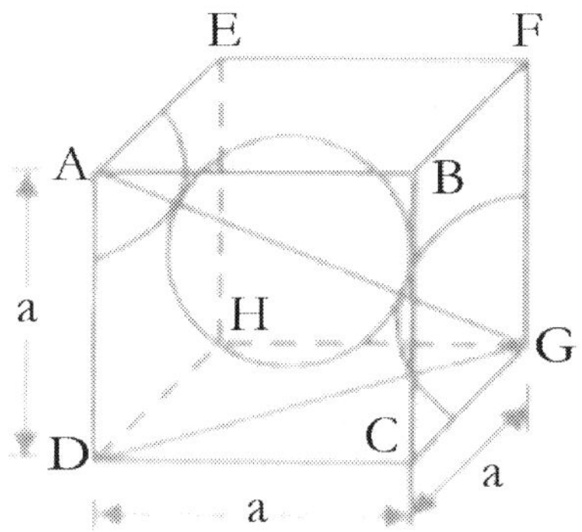

Effective number of atom in unit cell

$$Z_{BCC} = 2$$

Packing fraction

$$PF_{BCC} = 0.68$$

Example

Face Centered Cubic (FCC)

The distance between adjacent atoms

$$d_{FCC} = 2r = \frac{a}{\sqrt{2}}$$

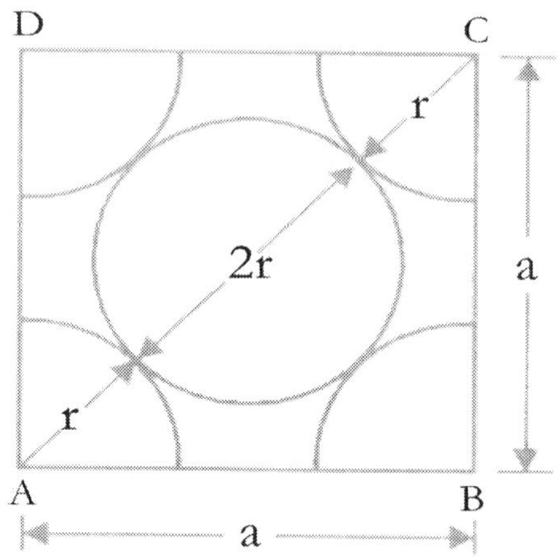

Effective number of atom in a unit cell

$$Z_{FCC} = 4$$

Packing fraction

$$PF_{FCC} = 0.7404$$

Example =NaCl, Cu, Silver, Gold, Al, Ni……

$$\text{Miller index} = \frac{\text{edge length}}{\text{Length of intercept}}$$

A Handbook on Electronics and Communications Engineering

11

Measurement and Instrumentation

CONTENTS

1 Characteristic of Instruments and Measurement Systems

Measurement is a process of mapping actually occurring variables into equivalent values. Deviations from perfect measurement mappings are called errors: what we get as the result of measurement is not exactly what is being measured.

Instrument systems are usually built up from a serial linkage of distinguishable building blocks. The actual physical assembly may not appear to be so but it can be broken down into a representative diagram of connected blocks.

Static Error

Static error is caused by any reason and which stays unchanged in value during the operation of the system. This error should be measured under static conditions.

$$\delta A = A_m - A_t$$

Where, A_m = Measured value of quantity or actual value

A_t = True value of quantity or Nominal value

Relative Static Error

The relative static error is defined as the ratio of absolute static error to the true value of the quantity under measurement.

$$\epsilon_r = \frac{\delta A}{A_t}$$

Static correction

Static correction is a very mild, harmless electric stimulation that gets your pet's attention.

$$\delta C = A_t - A_m = -\delta A$$

Static sensitivity

Static sensitivity of instrument or measurement system is ratio of the magnitude of output signal or response to the magnitude of input signal or quantity being measured.

$$\text{static sensitivity} = \frac{\Delta q_0}{\Delta q_i}$$

Where, Δq_0 = Infinitesimal change in output

Δq_i = Infinitesimal change in input

- $$\text{Non-linearity} = \frac{\text{(Max. deviation of output from the idealized straight line)}}{\text{full scale deflection}} \times 100$$

Combination of Quantities with limiting Errors

When the two quantities are combined, each having limiting error, then it is necessary to calculate the overall limiting error.

Sum or Difference of More than Two Quantities

Let $X = \pm x_1 \quad \pm x_2 \quad \pm x_3 \quad \pm x_4$

$$\frac{\delta X}{X} = \pm\left(\frac{x_1}{X}\frac{\delta x_1}{x_1} + \frac{x_2}{X}\frac{\delta x_2}{x_2} + \frac{x_3}{X}\frac{\delta x_3}{x_3} + \frac{x_4}{X}\frac{\delta x_4}{x_4}\right)$$

where $\pm\,\delta x_1 = \text{Re lative increment in quantity } x_1$

$\pm\,\delta x_2 = \text{Re lative increment in quantity } x_2$

$\pm\,\delta x = \text{Re lative increment in quantity } X$

$\frac{\delta x_1}{x_1} = \text{Relative limiting error in quality } x_1$

$\frac{\delta x_2}{x_2} = \text{Relative limiting error in quality } x_2$

$\frac{\delta x}{X} = \text{Relative limiting error in quality } X$

Product or Quotient of More than Two Quantities

$$X = x_1x_2x_3 \text{ or } X = \frac{x_1}{x_2x_3} \text{ or } X = \frac{1}{x_1x_2x_3}$$

$$\frac{\delta X}{X} = \pm\left(\frac{\delta x_1}{x_1} + \frac{\delta x_2}{x_2} + \frac{\delta x_3}{x_3}\right)$$

Composite Factor

$$X = x_1^n . x_2^m \qquad \boxed{\frac{\delta X}{X} = \pm\left(n\frac{\delta x_1}{x_1} + m\frac{\delta x_2}{x_2}\right)}$$

Let

- Arithmetic Mean

$$\overline{X} = \frac{\sum x}{n} = \frac{x_1 + x_2 + \cdots x_n}{n}$$

Where, x_1, x_2, ……..,

x_n = Reading or samples

n=Number of reading

Deviation

Deviation is a measure of difference between the observed value of a variable and some other value, often that variable's mean.

$$d_n = x_n - \overline{x}$$

Note: Algebraic sum of deviation is zero.

- Average deviation

$$\overline{D} = \frac{\sum|d|}{n} = \frac{|d_1| + |d_2| + \cdots |d_n|}{n}$$

- Standard deviation

For n>20

$$S.D. = \sigma = \sqrt{\frac{\sum d^2}{n}} = \sqrt{\frac{d_1^2 + d_2^2 + \cdots d_n^2}{n}}$$

For n<20

$$S.D. = s = \sqrt{\frac{\sum d^2}{n-1}}$$

Variance

Variance is always non-negative: a small variance indicates that the data points tend to be very close to the mean.

Variance for n>20

$$V = \sigma^2 = \frac{\sum d^2}{n}$$

For n<20

$$V = s^2 = \frac{\sum d^2}{n-1}$$

Normal or Gaussian Curve of Error

The Gaussian curve error and complementary error function are frequently used in probability theory since the normalized Gaussian curve represents the probability distribution with standard deviation σ relative to the average of a random distribution.

For Infinite Number of Reading

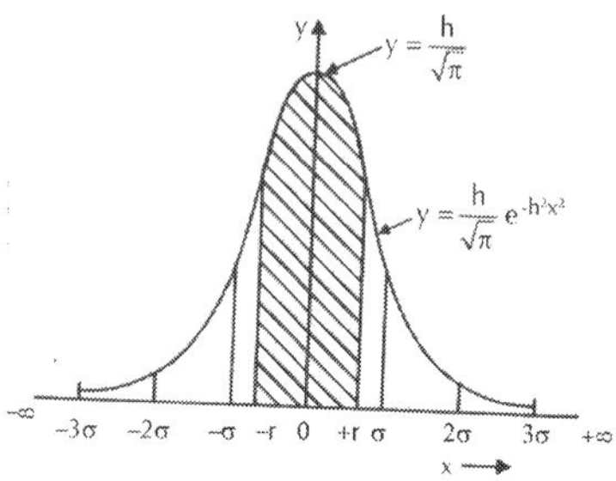

$$y = \frac{1}{\sigma\sqrt{2\pi}} \exp\left(-x^2 / 2\sigma^2\right)$$

Where x=magnitude of deviation from mean.

y=number of reading at any deviation x, (the probability of occurrences of deviation x)

σ=standard deviation

- Precision index

$$h = \frac{1}{\sigma\sqrt{2}}$$

Probable Error (P.E.)

$$r = \frac{0.4769}{h}$$

- Average deviation

$$\bar{D} = \frac{r}{0.8453} = \frac{1}{h\sqrt{\pi}}$$

- Standard deviation

$$\boxed{\sigma = \frac{r}{0.6745} = \frac{1}{h\sqrt{2}}}; \text{ P.E.} = r = 0.8453\,\bar{D} = 0.6745\sigma$$

For Finite Number of reading

- For n>20

$$\text{P.E.} = r = 0.6745\sqrt{\frac{\Sigma|d|^2}{n}}$$

- For n<20

$$\text{P.E.} = r = 0.6745\sqrt{\frac{\Sigma|d|^2}{n-1}}$$

- Standard deviation of the mean

$$\sigma_m = \frac{\sigma}{\sqrt{n}}$$

- Standard deviation

$$\sigma_\sigma = \frac{\sigma_m}{\sqrt{2}}$$

- Variance of combination of components

Let $X = f(x, x_2, xxx, n)$

$$V_x = \left(\frac{\partial X}{\partial x_1}\right)^2 V_{x_1} + \left(\frac{\partial X}{\partial x_2}\right)^2 V_{x_2} + \cdots + \left(\frac{\partial X}{\partial x_m}\right)^2 V_{x_n}$$

where $V_{x_1}, V_{x_2} ......V_{x_n}$ = Variance of $x_1, x_2, ...x_n$

- Standard Deviation of combination of Components

Let $X = f(x_1, x_2, ,........., x_n)$

$$\sigma_x = \sqrt{\left(\frac{\partial X}{\partial x_1}\right)^2 \sigma_{x_1}^2 + \left(\frac{\partial X}{\partial x_2}\right)^2 \sigma_{x_2}^2 + \cdots + \left(\frac{\partial X}{\partial x_n}\right)^2 V_{x_n}}$$

where $\sigma_{x_1}, \sigma_{x_2} ......\sigma_{x_n}$ = Standard deviation of $x_1, x_2, ...x_n$

- Probable Error of Combination of Components

Let $X = f(x_1, x_2, ,........., x_n)$

$$r_x = \sqrt{\left(\frac{\partial X}{\partial x_1}\right)^2 r_{x_1}^2 + \left(\frac{\partial X}{\partial x_2}\right)^2 r_{x_2}^2 + \cdots + \left(\frac{\partial X}{\partial x_m}\right)^2 r_{x_n}^2}$$

where $r_{x_1}, r_{x_2} ......r_{x_n}$ = Probable error of $x_1, x_2, ...x_n$

- Uncertainty of Combination of Components

Let $X = f(x_1, x_2, ,........., x_n)$

$$w_x = \sqrt{\left(\frac{\partial X}{\partial x_1}\right)^2 w_{x_1}^2 + \left(\frac{\partial X}{\partial x_2}\right)^2 w_{x_2}^2 + \cdots + \left(\frac{\partial X}{\partial x_n}\right)^2 w_{x_n}^2}$$

where $w_{x_1}, w_{x_2} ......w_{x_n}$ = Uncertainties of $x_1, x_2, ...x_n$

2 Circuit Component (Resistor, Inductors, Capacitors)

A component which is able to amplify or control voltages or currents without mechanical or other nonelectrical command or to switch currents or voltages without mechanical switches.

Frequency Error in Resistor

The frequency error on the resistor should not figure into the graph itself at all. Measure the slope of the graph to estimate the resistance, and then compare that with the value marked on the resistor. They are closer than the frequency errors.

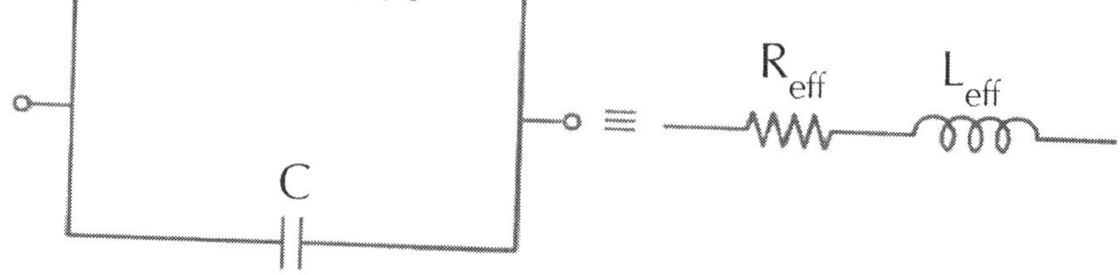

(Equivalent circuit of a resistor at low and medium frequencies)

Effective Resistance: The resistance to an alternating current, expressed as the ratio of the power dissipated to the square of the effective current.

$$R_{eff} = \frac{R}{1+\omega^2 C\left(CR^2 - 2L\right)}$$

Effective Inductance or Residual Inductance: The value to which the electrical resistance of a metal drops as the temperature is lowered to near absolute zero, caused by imperfections and impurities in the metal rather than by lattice vibrations.

$$L_{eff} = \frac{L - CR^2}{1+\omega^2 C\left(CR^2 - 2L\right)}$$

$$\tan\phi = \frac{X_{eff}}{R_{eff}} = \frac{\omega L_{eff}}{R_{eff}} = \frac{\omega\left(L - CR^2\right)}{R} = \omega\left(\frac{L}{R} - CR\right)$$

Where, Phase defect angle

Time Constant: The time constant, usually denoted by the Greek letter τ (tau), is the parameter characterizing the response to a step input of a first-order, linear time-invariant (LTI) system. The time constant is the main characteristic unit of a first-order LTI system.

$$\tau = \frac{L_{eff}}{R_{eff}} = \frac{L - CR^2}{R} = \frac{L}{R} - CR$$

- Condition for Resistance to remain independent of frequency

$$CR^2 = 2L$$

- Condition for Resistance to show no inductive effect is

$$CR^2 = L$$

- Effective resistance for zero effective inductance

$$R_{eff} = \frac{R}{1-\omega^2 LC}$$

- Quality factor or storage factor inductance

$$Q = \frac{\omega L}{R}$$

Frequency Error in Inductor

An inductor connected to a capacitor forms a tuned circuit, which acts as a resonator for oscillating current. Tuned circuits are widely used in radio frequency equipment such as radio transmitters and receivers, as narrow band pass filters to select a single frequency from a composite signal, and in electronic oscillators to generate sinusoidal signals.

A frequency error in inductor can effect on:

- Effective resistance

$$R_{eff} = \frac{R}{\left(1-\omega^2 LC\right)^2}$$

- Effective inductance

$$L_{eff} = L\left(1+\omega^2 CL\right)$$

Capacitor

A capacitor (originally known as a condenser) is a passive two-terminal electrical component used to store energy electrostatically in an electric field. The forms of practical capacitors vary widely, but all contain at least two electrical conductors (plates) separated by a dielectric (i.e. insulator).

Parallel Representation

Capacitors, like other electrical elements, can be connected to other elements either in series or in parallel. Sometimes it is useful to connect several capacitors in parallel in order to make a functional block such as the one in the figure. In such cases, it is important to know the equivalent capacitance of the parallel connection block.

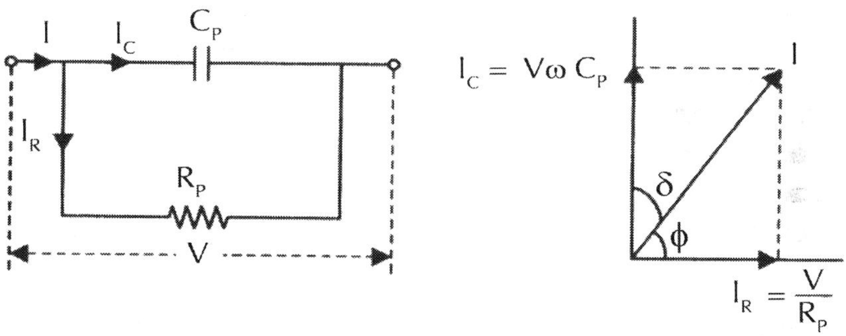

- Dielectric Loss

$$P_L = \omega C_P V^2$$

- Dissipation Factor

$$D = \tan\delta = \frac{1}{\omega C_P R_P}$$

Where, δ=loss angle of the capacitor

The value of δ is very small for capacitor

Series Representation

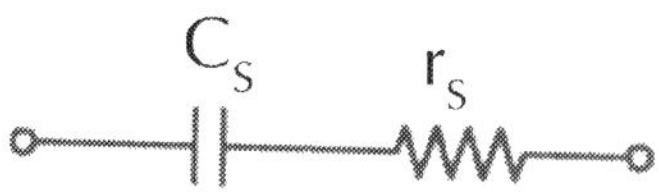

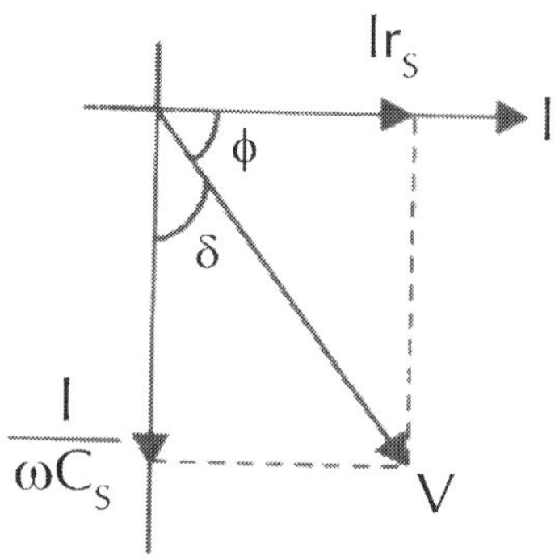

(Phasor diagram)

- Dielectric loss

$$P_L = \frac{I^2}{\omega C_s} \tan\delta$$

- Dissipation Factor

$$D = \tan\delta = \omega C_s r_s$$

Frequency Error in Capacitor

Effective Capacitance

$$C_{eff} = \frac{C}{1-\omega^2 LC}$$

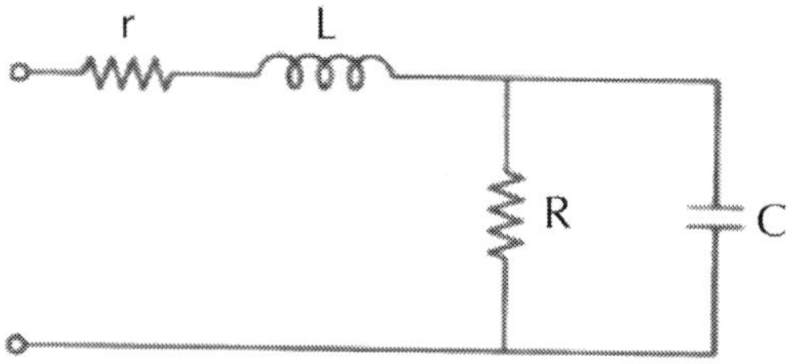

Equivalent Circuit of a Capacitor

For Medium Frequency

Effective Capacitance

$$C_{eff} = C\left(1+\omega^2 LC\right)$$

Effective series resistance

$$R_{eff} = r + \frac{R}{1+\omega^2 RC^2}$$

Loss angle

$$\tan\delta = \frac{1-\omega^2 LC}{\omega r + \dfrac{1}{\omega CR}}$$

For Low Frequency

Effective Capacitance

$$C_{eff} = C + \frac{1}{\omega^2 CR^2}$$

Effective series resistance

$$R_{eff} = \frac{R}{1 + \omega^2 C^2 R^2}$$

Loss angle (Low Frequency)

$$\tan \delta = \omega C r + \frac{1}{\omega C R}$$

3 Galvanometers

A galvanometer is a type of sensitive ammeter: an instrument for detecting electric current. It is an analog electromechanical actuator that produces a rotary deflection of some type of pointer in response to electric current through its coil in a magnetic field.

Galvanometer is the historical name given to a moving coil electric current detector. When a current is passed through a coil in a magnetic field, the coil experiences a torque proportional to the current. If the coil's movement is opposed by a coil spring, then the amount of deflection of a needle attached to the coil may be proportional to the current passing through the coil.

D' ARSONVAL Galvanometer

The most common type is the D'ARSONVAL galvanometer, in which the indicating system consists of a light coil of wire suspended from a metallic ribbon between the poles of a permanent magnet. The magnetic field produced by a current passing through the coil reacts with the magnetic field of the permanent magnet, producing a torque, or twisting force.

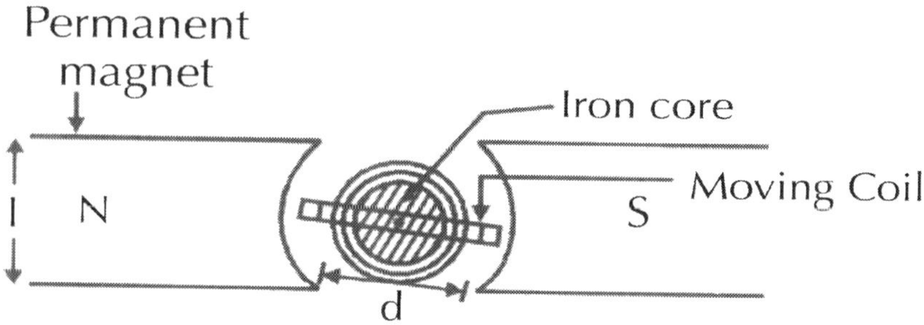

Deflecting Torque

$$T_d = BiNA = Gi$$

Where B= Flux density in air gap; Wb/m^2

I= current through moving coil; A

N= Number of turns in coil

A=ld= Area of coil; m^2

l, d= length of respectively vertical and horizontal side (width) of coil; m

G = Displacement constant of galvanometer

Controlling Torque

$$T_c = K\theta_f$$

Where K = Spring constant of suspension; Nm/rad

θ_f= final steady detection of moving coil; rad

- Final steady Deflection

$$\theta_f = \left(\frac{NBA}{K}\right)i = \left(\frac{G}{K}\right)i$$

Dynamic behavior of Galvanometers

Galvanometer are electromechanical electricity meters, which is proportional generate rotary mechanical motion to electricity. The principle is used among other things in moving coil instruments in combination with a pointer and a scale as indicating instrument. The time-dependent deflection angle of the galvanometer, which are changes over time; it is the important dynamic behavior of galvanometers.

Torque in Galvanometers

The torque on a current loop in a uniform magnetic field is used to measure electrical magnetic field is used to measure electrical currents. This current measuring device is called a moving coil galvanometer.

Inertial Torque

$$T_j = J\frac{d^2\theta}{dt^2}$$

Where J= moment of inertia of moving system about the axis of rotation; kg-m^2

θ= deflection at any time; rad

Damping Torque

$$T_D = D\frac{d\theta}{dt}$$

Where D= Damping constant

Controlling Torque

$$T_C = K\theta$$

Where, K= control constant

Deflecting Torque

$$T_d = Gi$$

Total resistance of galvanometer circuit for critical damping

$$R = \frac{G^2}{2\sqrt{KJ}}$$

External series resistance required for critical damping

$$R_e = R - R_g = \frac{G^2}{2\sqrt{KJ}} - R_g$$

Where, R_g = resistance of galvanometer

Notes:

- If $D^2 < 4kJ$, galvanometer is underdamped
- If $D^2 = 4kJ$, galvanometer is critically damped
- If $D^2 > 4kJ$, galvanometer is overdamped

Sensitivity

Sensitivity of galvanometer is defined as the current in micro ampere required to consume one millimeter deflection on a scale placed 1m away from the mirror.

Current Sensitivity

Current sensitivity of a galvanometer is defined as the deflection produced in the galvanometer when a unit current flows through it.

$$\boxed{S_i = \frac{\theta_f}{i} = \frac{G}{K}} \text{ red / A;} \quad \boxed{S_i = \frac{d}{i \times 10^6}} \text{ scale divisions / } \mu\text{A}$$

$$\boxed{S_i = \frac{2000G}{K \times 10^6}} = \text{mm / } \mu\text{A}$$

Voltage sensitivity

Voltage sensitivity of a galvanometer is defined as the deflection produced in the galvanometer when a unit voltage is applied across two terminals.

$$\boxed{S_v = \frac{d}{iR_g \times 10^6}} \text{ scale division / } \mu\text{V}$$

Megohm Sensitivity

It is define as the number of megohms required in series with the galvanometer to produce one scale division deflection. Numerically it is equal to current sensitivity.

$$\boxed{S_o = \frac{d}{i \times 10^{-6}}} \text{ MW / Scale division}$$

Note:

Sensitivity Galvanometer is one which produces a large deflection for a small current.

4 Analog Ammeters, Voltmeters and Ohmmeters

Analog ammeters are metered instruments that measure current flow in amperes. They display current levels on a dial, usually with a moving pointer or needle. A voltmeter is placed in parallel with a circuit element to measure the voltage drop across it and must be designed to draw very little current from the circuit so that it does not appreciably change the circuit it is measuring.

Ohmmeters are electrical devices used for measuring the resistance of a given conductor. This measuring instrument works on the basis of Ohm's Law (V=IR), which applies to electrical circuits.

Permanent Magnet Moving Coil (PMMC)

It is an ammeter or other electrical instrument in which a small coil of wire, supported on jeweled bearings between the poles of a permanent magnet, rotates when current is carried to it through spiral springs which also exert a restoring torque on the coil; the position of the coil is indicated by an attached pointer.

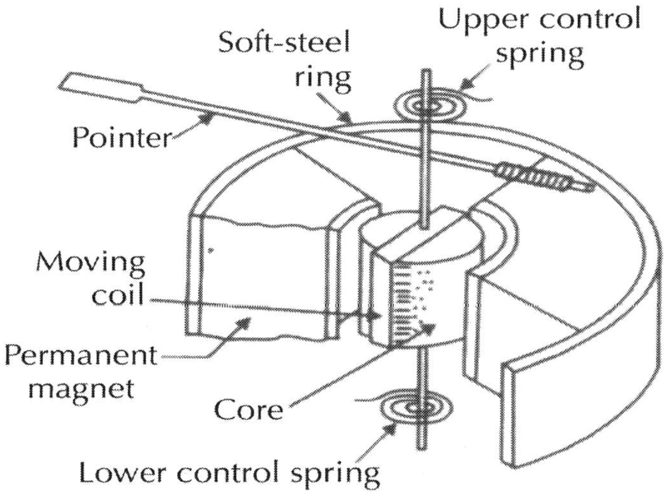

Final steady deflection

$$\boxed{\theta = \left(\frac{G}{K}\right) I}$$

Ammeter Shunts

An ammeter shunt allows the measurement of current values too large to be directly measured by a particular ammeter. In this case the shunt, a manganin resistor of accurately known resistance, is placed in series with the load so that all of the current to be measured will flow through it.

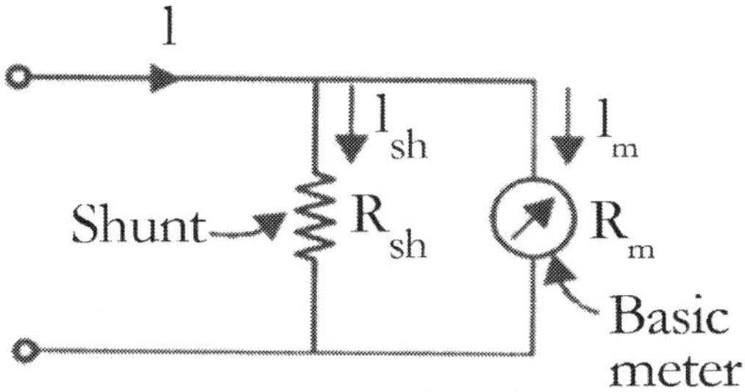

Basic ammeter circuit

$$I_{sh}R_{sh} = I_m R_m$$

$$I = \left(1 + \frac{R_m}{R_{sh}}\right) I_m$$

Where I = Current to be measured;

$I_m = I_{fs}$ = full scale deflection current of movement; A

R_m = Internal resistance of movement; Ω

R_{sh} = resistance of the shunt; Ω

Shunt resistance

$$\boxed{R_{sh} = \frac{R_m}{m - 1}}$$

Where

$$m = \frac{I}{I_m} = \left(1 + \frac{R_m}{R_{sh}}\right)$$

= multiplying power of shunt.

Universal or Ayrton Shunt

Ayrton shunt or universal shunt is a high-resistance shunt used in galvanometers to increase their range without changing the damping. Also it will eliminate the possibility of having a meter without a shunt which is a serious concern in make-before-break switches. (Multi range ammeter using universal shunt).

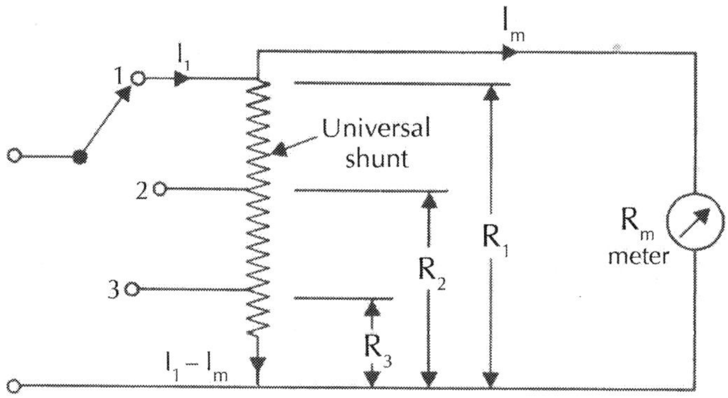

For switch at a position 1

$$R_1 = \frac{R_m}{(m-1)}$$

For switch at a position 2

$$R_2 = \frac{(R_1 + R_m)}{m_2}$$

For switch at a position 3

$$R_3 = \frac{(R_1 + R_m)}{m_3}$$

Where, $$m_1 = \frac{I_1}{I_m}, m_2 = \frac{I_2}{I_m}, m_3 = \frac{I_3}{I_m},$$

Voltmeter Multiplier

Voltmeter multiplier is a high non-inductive resistance connected in series with the voltmeter coil and is used for increasing the range of the voltmeter.

A voltmeter connected in series with the high non-inductive resistance (voltmeter multiplier) for measurement of high voltage is shown in fig.

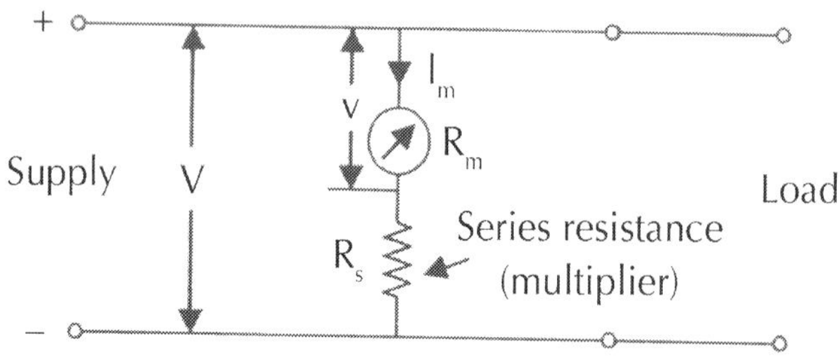

Multiplying factor for multiplier

$$m = \frac{V}{v} = 1 + \frac{R_s}{R_m}$$

Resistance of multiplier

$$R_s = (m-1)R_m$$

Where, R_s = Multiplier resistance

R_m = Internal resistance of meter

Potential Divider Arrangement

A potential divider is a passive linear circuit that produces an output voltage (V_{out}) that is a fraction of its input voltage (V_{in}). Potential dividers usually consist of two or more resistors arranged in series across a power supply.

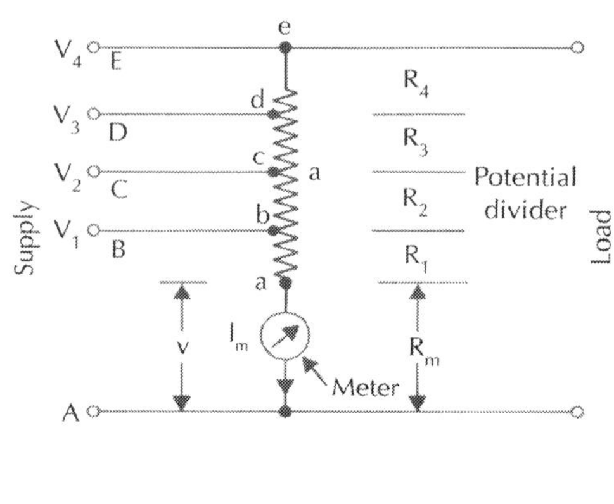

$$R_1 = (m_1 - 1)R_m$$
$$R_2 = (m_2 - m_1)R_m$$
$$R_3 = (m_3 - m_2)R_m$$
$$R_4 = (m_4 - m_3)R_m$$

Where, R_1=Resistance between point a and b
R_2=Resistance between point b and c
R_3=Resistance between point c and d
R_4=Resistance between point d and e

Sensitivity of a Voltmeter

The sensitivity of a voltmeter is given in ohms per volt. It is determined by dividing the sum of the resistance of the meter, plus the series resistance, by the full-scale reading in volts. In equation form, sensitivity is expressed as follows:

$$S_v = \frac{1}{I_{fs}} = \frac{1}{I_m}\Omega / V$$

Ohmmeter

An ohm meter uses a small electrical current to check the continuity of a circuit, also on different settings it can also tell how much resistance is in a circuit.

Ohmmeters typically use an ammeter to measure current through the material after it has been given some set voltage by the ohmmeter; the direct-current resistance of the sample can then be directly determined through Ohm's law.

Series type Ohmmeter

It consists of basic D'ARSONVAL movement connected in parallel with a shunting resistor R_2. This parallel circuit is in series with resistance R_1 and a battery of emf E. the series circuit is connected to the terminals A and B of unknown resistor R_x.

A circuit of series type ohmmeter is shown in figure.

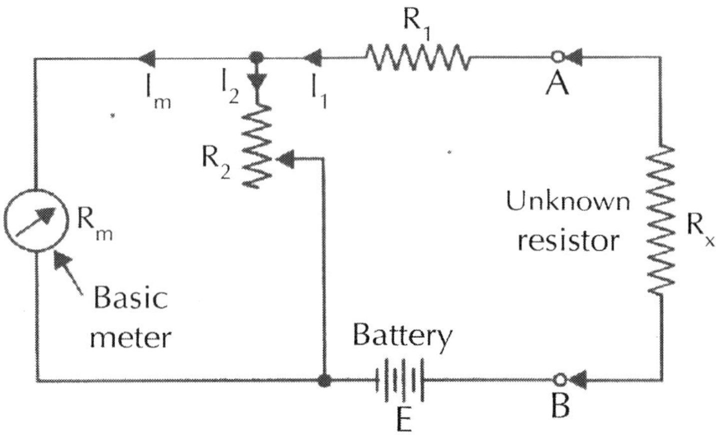

For the figure,

R_1 = current limiting resistor,

R2 = zero adjusting resistor,

E = emf of internal battery,

R_m = internal resistance of D'ARSONVAL movement,

Half Scale Resistance

$$R_h = R_1 + \frac{R_2 R_m}{R_2 + R_m}$$

Meter Current

$$I_m = \frac{ER_2}{(R_h + R_x)(R_2 + R_m)}$$

Full scale deflection current

$$I_{fs} = \frac{ER_2}{R_h(R_2 + R_m)}$$

Friction of full scale reading

$$s = \frac{I_m}{I_{fs}} = \frac{R_h}{R_x + R_h}$$

Shunt Type Ohmmeter

Shunt type Ohmmeter is used to measure small values of resistance. In this type of Ohmmeter, the movement mechanism is connected parallel to the unknown resistance R_x. In this circuit, it is compulsory to use a switch, if not; the current will always flow in the movement mechanism.

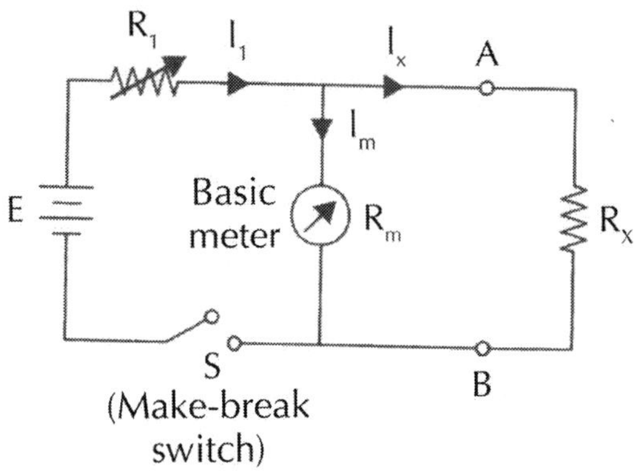

Half Scale reading of the meter

$$I_h = 0.5\, I_{fs} = \frac{ER_h}{R_1 R_m + R_h(R_1 + R_m)}$$

Half Scale reading of unknown resistance R_x is

$$R_h = \frac{R_1 R_m}{R_1 + R_m}$$

Where, R_m = Internal resistance of meter

R_1 = Adjustable resistance (shown in figure)

E= Supply Voltage

Ratio meter

A meter that measures the quotient of two electrical quantities; the deflection of the meter pointer is proportional to the ratio of the currents flowing through two coils.

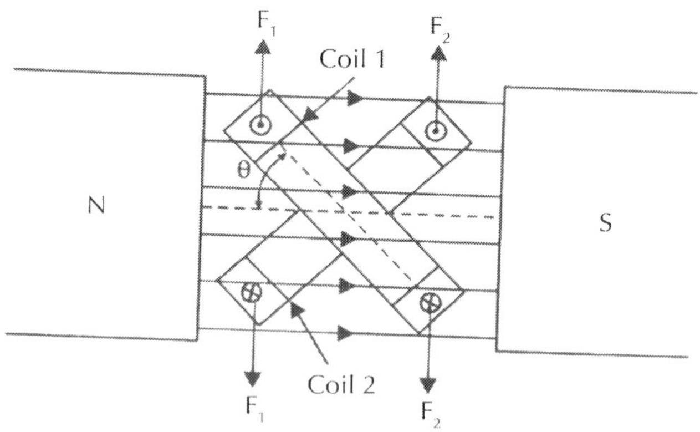

Deflecting torque acting on coil 1

$$T_{d1} = N_1 B l_1 d_1 I_1 \cos\theta$$

Deflecting torque acting on coil 1

$$T_{d2} = N_2 B l_2 d_2 I_2 \cos\theta$$

Where, I_1, I_2= current in coil 1 and 2

N_1, N_2= Numbers of turns in coil 1 and 2

l_1, l_2= length of coil 1 and 2

d1 d2= width of coil 1 and 2

B= Flux density of magnetic field

Deflection of Equilibrium

$$\theta = k\left(\frac{I_1}{I_2}\right)$$

Moving Iron Instrument

This instrument is one of the most primitive forms of measuring and relay instrument. Moving iron instruments are of mainly two types. Attraction type and repulsion type instrument. Whenever a piece of iron is placed nearer to a magnet it would be attracted by the magnet. The force of this attraction depends upon the strength said magnetic field. If the magnet is electromagnet is electromagnet then the magnetic field strength can easily be increased or decreased by increasing or decreasing current through its coil. Accordingly the attraction force acting on the piece of iron would also be increased and decreased.

Deflection Torque

$$T_d = \frac{1}{2} I^2 \frac{dL}{d\theta}$$

Deflection

$$\theta = \frac{1}{2} \frac{I^2}{K} \frac{dL}{d\theta}$$

Note:

For linear scale

$$\theta \cdot \frac{dL}{d\theta} = \text{cons tan t}$$

Shunt for moving Iron Instrument

The use of a shunt for moving iron instruments also introduces frequency errors. The inductive reactance of the shunt is low and the inductive reactance of the coil is relatively high. The impedance of the shunt remains nearly unchanged over a range of frequencies.

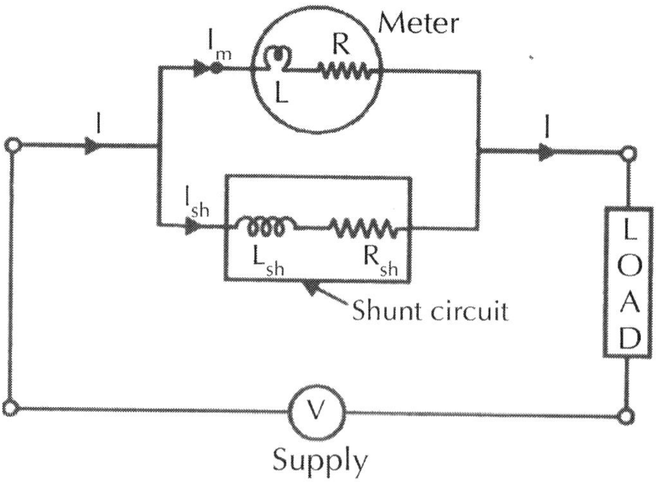

$$\frac{I_{sh}}{I_m} = \frac{R_{sh}}{R} \frac{\sqrt{1+(\omega L / R)^2}}{\sqrt{1+(\omega L_{sh} / R_{sh})^2}}$$

Note:

In order to the division of current through two parallel branches meter and the shunt shall remain the same for all frequencies, L_{Sh}/R_{Sh} must be equal to L/R i.e. time constants of meter and shunt should be same.

Multiplier for moving Iron Instrument

Instruments are manufactured that combine a moving-coil meter with a number of shunts and series multipliers, to provide a range of readings on a single scale graduated to read current and voltage.

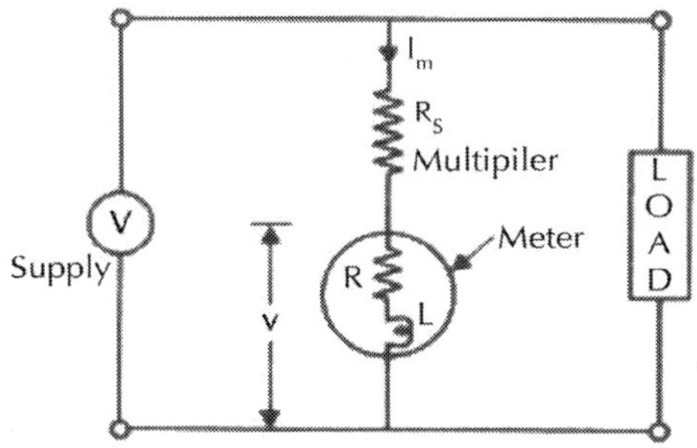

Voltage multiplier power

$$m = \frac{V}{v} = \frac{\sqrt{(R + R_s)^2 + \omega^2 L^2}}{\sqrt{R^2 + \omega^2 L^2}}$$

Error in Moving Iron Instrument

Error due to friction is quite small as torque-weight ratio is high in moving coil instruments. Stray fields cause relatively low values of magnetizing force produced by the coil. Efficient magnetic screening is essential to reduce this effect. Error due to variation of frequency causes change of reactance of the coil and also changes the eddy currents induced in neighboring metal.

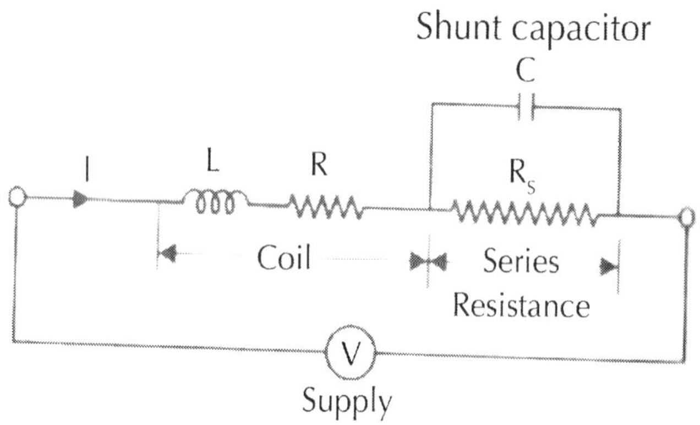

Shunt Capacitance

$$C = 0.41\frac{L}{R_s^2}$$

Eddy Current

- When ω is small

$$I'_e = \frac{\omega^2 M L_e I}{R_e^2}$$

- When ω is large

$$I'_e = \frac{MI}{L_e}$$

Where, R_e, L_e = resistance and inductance of eddy current path

5 Instrument Transformers

Instrument transformers are high accuracy class electrical devices used to isolate or transform voltage or current levels. The most common usage of instrument transformers is to operate instruments or metering from high voltage or high current circuits, safely isolating secondary control circuitry from the high voltages or currents.

Ratio of Instrument Transformers

It might appear that the extension of range could be conveniently done by the use of shunts for currents and multiplier for voltage measurement, as in DC.

But this method is suitable only for small values of current and voltage.

- Difficult to achieve accuracy with a shunt on AC
- Capability of having shunt of large range is not possible
- The power consumed by multipliers become large as the voltage increases
- The measuring circuit is not isolated electrically from the power circuit

Transformation Ratio (R): It is the ratio of magnitude of the primary phasor to secondary phasor.

$$R = \frac{|\text{primary phasor}|}{|\text{secondary phasor}|}$$

For C.T. (Current Transformer)

- Transformers used for current measurement
- Steps down the current to the level of ammeter.

$$R = \frac{\text{primary winding current}}{\text{secondary winding current}}$$

For P.T. (Potential Transformer)

- Transformers used for voltage (Potential) measurement
- Steps down the voltage to the level of voltmeter.

$$R = \frac{\text{primary winding voltage}}{\text{secondary winding voltage}}$$

Nominal Ration (K_n): It is the ratio of rated primary winding current (or voltage) to the rated secondary winding current (or voltage).

For C.T.

$$K_n = \frac{\text{rated primary winding current}}{\text{rated secondary winding current}}$$

For P.T.

$$K_n = \frac{\text{rated primary winding voltage}}{\text{rated secondary winding voltage}}$$

Turn Ratio (n)

For C.T.

$$n = \frac{\text{number of turns of secondary winding}}{\text{number of turns of primary winding}}$$

For P.T.

$$n = \frac{\text{number of turns of primary winding}}{\text{number of turns of secondary winding}}$$

Ratio Correction Factor (RCF)

$$RCF = \frac{R}{K_n}$$

Note:

The ratio Marked on the transformers is their nominal ratio.

Current Transformers

The Current Transformer (C.T.), is a type of "instrument transformer" that is designed to produce an alternating current in its secondary winding which is proportional to the current being measured in its primary.

Current transformers reduce high voltage currents to a much lower value and provide a convenient way of safely monitoring the actual electrical current flowing in an AC transmission line using a standard ammeter. The principal of operation of a current transformer is no different from that of an ordinary transformer.

Equivalent Circuit

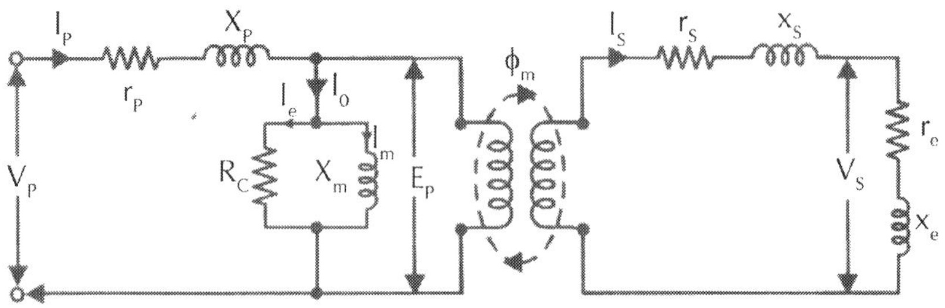

where r_s , x_s = resistance, resistance of secondary winding

r_e , x_e =resistance, resistance of external burden

E_p , E_s =priamary, secondary winding induced voltage

N_p , N_s =number of primary, secondry winding turns

Ip, Is= primary, secondary winding current

f=working flux of transformer

è=phase angle of transformer

ä=angle between secondary winding induced voltage and secondary winding current

Ä=phase angle of secondary winding load circuit

I_0=exciting current

I_m =magnetizing component of exciting current

Ie=loss component of exciting current

á=angle between exciting current and flux

Phasor Diagram

A phasor is a vector that has an arrow head at one end which signifies partly the maximum value of the vector quantity (V or I) and partly the end of the vector that rotates.

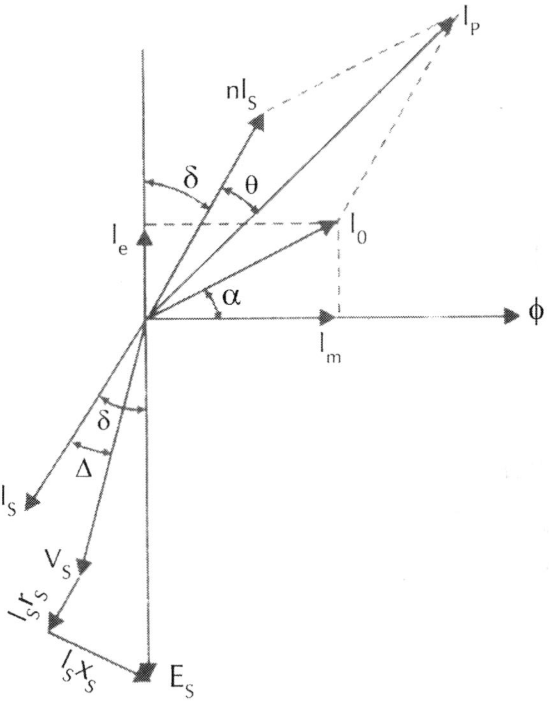

Transformation Ratio

$$R = \frac{I_p}{I_s} \simeq n + \frac{I_o}{I_s}\sin(\delta + \alpha)$$

$$R \simeq n + \frac{(I_m \sin\delta + I_e \cos\delta)}{I_s}$$

$$R \simeq n + \frac{I_e}{I_s} \simeq n\left(1 + \frac{I_e}{I_p}\right)$$

Where

$$I_m = I_o \cos\alpha$$
$$I_e = I_o \sin\alpha$$

Phase Angle

$$\boxed{\theta \simeq \frac{180}{\pi}\left(\frac{I_m \cos\delta - I_e \sin\delta}{nI_s}\right)} \text{ degree}$$

$$\boxed{\theta \simeq \frac{180}{\pi}\frac{I_m}{I_e}} \text{ degree}$$

Ratio Error

$$\text{Ratio error} = \frac{\text{no min al ratio}\left(K_n\right) - \text{actual ratio}\left(R\right)}{\text{actual ratio}\left(R\right)}$$

Potential Transformers

Potential Transformer is designed for monitoring single-phase and three-phase power line voltages in power metering applications. A Potential Transformer is a special type of transformer that allows meters to take readings from electrical service connections with higher voltage (potential) than the meter is normally capable of handling without at potential transformer.

Actual Transformation (voltage) ratio

$$R = n + \frac{\frac{I_s}{n}\left[R_p \cos\Delta + X_p \sin\Delta\right] + I_e r_p + I_m x_p}{V_s}$$

6 Measurement of Power and Wattmeter's

Phase Angle

$$\theta = \frac{I_s}{V_s}\left(X_s \cos\Delta - R_s \sin\Delta\right) + \frac{I_e x_p - I_m r_p}{nV_s} \text{ rad.}$$

Power

Power electronics is the application of solid-state electronics to the control and conversion of electric power. Electric power is the rate at which electric energy is transferred by an electric circuit. The SI unit of power is the watt, one joule per second.

D.C. Circuit

Direct current (DC) circuits basically consist of a loop of conducting wire through which an electric current flows.

Ammeter connected between load and volt meter

- Power consumed by load

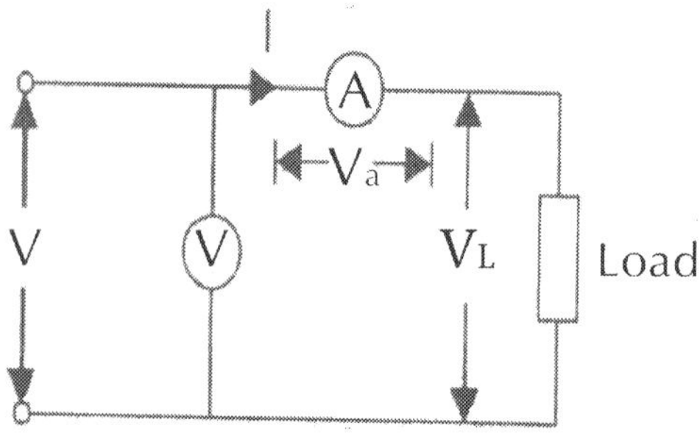

$$p = VI - I^2R_a$$

Where, V=Voltage across voltmeter

I=Current through ammeter

R_a=Resistance of ammeter

Volt meter connected between load and ammeter

- Power consumed by load

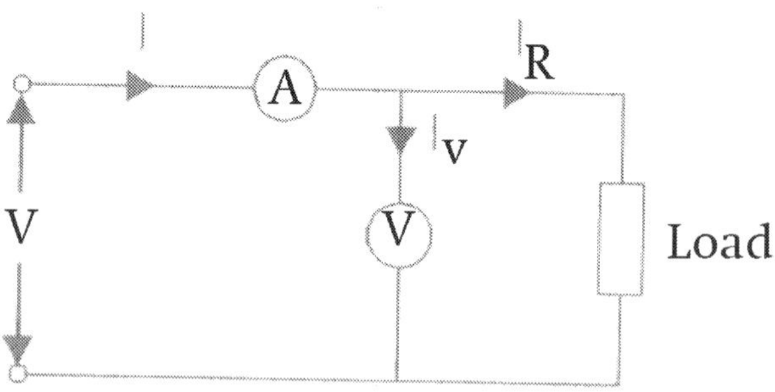

$$P = VI - \frac{V^2}{R_V}$$

Where, V=Voltage across voltmeter

I=current through ammeter

R_V=Resistance of voltmeter

A.C. Circuit

Alternating current (AC), the flow of electric charge periodically reverses direction, whereas in direct current the flow of electric charge is only in one direction.

- Instantaneous power

$$p = vi = V_m I_m \sin\omega t \cdot \sin(\omega t - \phi)$$

Where, $v = V_m \sin \omega t$

$i = I_m \sin(\omega t - \phi)$

- Average power

$P = VI\cos\phi$

Where, V, I=Rms values of voltage and current

$\cos\phi$ =Power factor of the load

Electrodynamometer wattmeter

An electrodynamometer is an instrument used for measuring the electric power. The same current passes through two concentric coils placed at right angles to each other, the resulting torque depends on the square of the current.

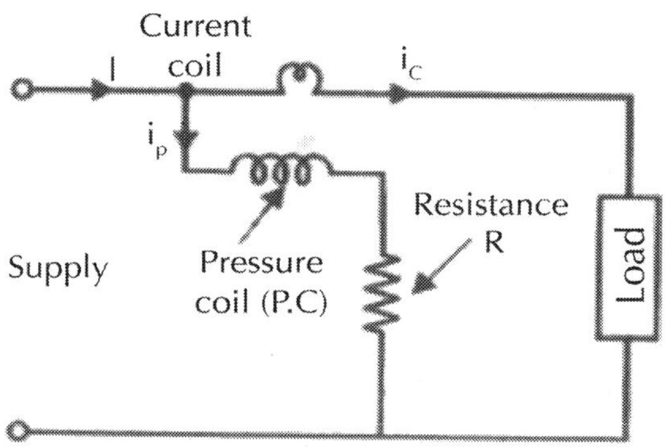

Instantaneous torque

$$T_i = i_1 i_2 \left(\frac{dM}{d\theta}\right)$$ Where i1, i2=instantaneous value of current in pressure and current coils

Deflecting torque

$$T_d = \frac{VI}{R_p} \cos\phi \cdot \frac{dM}{d\theta}$$ Where, RP=resistance of a pressure coil circuit

Controlling torque

$T_c = K\theta$ Where, k=spring constant

θ=final steady deflection

Deflection

$$\theta = \left(k_1 \frac{dM}{d\theta}\right) P$$ Where, P = power being measured

$= VI \cos\phi$

$$k_1 = \frac{1}{R_p k}$$

Errors in Electrodynamometer Wattmeter's

The wattmeter will read high when the load power factor is lagging ,as in that case the effect of pressure coil inductance is to reduce the phase angle between load current and pressure coil current . Hence the wattmeter will read high. This is very serious error.

Correction factor (k)

The correction factor is a factor by which the actual wattmeter reading is multiplied to get the true power

For lagging power factor $k = \dfrac{\cos\phi}{\cos\beta\cos(\phi-\beta)}$

Power leading power factor $k = \dfrac{\cos\phi}{\cos\beta\cos(\phi+\beta)}$

Where ϕ =Angle between current coil and pressure coil

β= Angle between current and voltage of pressure coil

True power =Correction factor x actual wattmeter reading

For β very small

Actual wattmeter reading = true power $(1+\tan\phi\ \tan\beta)$

Error = $\tan\phi\ \tan\beta$ x true power =VI $\sin\phi\ \tan\beta$

%error = $\tan\phi\ \tan\beta$

True power = VI $\cos\phi$ where, V=Voltage applied to pressure coil

I=Current in current coil

Power in Poly-Phase Systems

A poly-phase system is a means of distributing alternating-current electrical power. Poly-phase systems have three or more energized electrical conductors carrying alternating currents with a definite time offset between the voltage waves in each conductor.

Blondel's Theorem

If a network is supplied through n conductors, the total power measured by assuming the reading of n wattmeters so arranged that current element of a wattmeter is in each line and the corresponding voltage element is connected between that line and a common point, the common point is located on one of the lines, then the power may measure by (n -1) Wattmeters.

Two Wattmeter Method

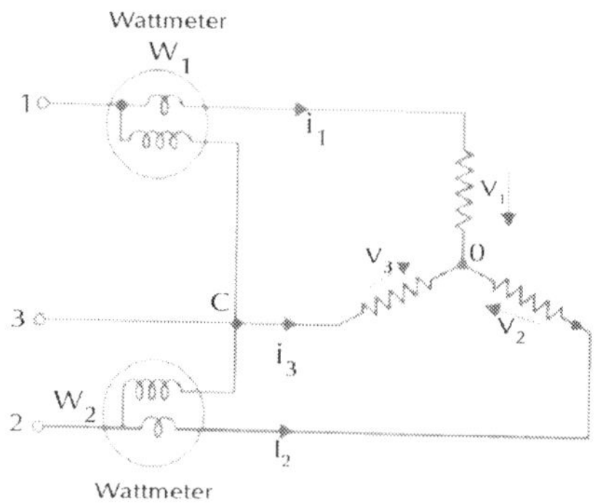

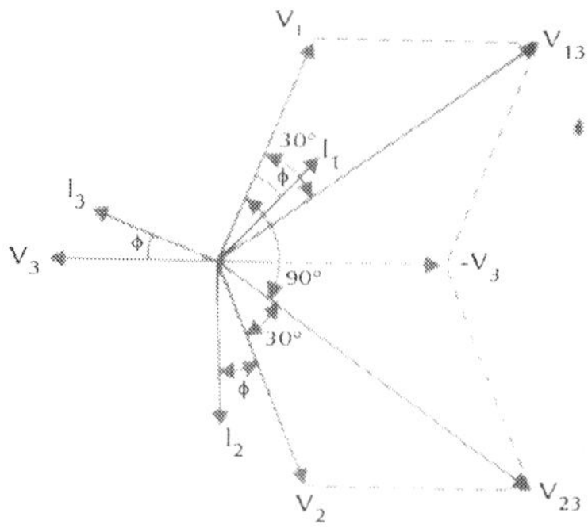

- Reading of P_1 wattmeter $P_1 = \sqrt{3}\,VI\cos(30° - \phi)$
- Reading of P_2 wattmeter $P_2 = \sqrt{3}\,VI\cos(30° + \phi)$
- Total power consumed by load $P = P_1 + P_2$
- Power factor

7 Measurement of Resistance

$$\cos\phi = \cos\left[\tan^{-1}\sqrt{3}\frac{P_1 - P_2}{P_1 + P_2}\right];$$

$$\tan\phi = \sqrt{3}\frac{P_1 - P_2}{P_1 + P_2}$$

Where ϕ =Angle between phase current and phase voltage.

Resistance

- Low resistance: All resistance of the order 1 Ω and below.
- Medium resistance: Resistance from 1 Ω upwards to about 0.1 M Ω.
- High resistance: All resistance of order 0.1 M Ω and above..

Measurement of Medium Resistance

- Ammeter voltmeter method

$$R_m = \frac{\text{voltmeter reading}}{\text{ammeter reading}} = \frac{V}{I}$$

Where, R_m =measured value of resistance

Circuit for Higher Resistance

- True value of resistance

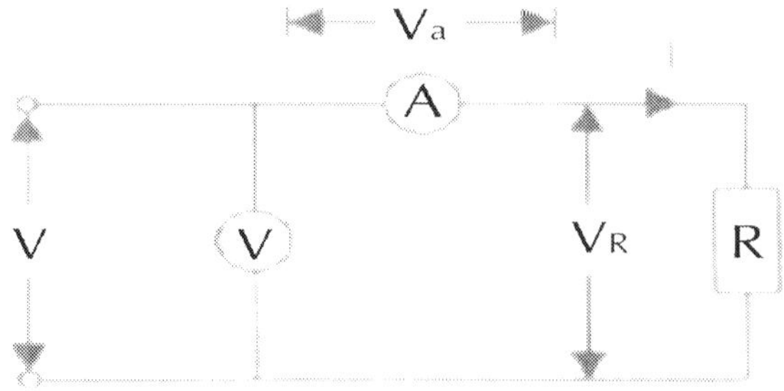

$R = R_{m1} - R_a$

$$R_m = R_{m1}\left(1 - \frac{R_a}{R_{m1}}\right)$$

Where, R_{m1}=Measured value of resistance

R_a = Resistance of ammeter

Relative error $\epsilon_r = \dfrac{R_{m1} - R}{R} = \dfrac{R_a}{R}$

Circuit for Higher Resistance

Resistance is the friction in an electrical circuit that controls the flow of current.

- True value of resistance

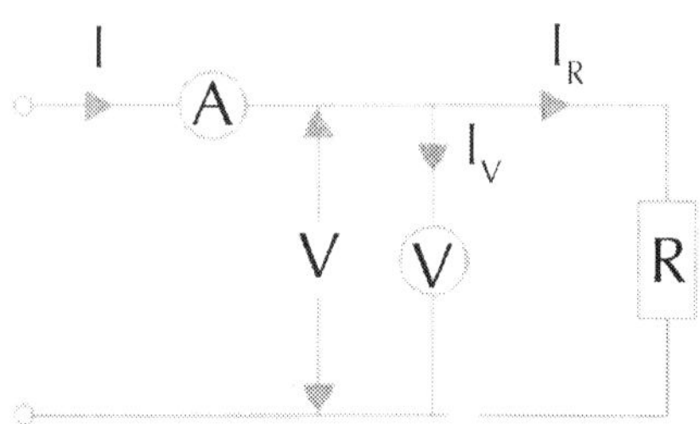

$$R = \frac{R_{m2}R_v}{R_v - R_{m2}}$$

Where, Rm2=Measured value of resistance

Rv=Resistance of voltmeter

For RV>>Rm2

$$R = R_{m2}\left(1 + \frac{R_{m2}}{R_v}\right)$$

- Relative error

$$\in_r = \frac{R_{m2} - R}{R} = \frac{R^2_{m2}}{R_v}$$

- Approximate relative error

$$\in = -\frac{R}{R_v}$$

Note:

Relative error for above two cases is equal when true value of resistance,

$$R = \sqrt{R_a R_v}$$

Wheatstone bridge

The Wheatstone bridge circuit is nothing more than two simple series-parallel arrangements of resistances connected between a voltage supply terminal and ground producing zero voltage difference between the two parallel branches when balanced.

At Balance

$$R = S\frac{P}{Q}$$

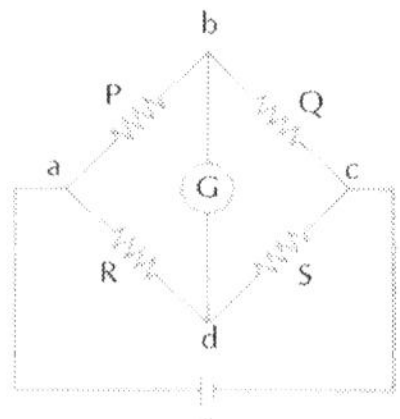

- Sensitivity of Wheatstone bridge

The relation for sensitivity of Wheatstone bridge using transfer function or any other methods.

$$S_B = \frac{\theta}{\Delta R / R} = \frac{S_v ESR}{(R+S)^2}$$

$$S_B = \frac{S_v E}{\frac{P}{Q} + 2 + \frac{Q}{P}}$$

Where, Sv=Voltage sensitivity of galvanometer

E= Bridge voltage

P, Q=Branch resistance

For a bridge with equal arms $S_B = \frac{S_v E}{4}$

Equivalent Circuit of Wheatstone Bridge

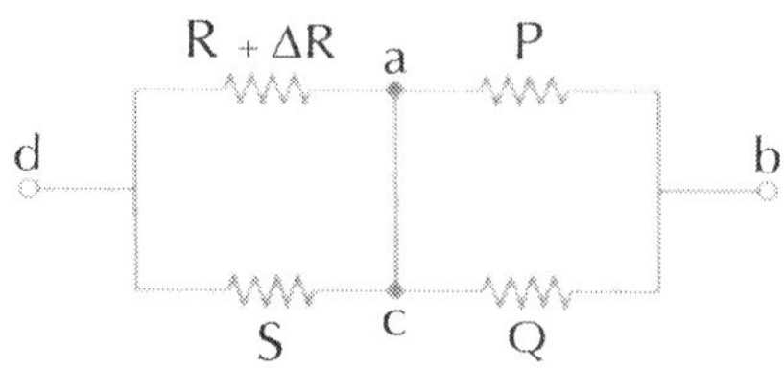

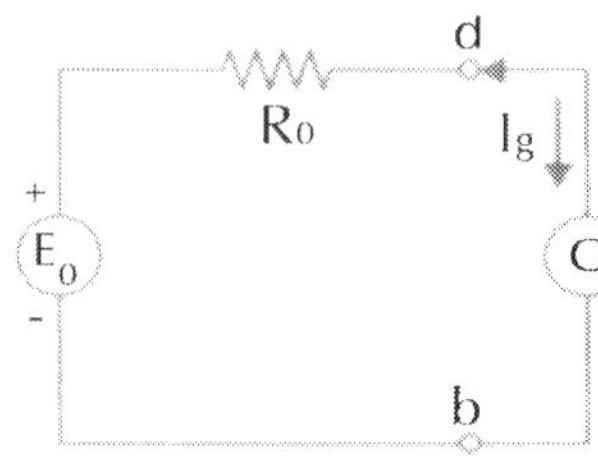

Galvanometer current $I_g = \frac{E_0}{R_0 + G}$

Where E0 = thevenin's or open circuit voltage appearing between terminals b and d with galvanometer circuit open circuited.

G = Resistance of the galvanometer circuit open deflection of the galvanometer for a small change in resistance in the unknown arm.

$$E_0 = E\left[\frac{R+\Delta R}{2R+\Delta R} - \frac{1}{2}\right] \approx E\left(\frac{\Delta R}{4R}\right) \text{ as } \Delta R << R$$

ΔR =Change in resistance R

- Thevenin equivalent resistance of bridge $R_0 = \frac{RS}{R+S} + \frac{PQ}{P+Q}$
- Galvanometer deflection $\theta = \frac{S_v ES\Delta R}{(R+S)^2} = \frac{S_i ES\Delta R}{(R_0+G)(R+S)^2}$

Where, Si=Current sensitively of galvanometer

- Bridge sensitivity $S_B = \frac{\theta}{\Delta R / R} = \frac{S_i ESR}{(R_0+G)(R+S)^2}$

Measurement of Low Resistance

It is used for the measurement of low resistances. The Kelvin double bridge is the modification of the Wheatstone bridge and provides greatly increased accuracy in measurement of low value resistance.

Kelvin Double Bridge Method

The Kelvin double bridge incorporates the idea of a second set of ratio arms-hence the name double bridge-and the use of four terminal resistors for the low resistance arms.

For zero galvanometer deflection

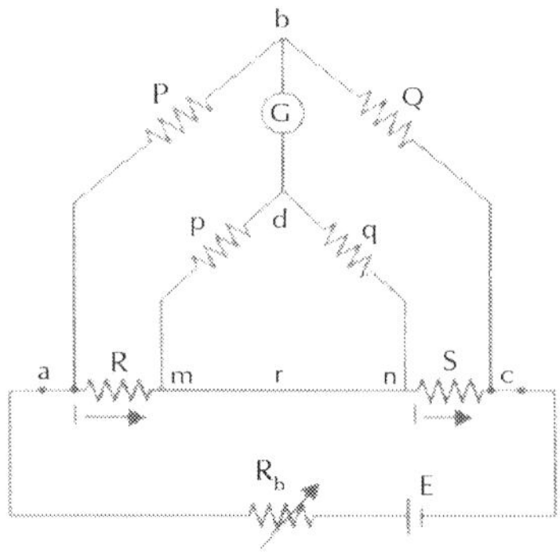

$$E_{ab} = E_{amd}$$

$$R = \frac{P}{Q} \cdot S + \frac{qr}{p + qr + r}\left[\frac{P}{Q} - \frac{p}{q}\right]$$

Note:

- If $\frac{P}{Q} = \frac{p}{q}$

$$R = \frac{P}{Q}.S$$

Accuracies by Kelvin double bridge method

- From 1000 μΩ to 1.0 Ω: 0.05%
- From 100 μΩ to 1000 μΩ: 0.5% to 0.05%
- From 10 μΩ to 100 μΩ: 0.5%to 0.2

Measurement of High Resistance

- Loss of charge method

8 A. C. Bridges

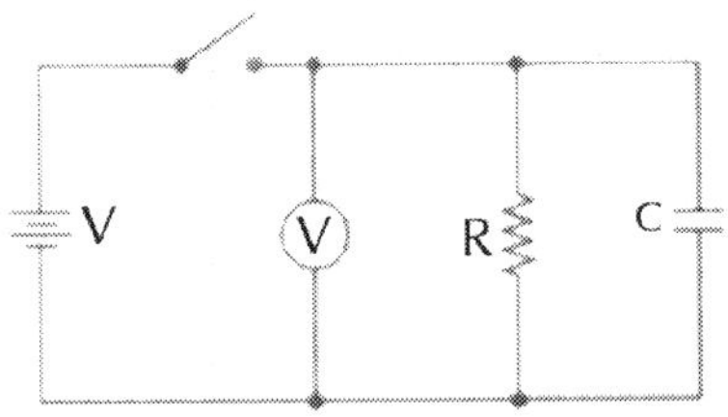

$$R = \frac{0.4343t}{C \log_{10} \frac{V}{V-e}}$$

AC Bridges consist of a source, balance detector and four arms. In AC bridges, all the four arms consists of impedance. The AC bridges are formed by replacing the DC battery with an AC source and galvanometer by detector of Wheatstone bridge. They are highly useful to find out inductance, capacitance, storage factor, dissipation factor.

Used to measure self-inductance, Mutual inductance, capacitance, and frequency

General equation for bridge balance

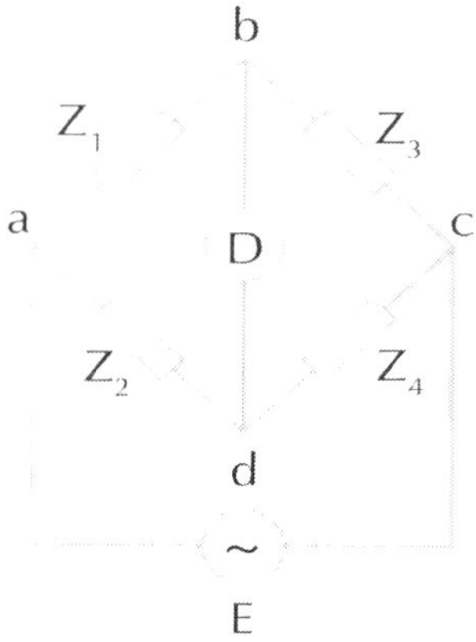

$$\bar{Z}_1\bar{Z}_4 = \bar{Z}_2\bar{Z}_3$$

Magnitude condition

$$|Z_1||Z_4| = |Z_2||Z_3|$$

Angle condition

$$\angle\theta_1 + \angle\theta_4 = \angle\theta_2 + \angle\theta_3$$

Note: Both the above condition must be satisfied for the bridge to be balanced.

Measurement of Self Inductance

Measurement of self-inductance is follows:

Maxwell's Inductance Bridge

The alternating current bridges are mainly designed to measure the electrical quantities such as inductance and capacitance. These AC bridges are the improved versions of Whetstone Bridge. These bridges in their basic form consist of normally four arms, an AC power source and a detector.

Let us now discuss Maxwell's inductor bridge. The figure shows the circuit diagram of Maxwell's inductor bridge.

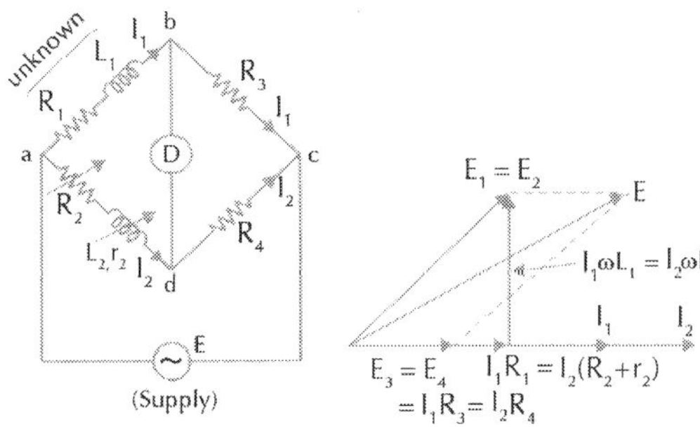

In this bridge the arms bc and cd are purely resistive while the phase balance depends on the arms ab and ad.

$$R_1 = \frac{R_3}{R_4}(R_2 + r_2)$$

$$L_1 = \frac{R_3}{R_4}.L_4$$

Where, L_1=Unknown inductance of resistance R_1

L_2=Variable inductance of fixed resistance r_2

R_2= Variable resistance connected in series with L_2

R_3, R_4=Known non-inductive resistance

Maxwell's Inductance Capacitance Bridge

In this Maxwell Bridge, the unknown inductor is measured by the standard variable capacitor. Circuit of this bridge is given below,

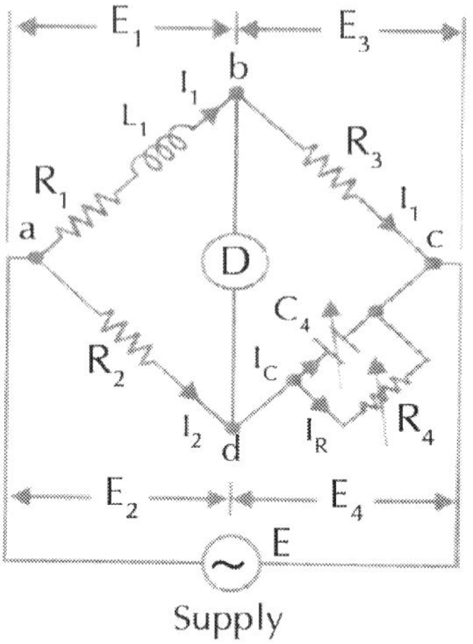

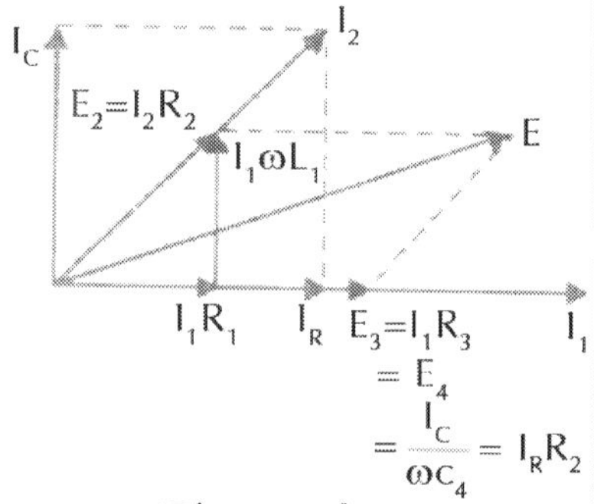

(Phasor diagram)

$$R_1 = \frac{R_2 R_3}{R_4} \quad \text{and} \quad L_1 = R_2 R_3 C_4$$

- Q-factor of the coil $$Q = \frac{\omega L_1}{R_1} = \omega C_4 R_4$$

Hay's Bridge

The Hay's bridge is modification of the Maxwell's bridge. The connection diagram of the Hay's bridge is shown in figure below.

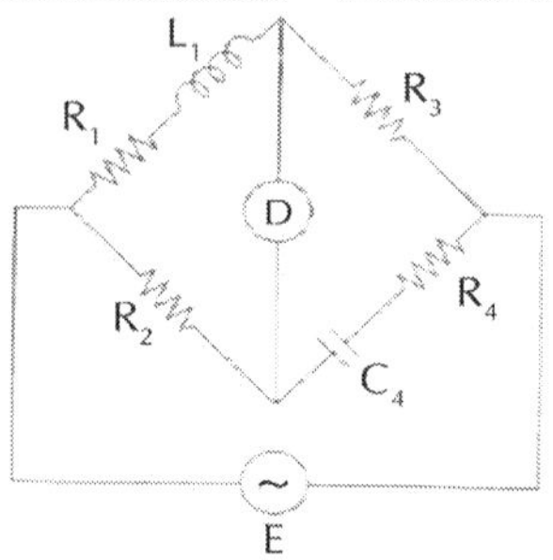

$$R_1 = \frac{\omega^2 R_2 R_3 R_4 C_4^2}{1+\omega^2 C_4^2 R_4^2} = \frac{R_2 R_3}{R_4}\left(\frac{1}{1+Q^2}\right)$$

$$L_1 = \frac{R_2 R_3 C_4}{1+\omega^2 C_4^2 R_4^2} = \frac{R_2 R_3 R_4}{1+\left(\frac{1}{Q}\right)^2}$$

For Q>10

$$L_1 = R_2 R_3 C_4$$

$$Q = \frac{\omega L_1}{R_1} = \frac{1}{\omega C_4 R_4}$$

Where, L1=Unknown-inductance having a resistance R1, R2, R3, R4=known non-inductive resistance

C_4= Standard capacitor

Anderson's Bridge

Anderson's bridge though we have Maxwell bridge and Hay's bridge to measure quality factor of the circuit. The main disadvantage of using Hay's bridge and Maxwell Bridge is that, they are unsuitable of measuring the low quality factor. However Hay's bridge and Maxwell Bridge are suitable for measuring accurately high and medium quality factor respectively. So, there is need of bridge which can measure low quality factor and this bridge is modified Maxwell's bridge and known as Anderson's bridge.

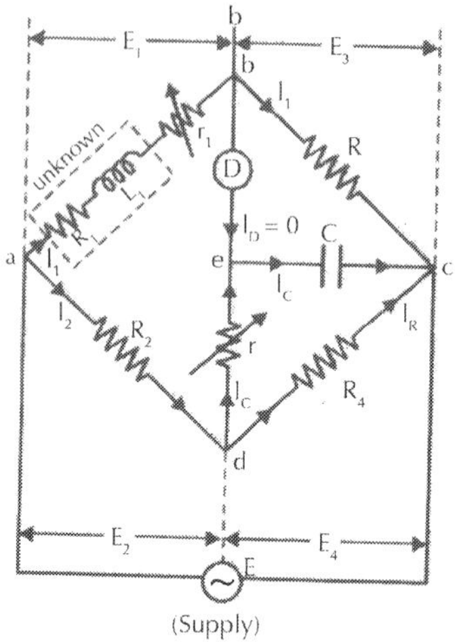

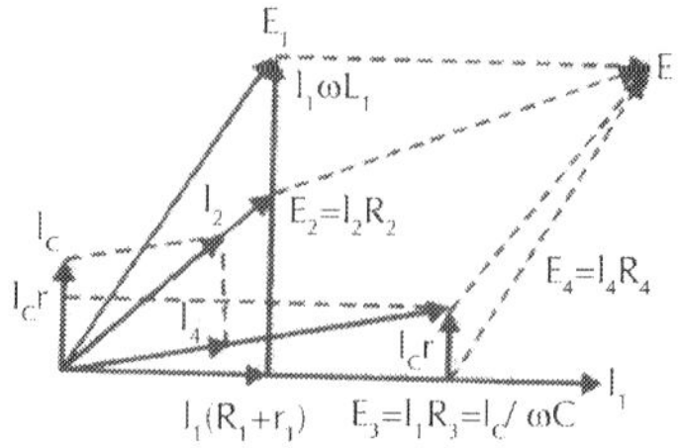

$$R_1 = \frac{R_2 R_3}{R_4} - r_1;\ L_1 = \frac{CR_3}{R_4}\left[r\left(R_4 + R_2\right) + R_2 R_4\right]$$

Where, L_1=self-inductance to be measured

R_1=Resistance of self-inductor

r_1=Resistance connected in series with seft-inductor

r_2, R_2, R_3, R_4=Known non-inductive resistances

C=Fixed standard capacitor

Owen's Bridge

An Owen's bridge circuit is given below.

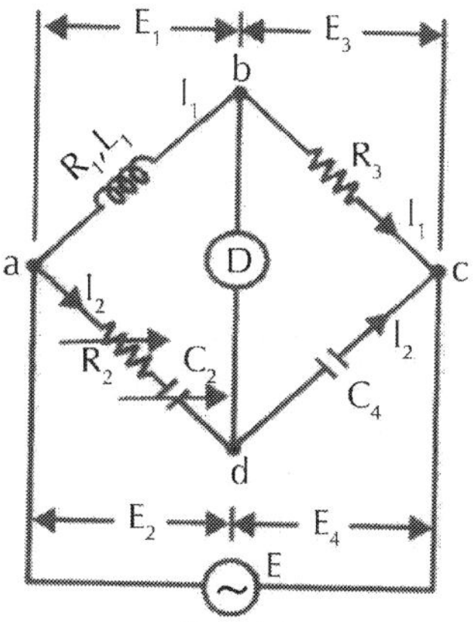

(Supply)

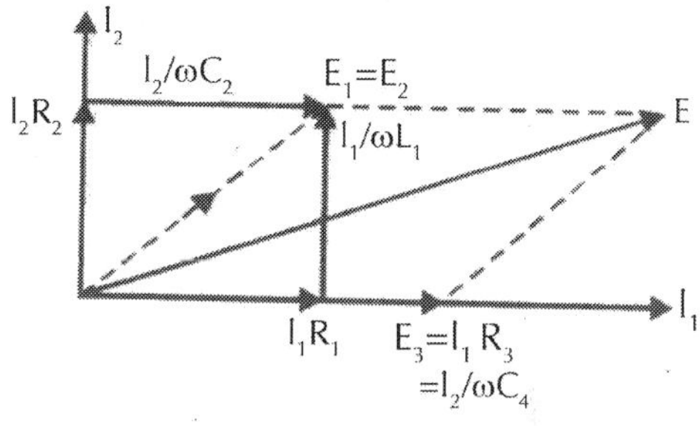

(Phasor diagram)

$$L_1 = R_2 R_3 C_4$$

$$R_1 = R_3 \frac{C_4}{C_2}$$

Where, L1=Unknown self-inductance of resistance R1

R2=Variable non-inductive resistance '

R3=Fixed non-inductive resistance

C3=Variable standard capacitor

C4=Fixed standard capacitor

Measurement of Incremental Inductance

- Incremental inductance

  $L_1 = R_2 R_3 C_4$

- Incremental permeability

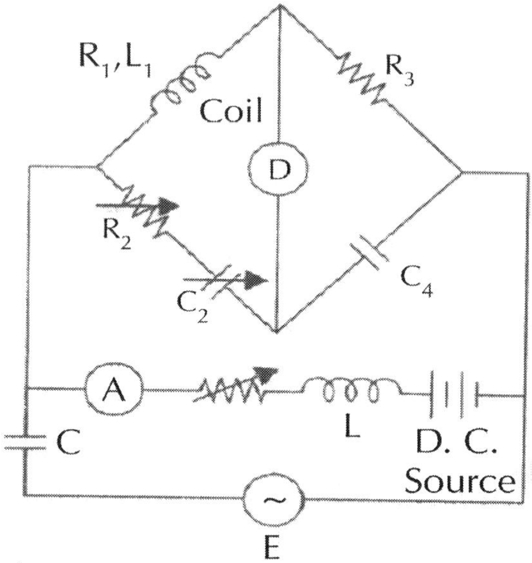

$$\mu = \frac{L_1 l}{N^2 A}$$

Where, N=Number of turns

A=Area of flux path

l=Length flux path

R_2=Variable non-inductive resistance

R_3=Fixed non-inductive resistance

C^4=Fixed standard capacitor

Measurement of Capacitance

De Sauty's Bridge

This bridge provide us the most suitable method for comparing the two values of capacitor if we neglect dielectric losses in the bridge circuit. The circuit of De Sauty's bridge is shown below.

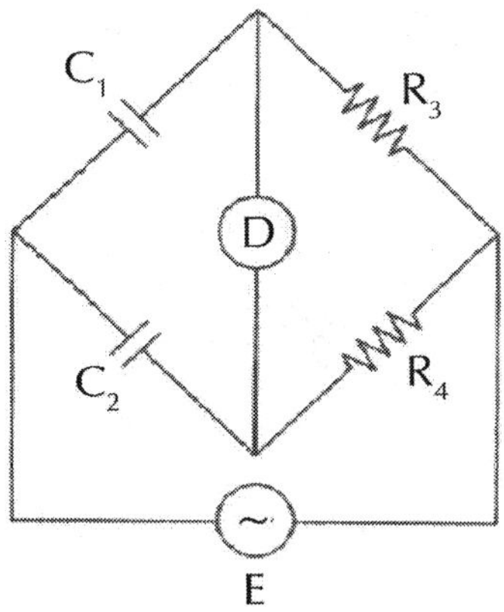

$$C_1 = \frac{C_1 R_4}{R_3}$$

Where, C_1=Capacitor whose

Capacitance to be measured

C_2=A standard capacitor

R_3, R_4= Non-inductive resistance

Schering Bridge

This bridge is used to measure to the capacitance of the capacitor, dissipation factor and measurement of relative permittivity. Let us consider the circuit of Schering Bridge as shown below:

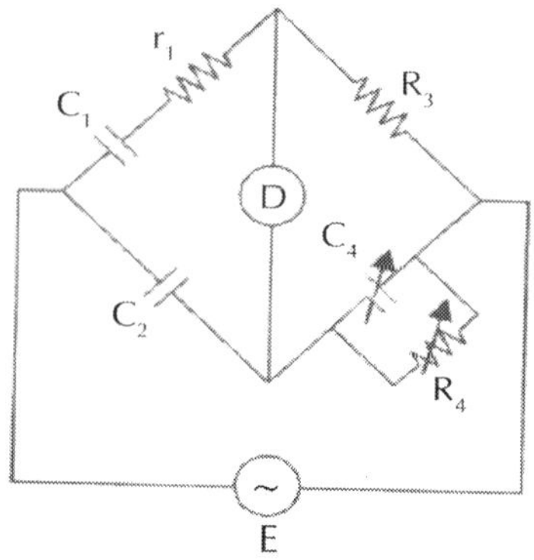

$$r_1 = \frac{R_3 C_4}{C_2}$$

$$C_1 = C_1\left(\frac{R_4}{R_3}\right)$$

Dissipation factor

$D_1 = \omega C_1 r_1 = \omega C_4 R_4$

9 Magnetic Measurements

Where, C_1= Capacitor whose capacitance is to be determined

r, = A series resistance representing the loss in the capacitor C_1

C_2 = A standard loss-free capacitor

R_3 = A non-inductive resistance C_4 = A variable capacitor

R_4 = A variable non-inductive resistance in parallel with variable capacitor C_4

Measurement of Frequency

Measurement of frequency are follows:

Wien's Bridge

Wien Bridge has a series RC combination in one and a parallel combination in the adjoining arm. Wien's bridge

Frequency

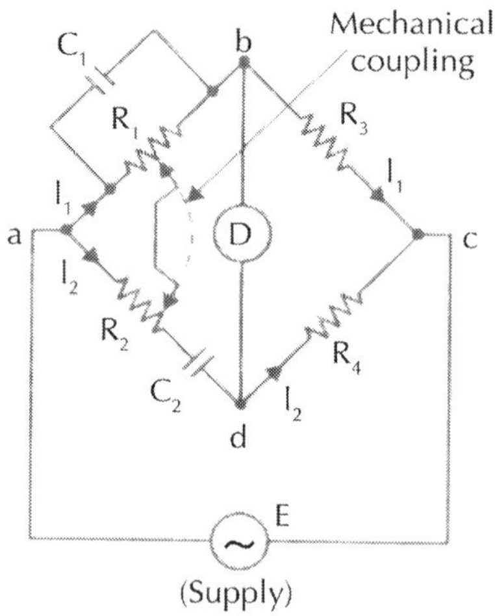

$$f = \frac{1}{2\pi\sqrt{R_1R_2C_1C_2}} \quad \text{Hz}$$

For R1=R2=R and C1=C2=C

$$f = \frac{1}{2\pi RC}$$

Electrical circuits and their components have characteristics that affect the ability of the circuit to conduct a current and generate a magnetic field. Electrical measurements of these characteristics are often determined by calculations based on measurable quantities of the circuit, like voltage, current, and resistance.

Flux Density

The amount of magnetic field flux concentrated in a given area.

$$B = \frac{\phi}{A_s} = \frac{Rk_q\theta_1}{2NA_s}$$

Where, ϕ=Flux linking search coil

As=Cross-section area of specimen

R=Resistance of ballistic galvanometer circuit

$K_q\,\theta_1$=Charge indicated by ballistic galvanometer

N=Number of turns in the search coil

Hysteresis loss per unit volume

Hysteresis loss in transformer can be explained in different ways. We will discuss two of them, one is physical explanation and the other is mathematical explanation.

$$p_h = \eta f\, B_m^k$$

Where, η=Hysteresis coefficient

f=Frequency; Hz

B_m=Maximum flux density; Wb/M^2

K=Steinmetz coefficient

Note: The value of k varies 1.6 to 2.

Eddy current loss per unit volume for laminations

$$p_e = \frac{4k_f^2 f^2 B_m^2 t^2}{3\rho}$$

Where, k_f=from factor

t=Thickness of laminations; m

ρ=Resistivity of material; Ω-m

Total iron loss per unit volume

$$p_i = p_h + p_e = \eta f B_m^k + \frac{4k_f^2 f^2 B_m^2 t^2}{3\rho}$$

- Maximum flux density

$$B_m = \frac{E_2}{4k_f f A_s N_2}$$

For sinusoidal supply

$$\frac{E_2}{4.44 f A_s N_2}$$

Where, E2=Voltage induced in secondary winding

$E_2 = 4K_f\ f\ \phi_m\ N_2$

K_f=from factor (=1.11 for sinusoidal supply)

f=Frequency

A_s=cross-sectional area of specimen

N_2=Number of turns in secondary winding

10 Cathode Ray Oscilloscope

ϕ_m =Maximum flux linkage

Average current through vacuum tube diode

$$I_{av} = \frac{E_{av}}{2R} = \frac{E_{rms}}{2 \times 1.11 \times R} = 0.45 \frac{E_{rms}}{R}$$

Where, E_{rms}=rms value of applied voltage

E_{av}=Average value of applied voltage

R=Load resistance

Difference Amplifier Type of Electronic Voltmeter

A voltmeter is an instrument used to measure voltage or electrical potential difference between two points in basic electric circuits. Analog voltmeters move a pointer across a scale in proportional to the voltage of the circuit.

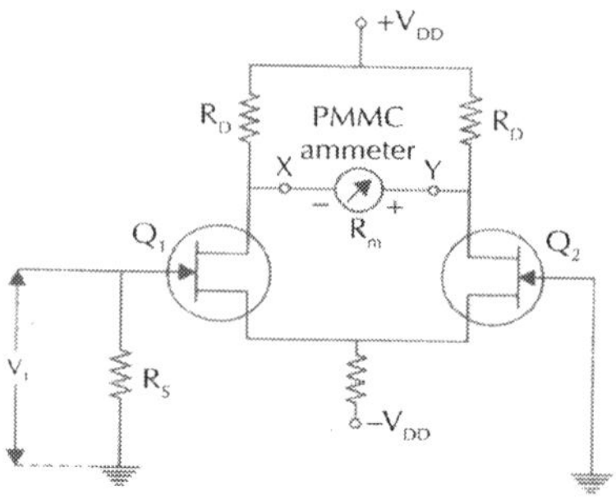

Thevenin's voltage across terminal X-Y

$$V_{th} = g_m \left(\frac{r_d R_D}{r_d + R_D} \right) v_1$$

Thevenin's resistance looking into terminals X-Y

$$R_{Th} = \frac{2 r_d R_D}{r_d + R_D}$$

Current through ammeter

$$i = \frac{V_{Th}}{R_{Th} + R_m} = \frac{g_m r_d R_D / (r_d + R_D)}{2 r_d R_D / (r_d + R_D) + R_m} . v_1$$

When RD<<rd

$$i = \frac{g_m R_D}{2R_D + R_m} . v_1$$

Electrostatic Deflection

Electrostatic deflection refers to a technique for modifying the path of a stream of charged particles by the use of an electric field applied transverse to the path of the particles.

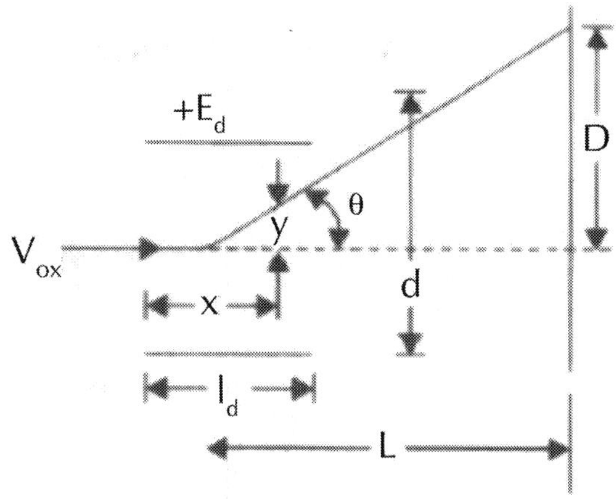

$$y = \frac{1}{2}\frac{eE_y}{mv_{ox}^2}x^2$$

Where, y=Displacement y-direction; m

e=Charge of an electron; Coulomb

E_y=Electric field intensity in y-direction; V/m

11 High Frequency Measurement

M=Mass of electron; kg

V_{ox}=Velocity of electron when entering the field of deflecting plates; m/s

X=Displacement in X direction; m

Deflection

The deflection distance of a member under a load is directly related to the slope of the deflected shape of the member under that load and can be calculated by integrating the function that mathematically describes the slope of the member under that load.

$$D = \frac{Ll_d E_d}{2dE_a}$$

Where, L=Distance between screen and the Centre of deflecting plates; m

l_d=Length of deflecting plates; m

E_d=Potential between deflecting plates; V

d=Distance between deflecting plates; m

E_a=voltage of pre-accelerating anode; V

$$\text{Deflection sensitivity } S = \frac{D}{E_d} = \frac{Lld}{2dE_a} m/v$$

$$\text{Deflection factor } G = \frac{1}{2} = \frac{2dE_a}{Lld} v/m$$

Sawtooth Generator

The part of the circuit consisting of the capacitor C, transistor, zener diode and the resistors form a constant current source to charge the capacitor. Initially assume the capacitor is fully discharged.

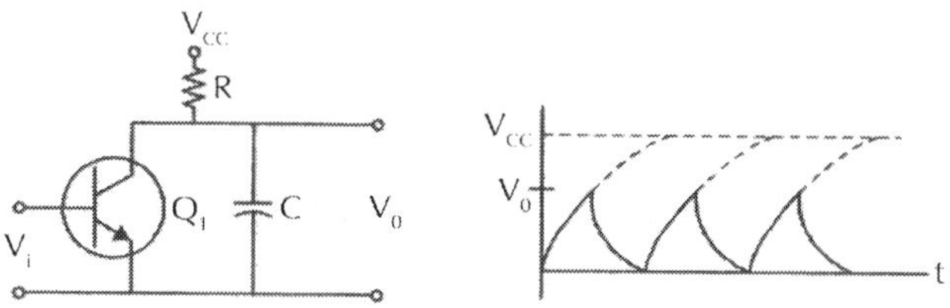

$$V_0 = V_{cc}\left[1 - \exp(-t/RC)\right]$$

Where, V0=Instantaneous voltage across the capacitor at time t; V

- Rising time of vertical amplifier

$$t_r = \frac{0.35}{BW}$$

Where, BW=Bandwidth

$$\frac{f_y}{f_x} = \frac{(\text{number of intersections of the horizontal line with the curve})}{(\text{number of intersection of the vertical line with the curve})}$$

Where, fy=Frequency of signal applied to Y plates

f_x= Frequency of signal applied to X plates

Q-Meter

A high value of Q is always desirable as it means high inductive reactance and low resistance. A low value of Q indicates that the resistance component is relatively high and so there is a comparatively large loss of power.

12 Transducers

Measurement of Q

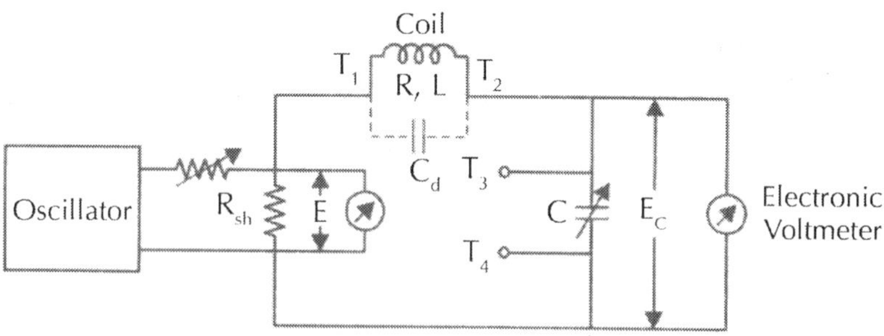

- Resonant frequency of Q-Meter $f_0 = \dfrac{1}{2\pi\sqrt{LC}}$
- Measured value of Q $Q_m = \dfrac{\omega_0 L}{R + R_{sh}}$
- True value of Q $Q_m = \dfrac{\omega_0 L}{R}; Q_t = Q_m\left(1 + \dfrac{R_{sh}}{R}\right) = Q_m\left(1 + \dfrac{C_d}{C}\right)$

Where, R=Resistance of coil

L=Inductance of coil

R_{sh}=Shunt Resistance

C=Tuning capacitance

C_d=Distributed or self-capacitance

Measurement of Inductance $L = \frac{1}{4\pi^2 f_o^2 C}$

Measurement of Effective Resistance

The resistance to an alternating current, expressed as the ratio of the power dissipated to the square of the effective current.

- Effective resistance $R = \frac{\omega_o L}{Q_t}$

Measurement of Self-capacitance

In electrical circuits, the term capacitance is usually a shorthand for the mutual capacitance between two adjacent conductors, such as the two plates of a capacitor. However, for an isolated conductor there also exists a property called self-capacitance.

Resonance frequency

$$f_1 = \frac{1}{2\pi\sqrt{L(C_1 + C_d)}};\ f_2 = \frac{1}{2\pi\sqrt{L(C_2 + C_d)}}$$

Where $f_2=2f_1$

- Distributed or self-capacitance

$$C_d = \frac{C_1 - 4C_2}{3}$$

Where, C1=Tuning capacitance at frequency f1

C_2=Tuning capacitance at frequency f_2

Strain Gauge

A strain gauge is a device that is used to measure the strain that occurs in an object. The device is still being used in many electronic circuits mainly as the principle sensing element for sensors like torque sensors, pressure sensors, load cells

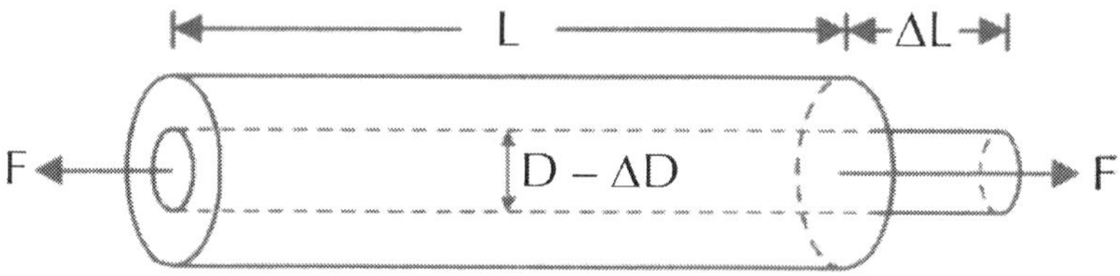

Gauge factor of strain gauge

$$G_f = 1 + 2v + \frac{\Delta\rho/\rho}{\in}$$

Where, $\frac{\Delta\rho}{\rho}$ = Per unit change in resistivity

v = Poisson's ratio

$\in$ = Strain

For $\frac{\Delta\rho}{\rho} \to 0$ $\quad G_f \approx 1 + 2v$

Poisson's ratio

Poisson's ratio is the ratio of transverse contraction strain to longitudinal extension strain in the direction of stretching force. Tensile deformation is considered positive and compressive deformation is considered negative.

$$\text{Possion's ratio}\,(v) = \frac{\text{lateral strain}}{\text{longitudinal strain}} = \frac{-\partial D/D}{\partial L/L}$$

- Strain

$$\text{Strain}\,(\in) = \frac{\Delta L}{L}$$

Where, $\frac{\Delta L}{L}$ = Per unit change in length

Thermistor

Thermistors are thermally sensitive resistors whose prime function is to exhibit a large, predictable and precise change in electrical resistance when subjected to a corresponding change in body temperature.

- Resistance of thermistor $R_{T_1} = R_{T_2} \exp\left[\beta\left(\frac{1}{T_1} - \frac{1}{T_2}\right)\right]$

Where, RT1=Resistance of thermistor at absolute temperature T1; $^\circ k$

RT2= Resistance of thermistor at absolute temperature T2; $^\circ k$

β=A constant depending upon the material of thermistor

- Steinhart-Hart equation

The Steinhart–Hart equation is a model of the resistance of a semiconductor at different temperatures. The equation is:

$$\frac{1}{T} = A + B\ln R + C(\ln R)^3$$

Where, T=Temperature; $^\circ k$

R=Resistance of thermistor; Ω

A,B, C=Curve fitting constant

- Thermistor resistance

$$R_T = aR_o \exp(b/T)$$

Where, R_T, R_c=Resistance of thermistor at temperature T$^\circ k$ and ice point respectively

Thermocouple

A thermocouple is a temperature-measuring device consisting of two dissimilar conductors that contact each other at one or more spots, where a temperature differential is experienced by the different conductors.

- E.M.F. produce in a thermocouple $E = a(\Delta\theta) + b(\Delta\theta)^2$

Where, Δθ=Difference in temperature between the hot thermocouple junction and the reference junction of thermocouple; $^\circ C$

a, b=Constant

LVDT

The linear variable differential transformer is a type of electrical transformer used for measuring linear displacement. A counterpart to this device that is used for measuring rotary displacement is called a rotary variable differential transformer.

- Sensitivity of LVDT $\text{Sensitivity} = \dfrac{\text{output voltage}}{\text{displacement}}$

Capacitive Transducers

The capacitive transducer is used extensively for the measurement of displacement, pressure etc. Let us see the principle of working of capacitive transducer or sensor also called as variable capacitance transducer.

Capacitance

The capacitive transducer is nothing but the capacitor with variable capacitance.

- Capacitance of parallel plate capacitor

$C = \dfrac{\in A}{d} = \dfrac{\in xw}{d}$ Where, A=Overlapping area of plates

x=Length of overlapping part of plates

w=Width of overlapping part of plates

d=Distance between two plates

$\in$=Permittivity of medium

- Capacitance of cylindrical capacitor $\boxed{C = \dfrac{2\pi \in x}{\log_e\left(D_2 / D_1\right)}}$ F
- Capacitance at angular displacement θ

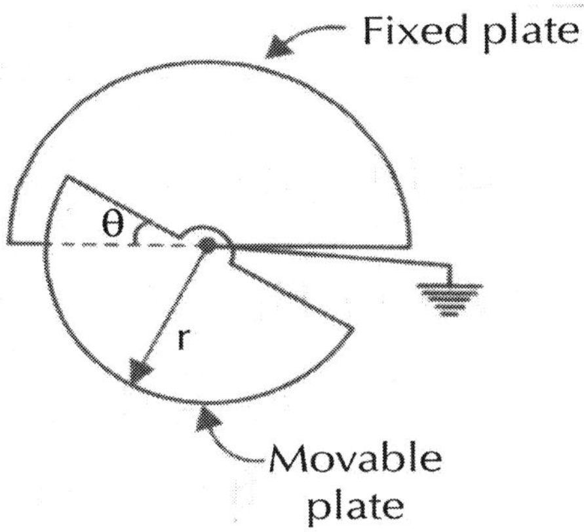

$$C = \frac{\in \theta r^2}{2d}$$

Where r=Radius of semicircular plate

θ=Angular displacement in radian

Sensitivity

- Sensitivity of parallel palate capacitive transducer

$$S = \frac{\partial C}{\partial x} = \in \frac{w}{d}$$

Where, x=Length of overlapping part of cylinders; m

D2=Inner diameter of outer cylindrical electrode; m

D1=Outer diameter of inner cylindrical electrodes; m

- Sensitivity of cylindrical capacitive transducer

$$S = \frac{2\pi \in}{\log_e (D_2 / D_1)} \, f/m$$

Sensitivity of variable capacitive transducer

$$S = \frac{\in r^2}{2d}$$

Piezo-Electric Transducer

A piezoelectric transducer is a device that transforms one type of energy to another by taking advantage of the piezoelectric properties of certain crystals or other materials.

- Voltage sensitivity of crystal

$$\boxed{g = \frac{\text{Electric field}}{\text{Stress}} = \frac{\in}{P}} \text{ Vm / N}$$

Where, P=Pressure or stress; N/m2

- Charge sensitivity

$$\boxed{d = \in_r \in_0 g} \text{ C / N}$$

- Output voltage

$$E_o = gtp$$

Where, t= thickness of crystal; m

A Handbook on Electronics and Communications Engineering

12

Microwave Engineering

CONTENTS

1 Basics

Microwave engineering pertains to the study and design of microwave circuits, components, and systems. Fundamental principles are applied to analysis, design and measurement techniques in this field.

- Microwave range is (10^9Hz to 10^{13} Hz) for all applications.
- Microwave are used for terrestrial range and cannot be used for ground wave communication.

For microwave frequencies $\frac{\sigma}{\omega \in} << 1$ ground behaves as dielectric cannot reflect back the em waves at microwave frequencies so frequencies cannot be used for ground wave communication.

IEEE bands for microwave

L	$1-2$ GHz
S	$(2-4)$ GHz
C	$4-8$ GHz
X	$8-12$ GHz
Ku	$12-18$ GHz
K	$18-27$ GHz
Ka	$27-40$ GHz
	> GHz Mi lim eter bands.

- At low frequency more V-I structure and less of the E-H format and a thick conductor is used for large skin depth to guide electromagnetic wave.
- As the frequency increases the energy transfer is more in E-H format and hence a thin conductor is sufficient to guide the wave.
- EM waves in common are dispersive in nature and have spherical wave front.
- Dispersive nature of EM wave has advantage of that it can be used one to rnultipoint communication.
- EM waves strength decreases with increase in distance of propagation due to dispersive nature.

2 Waveguide and Hybrid Circuit

A hybrid integrated circuit, hybrid microcircuit, or simply hybrid is a miniaturized electronic circuit constructed of individual devices, such as semiconductor devices (e.g. transistors and diodes) and passive components bonded to a substrate or printed circuit board (PCB).

Rectangular Waveguide

Waveguide may refer to any linear structure that conveys electromagnetic waves between its endpoints.

Rectangular waveguides are one of the earliest type of the transmission lines. A lot of components such as isolators, detectors, attenuators, couplers and slotted lines are available for various standard waveguide bands between 1 GHz to above 220 GHz.

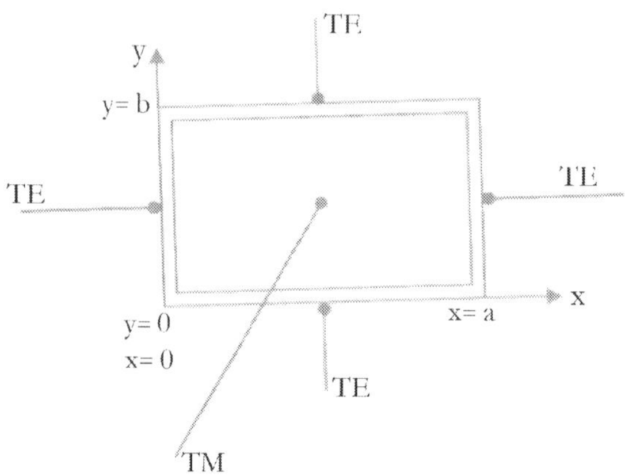

Helmholtz equation is given by

$$\nabla^2 E = \gamma^2 E$$

$$\nabla^2 H = \gamma^2 E$$

Where, γ →propagation constant of EM wave.

E, H → are electric and magnetic fields.

- There are four conductors at x=0, x=a, y=0 and y=b, so there will be four boundary conditions which will be used solving Helmholtz equation.

Boundary condition at a conductor is

$$E_{\tan g} = 0$$

$$H_{norm} = 0$$

Where,

E_{tang} → tangential component of electric component of electric field on conductor surface

H_{norm} → normal component of magnetic field at conductor surface

Propagation constant $\bar{\gamma}$ is given by

$$\bar{\gamma} = \sqrt{\left(\frac{m\pi}{a}\right)^2 + \left(\frac{n\pi}{b}\right)^2 - \omega^2 \mu_0 \in_0}$$

Where,

$\bar{\gamma}$ →propagation constant along guide axis.

$\left(\frac{m\pi}{b}\right)^2$ → Propagation constant along x-axis

γ Propagation constant along y-axis.

m, n→ are integers.

If $\bar{\gamma}$ will be real quantity then wave will not propagate through waveguide rather it attenuates.

If $\bar{\gamma}$ will be imaginary quantity then wave will propagate through waveguide.

Cutoff frequency f_c is given by

$$f_c = \frac{c}{2\pi}\sqrt{\left(\frac{m\pi}{a}\right)^2 + \left(\frac{n\pi}{b}\right)^2} \rightarrow \text{for free space}$$

- If frequency for a particular mode will be greater than cutoff frequency then wave will travel through wave guide
- If frequency will be equal to cutoff frequency then wave will oscillate between the walls of waveguide.
- If frequency will be less than cutoff frequency then wave will not propagate through waveguide.
- Number m and n stands for the number of half cycles completed by the wave between the guide walls or number of maximum between guide walls or number π-phase completions.

$$k_c = \sqrt{\left(\frac{m\pi}{a}\right)^2 + \left(\frac{n\pi}{b}\right)^2} \text{ is called the cutoff wave number}$$

Cutoff frequency for waveguide where medium is perfect dielectric is

given by $f_c = \dfrac{c}{2\pi\sqrt{\mu_r \in_r}}\sqrt{\left(\dfrac{m\pi}{a}\right)^2 + \left(\dfrac{n\pi}{b}\right)^2}$

Phase Velocity (v_p)

The phase velocity of a wave is the rate at which the phase of the wave propagates in space. This is the velocity at which the phase of any one frequency component of the wave travels. For such a component, any given phase of the wave will appear to travel at the phase velocity.

Phase velocity V_p is given by $$V_p = \frac{c}{\left(\sqrt{1-\left(\frac{f_c}{f}\right)^2}\right)\sqrt{\mu_r \epsilon_r}}$$

$V_p > c$ for waveguide medium to be free space.

Phase velocity is actual phase rate change which is greater than C.

$$\cos\theta = \sqrt{1-\left(\frac{f_c}{f}\right)^2}$$ where, $\bar{\gamma}$ →is called wave angle or tilt angle.

Effectively the wave is travelling along the guide axis at a rate of $\bar{\gamma}$ hence if the wave angle is θ the wave length along guide axis appears to be larger than the free space wavelength.

The phase shift per unit forward distance travelled is greater than that of straight line path in zig-zag path hence $V_p > C$.

Group Velocity

The group velocity of a wave is the velocity with which the overall shape of the waves' amplitudes known as the modulation or envelope of the wave propagates through space.

- Group velocity v_g is the velocity by which group of wave travel and is given by

$$v_g = \frac{c}{\sqrt{\mu_r \in_r}}\sqrt{1-\left(\frac{f_c}{f}\right)^2} \text{ or } v_g = c\cos\theta$$ for waveguide medium to be free space.

Group velocity signifies the physical rate of energy transfer along be free space

TE/TM Modes

A transverse mode of a beam of electromagnetic radiation is a particular electromagnetic field pattern of radiation measured in a plane perpendicular to the propagation direction of the beam.

A mode refers to a feed mechanism involving the number of connections or feed points and position of feed points

Transverse Magnetic Mode (TM mode)

This is called TM mode because magnetic field is perpendicular to the direction of wave propagation that is for TM mode H_x and H_y exist but z will be zero.

TM Wave Solution

- For TM waves (H_z=0)

$$E_z = E(x,y,z,t)_z = E_{z0}\sin\left(\frac{m\pi x}{a}\right)\sin\left(\frac{n\pi y}{b}\right)e^{-\gamma z}e^{j\omega t}\hat{a}_z$$

$$E_x = E(x,y,z,t)_x = E_{x0}\cos\left(\frac{m\pi x}{a}\right)\sin\left(\frac{n\pi y}{b}\right)e^{-\gamma z}e^{j\omega t}\hat{a}_x$$

$$E_y = E(x,y,z,t)_y = E_{y0}\sin\left(\frac{m\pi x}{a}\right)\cos\left(\frac{n\pi y}{b}\right)e^{-\gamma z}e^{j\omega t}\hat{a}_y$$

$$H_x = H(x,y,z,t)_x = H_{x0}\sin\left(\frac{m\pi x}{a}\right)\cos\left(\frac{n\pi y}{b}\right)e^{-\gamma z}e^{j\omega t}\hat{a}_x$$

$$H_y = H(x,y,z,t)_y = H_{y0}\cos\left(\frac{m\pi x}{a}\right)\sin\left(\frac{n\pi y}{b}\right)e^{-\gamma z}e^{j\omega t}\hat{a}_y$$

- If an of the m and n will be zero then E_z will be zero and wave will not exist so m and n cannot be zero
- TM_{0n} and TM_{m0} mode does not exist inside a rectangular wave guide.
- The modes which does not exist inside a wave guide are called

Evanescent modes

- TM_{0n} and TM_{m0} mode are evanescent mode for rectangular wave guide.
- Phase constant β_g inside waveguide for TM modes is
- The wave will only propagate down the waveguide if the excitation frequency is larger than the cut-on frequency.

$$\beta_g = \omega\sqrt{\mu \in}\sqrt{1-\left(\frac{f_c}{f}\right)^2}$$

Guide wavelength λ_g is given by

$$\lambda_g = \frac{\lambda}{\sqrt{1-\left(\frac{f_c}{f}\right)^2}} \text{ where } \lambda = \frac{1}{\sqrt{\mu \in .f}} = \frac{C}{\sqrt{\mu_r \in_r}} \cdot \frac{1}{f}$$

Where μ_r and $\in_r$ are relative permittivity and relative dielectric constants of medium of waveguide?

Group velocity v_g is given by $v_g = \frac{C}{\sqrt{\mu_r \in_r}} \times \sqrt{1-\left(\frac{f_c}{f}\right)^2}$

Wave impedance inside the wave guide is $z_g = \sqrt{\frac{\mu}{\in}}\sqrt{1-\left(\frac{f_c}{f}\right)^2}$

TE Mode

- These are called TE mode as electric field is perpendicular to the direction of propagation of wave.
- A transverse mode of a beam of electromagnetic radiation is a particular electromagnetic field pattern of radiation measured in a plane perpendicular to the propagation direction of the beam.

TE Wave Solution (E_z=0)

For TE modes E_z=0

$$E_x = E_{x0} \cos\left(\frac{m\pi x}{a}\right)\sin\left(\frac{n\pi y}{b}\right)e^{-\bar{\gamma}z}e^{j\omega t}\hat{a}_x$$

$$E_y = E_{y0} \sin\left(\frac{m\pi x}{a}\right)\cos\left(\frac{n\pi y}{b}\right)e^{-\bar{\gamma}z}e^{j\omega t}\hat{a}_y$$

$$E_z = 0$$

$$H_x = H_{x0} \sin\left(\frac{m\pi x}{a}\right)\cos\left(\frac{n\pi y}{b}\right)e^{-\bar{\gamma}z}e^{j\omega t}\hat{a}_x$$

$$H_y = H_{y0} \cos\left(\frac{m\pi x}{a}\right)\sin\left(\frac{n\pi y}{b}\right)e^{-\bar{\gamma}z}e^{j\omega t}\hat{a}_y$$

$$H_z = H_{z0} \cos\left(\frac{m\pi x}{a}\right)\cos\left(\frac{n\pi y}{b}\right)e^{-\bar{\gamma}z}e^{j\omega t}\hat{a}_z$$

- It is clear that if m or n any of them will be zero H_z will exist.
- So TE_{m0} and TE_{n0} modes exist inside rectangular waveguide.
- But both m and n cannot be zero at this same time.
- A mode having lowest cutoff frequency will be called dominant mode.
- TE10 mode is dominant mode inside a rectangular waveguide if a>b.
- TE01 mode is dominant mode inside a rectangular waveguide if b>a.
- TE00 made does not exist inside a rectangular waveguide.
- Cutoff frequency for TE mode is

Cutoff frequency is an essential parameter for any waveguide - it does not propagate signals below this frequency.

$$f_c = \frac{C}{2\pi\sqrt{\mu_r \in_r}}\sqrt{\left(\frac{m\pi}{a}\right)^2 + \left(\frac{m\pi}{b}\right)^2}$$

- Phase constant β_g inside wave guide for TE mode is given by

$$\beta_g = \omega\sqrt{\mu \in}\sqrt{1-\left(\frac{f_c}{f}\right)^2}$$

- Group velocity v_g inside waveguide for TE mode is

$$v_g = \frac{C}{\sqrt{\mu_r \in_r}}\sqrt{1-\left(\frac{f_c}{f}\right)^2}$$

Phase velocity v_p for TE mode is given by

$$v_p = \frac{C}{\sqrt{\mu_r \in_r}}\sqrt{1-\left(\frac{f_c}{f}\right)^2}$$

Characteristic impedance Z_g inside waveguide for TE mode is given by

$$z_g = \frac{\sqrt{\mu/\in}}{\sqrt{1-\left(\frac{f_c}{f}\right)^2}}$$

When two or more modes have the same cutoff frequency, then they are said to be degenerate modes.

- TE_{mn} and TM_{mn} models are always degenerate modes.
- If frequency of operation is less than the cutoff frequency of dominant mode then no modes travel through wave guide.
- If frequency of operation is less than cutoff frequency of a mode (other than dominant mode) then modes higher than that mode will not propagate through waveguide but modes lower than that mode will propagate through wave guide.

Some Modes Diagram

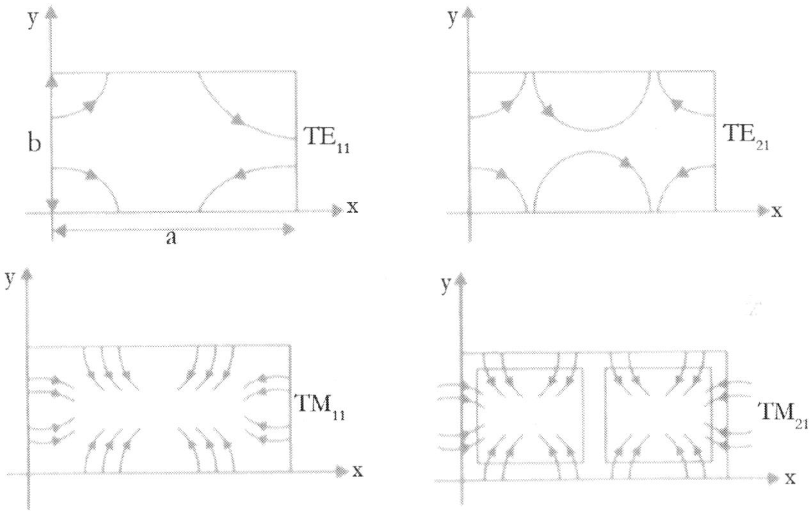

TEM (transverse electric magnetic) modes does not exist inside the waveguide because when both E_z and H_z will be zero then no component of E or H-field will inside waveguide.

Cylindrical Waveguide

Cylindrical waveguide which is simply a hollow tube with circular cross section of radius a and extending along the z-direction. There are two sets of modes, TE and TM modes, which can propagate in a cylindrical waveguide.

A waveguide is an electromagnetic feed line used in microwave communications, broadcasting, and radar installations. A waveguide consists of a rectangular or cylindrical metal tube or pipe. The electromagnetic field propagates lengthwise.

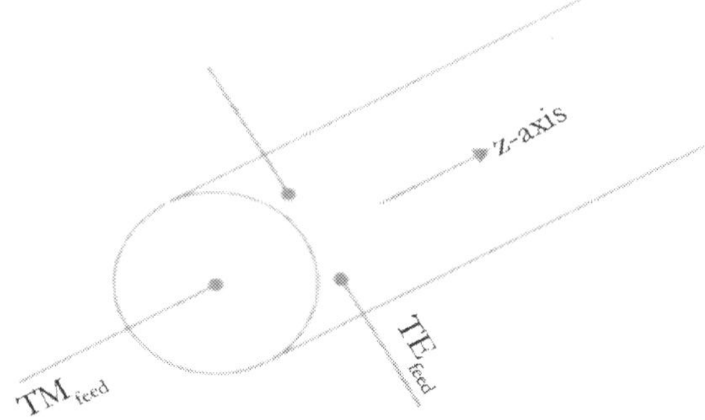

In case of cylindrical waveguide there is only one boundary condition that is at ρ=a, E_{tang}=0 and H_{norm}=0

$\bar{\gamma}$ is propagation constant along guide axis and β_ρ is phase constant in ρ direction then $\bar{\gamma}$ is given by

$$\bar{\gamma} = \sqrt{\beta_p^2 - \omega^2 \mu \in}$$

Where, $\beta \rightarrow \omega\sqrt{\mu \in}$ is phase constant in the medium inside wave guide when there is no wave guide.

Since at boundary, curvature of cylinder is involved so solution does not involve linear trigonometric functions (like, sin 0, cos 0) but solution will be of the type sin (sinθ) or, sin (cosθ), cos (sinθ), cos (cosθ) i.e. Bessel function

Solution along the guide walls is of the type $J_n(_p.p)$ where n is order of Bessel function.

TE Wave Solutions

For TE mode E_z=0

$$E_\rho = E_{\rho 0} J_n\left(\frac{X'_{np}\rho}{a}\right)\sin n\phi e^{-\bar{\gamma}z} e^{j\omega t}\hat{a}_\rho$$

$$E_\phi = E_{\phi 0} J'_n\left(\frac{X'_{np}\rho}{a}\right)\cos n\phi e^{-\bar{\gamma}z} e^{j\omega t}\hat{a}_\phi$$

$$H_\rho = H_{\rho 0} J_n\left(\frac{X'_{np}\rho}{a}\right)\cos n\phi e^{-\bar{\gamma}z} e^{j\omega t}\hat{a}_\rho$$

$$H_\phi = H_{\phi 0} J_n\left(\frac{X'_{np}\rho}{a}\right)\sin n\phi e^{-\bar{\gamma}z} e^{j\omega t}\hat{a}_\phi$$

$$H_z = H_{z0} J_n\left(\frac{X'_{np}\rho}{a}\right)\cos n\phi e^{-\bar{\gamma}z} e^{j\omega t}\hat{a}_z$$

- Mode propagation constant is given by

$$\bar{\gamma} = \sqrt{\left(\frac{X'_{np}}{a}\right)^2 - \omega^2\mu\in}$$

Where, X'_{np} →is solution of nth order derivation of Bessel function.

Cutoff frequency for TE mode is given by $f_c = \frac{C}{2\pi\sqrt{\mu_r \in_r}} \times \frac{X'_{np}}{a}$

Phase velocity V_p is given by

$$v_p = \frac{1}{\sqrt{\mu\in}\sqrt{1-\left(\frac{f_c}{f}\right)^2}}$$

Guide wavelength λ_g for TE mode is given by

$$\lambda_g = \frac{\lambda}{\sqrt{1-\left(\frac{f_c}{f}\right)^2}}$$

Where

$$\lambda = \frac{1}{\sqrt{\mu \in}.f}$$

Characteristic impedance Z_g inside waveguide for TE mode is given by

$$z_g = \frac{\sqrt{\frac{\mu}{\in}}}{\sqrt{1-\left(\frac{f_c}{f}\right)^2}}$$

Where, $\sqrt{\frac{\mu}{\in}}$ intrinsic impedance in an unbounded dielectric.

TM Wave Solution

- For TM waves H_z=0

$$E_\rho = E_{\rho 0} J_n' \left(\frac{X_{np}\rho}{a}\right) \cos n\phi e^{-\bar{\gamma} z} e^{j\omega t} \hat{a}_\rho$$

$$E_\phi = E_{\phi 0} J_n \left(\frac{X_{np}\rho}{a}\right) \sin n\phi e^{-\bar{\gamma} z} e^{j\omega t} \hat{a}_\phi$$

$$E_z = E_{z0} J_n \left(\frac{X_{np}\rho}{a}\right) \cos n\phi e^{-\bar{\gamma} z} e^{j\omega t} \hat{a}_z$$

$$H_\rho = H_{\rho 0} J_n \left(\frac{X_{np}\rho}{a}\right) \sin n\phi e^{-\bar{\gamma} z} e^{j\omega t} \hat{a}_\rho$$

$$H_\phi = H_{\phi 0} J_n' \left(\frac{X_{np}\rho}{a}\right) \cos n\phi e^{-\bar{\gamma} z} e^{j\omega t} \hat{a}_\phi$$

Where, X_{np} is solution of nth order Bessel function actually pth root of nth order Bessel function

- Propagation constant for TM mode is given by

$$\bar{\gamma} = \sqrt{\left(\frac{X_{np}}{a}\right)^2 - \omega^2 \mu \in}$$

Where, a- is radius of cylindrical waveguide.

Cutoff frequency for TM mode is given by $f_c = \dfrac{X_{np}}{2\pi a\sqrt{\mu \in}}$

Phase velocity vp for TM mode given by $v_p = \dfrac{1}{\sqrt{\mu \in}\sqrt{1-\left(\frac{f_c}{f}\right)^2}}$

Guide wavelength λ_g is given by

$$\lambda_g = \frac{\lambda}{\sqrt{1-\left(\frac{f_c}{f}\right)^2}} \text{; where, } \lambda = \frac{1}{f\sqrt{\mu \in}}$$

- Characteristic impedance Z_g for TM mode is given by

$$Z_g = \sqrt{\frac{\mu}{\in}}\sqrt{1-\left(\frac{f_c}{f}\right)^2}$$

Where, $\sqrt{\frac{\mu}{\in}}$ is intrinsic impedance of unbounded dielectric? l

- TEM mode does not exist inside cylindrical waveguide.
- Cutoff frequencies in case of cylindrical waveguide for TE and TM mode is different while it is same in case of rectangular waveguide.
- TM_{no} modes does not exist inside the waveguide called evanescent mode but TM_{Op} for p = 1, 2, 3 .... exist inside cylindrical waveguide.
- TEop modes exist inside waveguide but TE_{no} modes does not exist inside wave guide.
- TE_{11} mode is dominate mode inside cylindrical waveguide not TM_{01}.
- Dominant mode signifies the range of frequencies that waveguide can support, any frequency supported by a mode will be supported by dominant mode.
- In single mode operation mode used is dominant mode only.

TE and TM Field in Circular or Cylindrical Waveguide

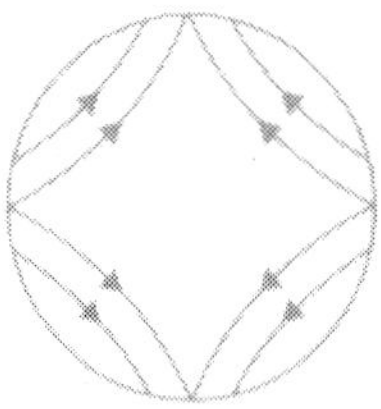

TE_{11}

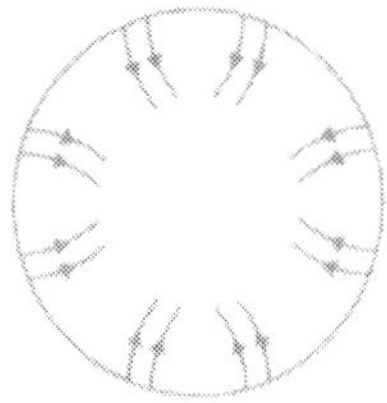

TM_{01}

Power Transmission in Rectangular Waveguide

Power transmitted through a rectangular waveguide is given by

$$P_{tr} = \oint \frac{1}{2}\left(E \times H^*\right).ds$$

Where

H* is complex conjugate of H.

- For a lossless dielectric, the time-average power flow through a rectangular guide is given by

$$P_{tr} = \frac{1}{2Z_g}\int |E|^2 \, da = \frac{Z_g}{2}\int |H|^2 \, da$$

$$\text{where, } Z_g = \frac{E_x}{H_y} = -\frac{E_y}{H_x}$$

$$|E|^2 = |E_x|^2 + |E_y|^2$$

$$|H|^2 = |H_x|^2 + |H_y|^2$$

- For TE_{mn} modes, the average power transmitted through a rectangular waveguide is given by

$$P_{tr} = \frac{\sqrt{1-\left(\frac{f_c}{f}\right)^2}}{2\sqrt{\frac{\mu}{\in}}} \int_0^b \int_0^a \left(|E_x|^2 + |E_y|^2\right) dxdy$$

- For TE_{mn} modes, the average power transmitted through a rectangular waveguide is given by

$$P_{tr} = \frac{1}{2\sqrt{\frac{\mu}{\in}}\sqrt{1-\left(\frac{f_c}{f}\right)^2}} \int_0^b \int_0^a \left(|E_x|^2 + |E_y|^2\right) dxdy$$

Power Transmission in Circular/Cylindrical Waveguide

The time-average power transmitted through a cylindrical waveguide is given by

$$P_{tr} = \frac{1}{2Z_g}\int_0^{2\pi}\int_0^a \left[\left|E_\phi\right|^2 + \left|E_\rho\right|^2 \right] \rho d_\rho d_\phi$$

where

$$Z_g = \frac{E_\rho}{H_\phi} = -\frac{E_\phi}{H_\rho} \text{ for particular given mode.}$$

a is radius of wave guide

- For TE_{np} modes, the average power transmitted through a rectangular waveguide is given by

$$P_{tr} = \frac{\sqrt{1-\left(\frac{f_c}{f}\right)^2}}{2\sqrt{\frac{\mu}{\in}}}\int_0^{2\pi}\int_0^a \left(\left|E_\rho\right|^2 + \left|E_\phi\right|^2 \right) \rho d\rho d\phi$$

- For TE_{np} modes, the average power transmitted through a rectangular waveguide is given by

$$P_{tr} = \frac{1}{2\sqrt{\frac{\mu}{\in}}\sqrt{1-\left(\frac{f_c}{f}\right)^2}}\int_0^{2\pi}\int_0^a \left(\left|H_\rho\right|^2 + \left|H_\phi\right|^2 \right) \rho d\rho d\phi$$

Cavity Resonators

Cavity resonator is a closed, confined conducting structure operating at $\bar{\gamma} = 0$ conditions. The wave does not travel along the guide axis but resonates between the walls of waveguide with E and H sustaining each other.

Cavity resonator is used to produced microwave frequencies or microwave signal by giving energy or power periodically.

It is a frequency selective resonant circuit whose i/p is a multiple frequency signal and o/p is a single frequency selected signal.

Rectangular Cavity Resonator

- For TE mode H_Z is given by

$$H_z = H_z \cos\left(\frac{m\pi x}{a}\right)\cos\left(\frac{m\pi y}{b}\right)\sin\left(\frac{p\pi z}{d}\right)$$

where a, b, d are dimensions of cavity.

m = 0, 1, 2.............., n = 0, 1, 2, 3.........; p = 1, 2, 3 ....

- For TE mode E_Z is given by

$$E_z = E_{z0} \sin\left(\frac{m\pi x}{a}\right)\sin\left(\frac{n\pi y}{b}\right)\cos\left(\frac{p\pi z}{d}\right)$$

m = 1, 2.............., n = 1, 2, 3.........; p = 0, 1, 2, 3 ....

- For a lossless dielectric and for both TE and TM modes resonant frequency is given by

$$f_r = \frac{1}{2\sqrt{\mu \in}}\sqrt{\left(\frac{m}{a}\right)^2 + \left(\frac{n}{b}\right)^2 + \left(\frac{p}{d}\right)^2}$$

For a > b < d, the dominant mode is TE_{101}.

Circular Cavity Resonator

A circular cavity resonator is a circular waveguide with two ends closed.

For TE mode H_Z, is given by

$$H_z = H_{z0} J_n\left(\frac{X'_{no}\rho}{a}\right)\cos(n\phi)\sin\left(\frac{q\pi z}{d}\right)$$

a is radius of cylinder

d is height of cylinder

q = 1, 2, 3

For TE mode resonant frequency f_r is given by

$$f_r = \frac{1}{2\pi\sqrt{\mu \in}}\sqrt{\left(\frac{X'_{np}}{a}\right)^2 + \left(\frac{q\pi}{a}\right)^2}$$

q = 1, 2, 3

For TE mode resonant E_Z is given by

$$E_z = E_{z0} J_n\left(\frac{X_{np}\rho}{a}\right)\cos(n\phi)\cos\left(\frac{q\pi z}{d}\right)$$

q = 0, 1, 2, 3

For TE mode resonant frequency f_r is given by

$$f_r = \frac{1}{2\pi\sqrt{\mu \in}}\sqrt{\left(\frac{X_{np}}{a}\right)^2 + \left(\frac{q\pi}{d}\right)^2}$$

Q - factor of a Cavity Rasonator

Quality factor Q is a measure of the frequency selectivity of a resonant or anti-resonant circuit and is given by

$$Q = 2\pi\frac{\text{max imum energy stored}}{\text{energy dissipated per cycle}} = \frac{\omega W}{P}$$

Where

W is the maximum stored energy.

P is the average power loss.

- Q of cavity resonator can be expressed approximately by

Where

$$Q = \frac{\omega\mu}{2R_s}$$

ω is operating frequency in rad/sec

R_S is surface resistance of the resonator.

- The resonant frequency and unloaded Q_0 of a cavity resonator is given by

$$f_0 = \frac{1}{2\pi\sqrt{LC}}; Q_0 = \frac{\omega_0 L}{R}$$

Where,

R, L, C are equivalent resistance, inductance and capacitance of cavity resonator.

- If the cavity is coupled by means of an ideal N : 1 transformer and a series inductance L_S to a generator having internal impedance Z_g, then coupling circuit and its equivalent are as

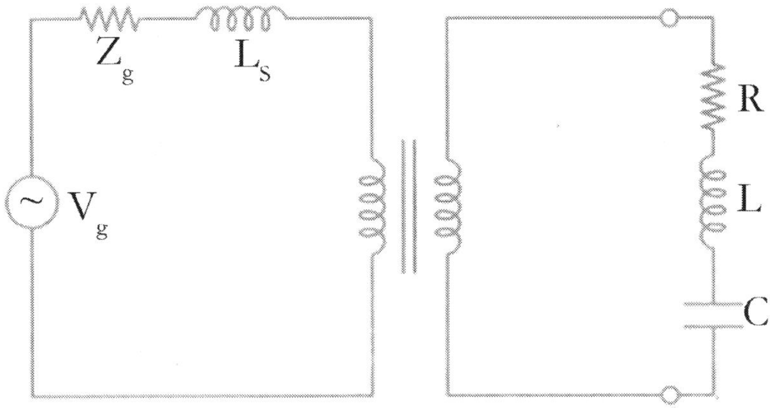

Coupling Circuit

Its equivalent circuit is the loaded Q_e of the system is given by

$$Q_e = \frac{\omega_0 L}{R + N^2 Z_g} \text{ for } \left|N^2 L_S\right| << \left|R + N^2 Z_g\right|$$

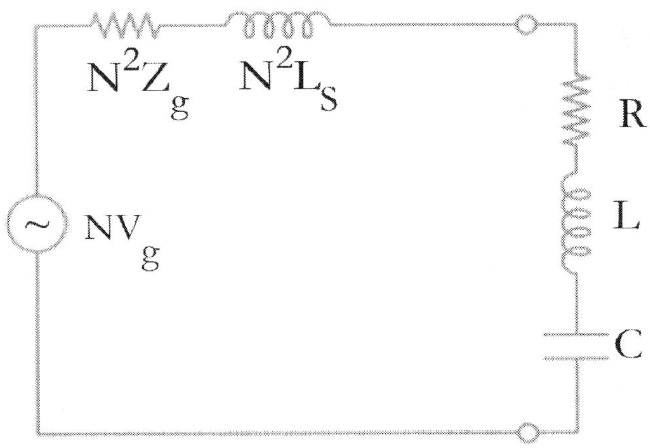

Coupling coefficient of the system is given by

$$K = \frac{N^2 Z_g}{R}$$

Value Q_e would become

$$Q_e = \frac{Q_0}{1+K}; \quad \frac{1}{Q_e} = \frac{1}{Q_0} + \frac{1}{Q_{ext}}$$

where

$$Q_{ext} = \frac{Q_0}{K} = \frac{\omega_0 L}{KR}$$

is the external Q.

If resonator is matched to generator then $Q_0 = Q_{ext}$ and $Q_1 = \frac{1}{2} Q_0 = \frac{1}{2} Q_{ext}$ and K = 1 and called critical coupling.

If K>1 then $Q_I = \frac{Q_0}{1+\rho}$ where ρ where ρ is standing wave ratio and K = ρ is called over coupling.

If K < 1, then $Q_0 = \frac{\rho}{\rho+1} Q_0$ and $K = \frac{1}{\rho}$ and is called under coupling.

Microwave Hybrid Circuits

The interconnection of two or more microwave devices may be regarded as a microwave junction. Commonly used microwave junction include such as waveguides, tees, directional coupler, circulator etc.

S- parameter of a network

S- parameter of 2 port network is

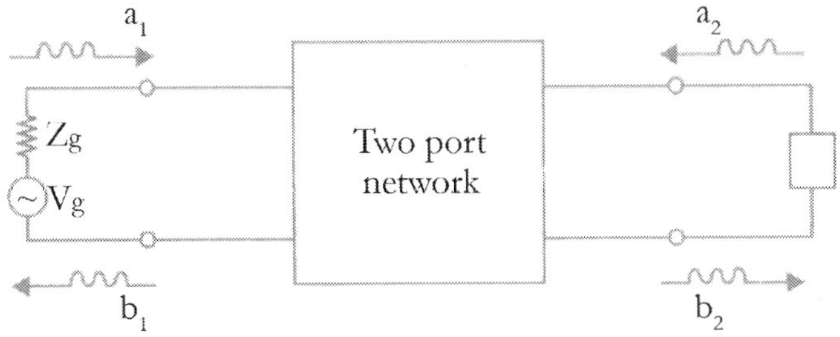

$b_1 = S_{11}a_1 + S_{12}a_2$

$b_2 = S_{21}a_1 + S_{22}a_2$

where,

a_1, a_2 is correspond to incident waveforms.

b_1, b_2 is correspond to reflected waveforms.

Tee Junction

A short circuit may always be placed in one of the arms of three port junction in such a way that no power can be transferred to the other two arms. It is impossible for a general three port junction of arbitrary symmetry to have matched impedance in all three arms.

E-plane Tee (Series tee)

E-plane tee is a waveguide tee in which axis of side arm is parallel to the E field of main waveguide and is as

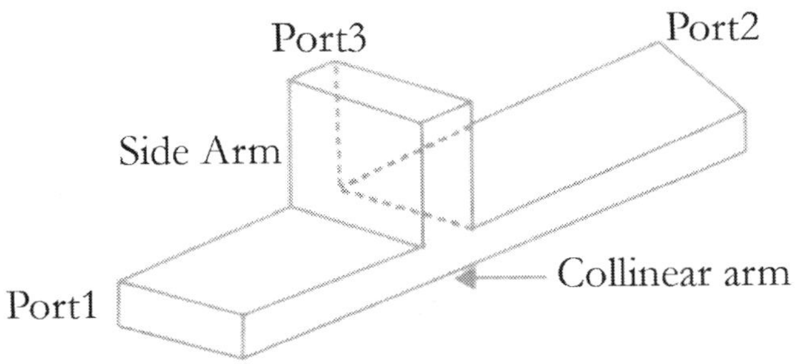

When the waves are fed into side arm (port 3) the wave appearing at port 1 and port 2 of collinear arm will be in opposite phase but, having same magnitude i,e.

$S_{13} = -S_{23}$ for E-plane tee

For a matched junction, the s-matrix is given by

$$S = \begin{bmatrix} 0 & S_{12} & S_{13} \\ S_{21} & 0 & S_{23} \\ S_{31} & S_{32} & 0 \end{bmatrix}$$

for the symmetry property of s-matrix

$S_{12}=S_{21}$, $S_{13} =S_{31}$, $S_{23}=S_{32}$

Unity property of s-matrix is

$$S_{21}.S_{21}^{*} + S_{31}S_{31}^{*} = 1$$

$$\boxed{\begin{array}{l} S_{12}.S_{12}^{*} + S_{32}S_{32}^{*} = 1 \\ S_{13}^{*}S_{13} + S_{23}S_{23}^{*} = 1 \end{array}}$$

For all matched junction.

S-matrix of E-plane Tee is Given By

$$S = \begin{bmatrix} S_{11} & S_{12} & S_{13} \\ S_{12} & S_{11} & -S_{13} \\ S_{13} & -S_{13} & S_{33} \end{bmatrix}$$

E-plane tee is also called series tee.

H-plane Tee

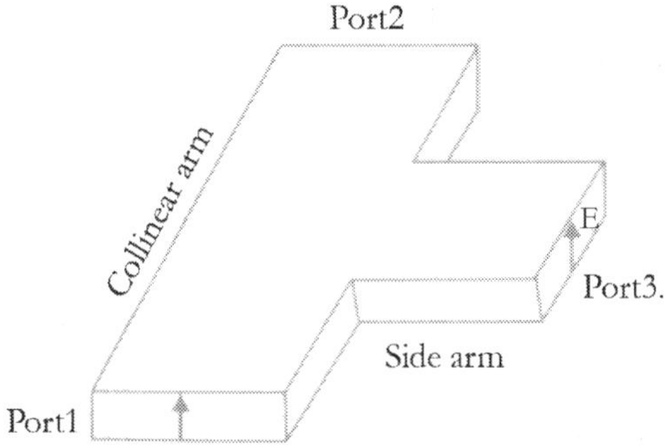

An H-plane tee is a waveguide tee in which the axis of the side arm is shunting the E-field or parallel to H field of the main waveguide.

For H-plane tee

$S_{13} = S_{23}$

s-matrix of H-plane tee is given by

$$S = \begin{bmatrix} S_{11} & S_{12} & S_{13} \\ S_{12} & S_{11} & S_{13} \\ S_{13} & S_{13} & S_{33} \end{bmatrix}$$

Hybrid Tee or Magic Tee

- Magic tee is a combination of H-plane tee and E-plane tee.
- s-matrix of hybrid tee is given by

$$S = \begin{bmatrix} 0 & 0 & S_{13} & S_{14} \\ 0 & 0 & -S_{13} & S_{24} \\ S_{31} & -S_{31} & 0 & 0 \\ S_{14} & S_{24} & 0 & 0 \end{bmatrix}$$

- Magic tee is used in Radar communication.
- Magic tee is used in coherent SSB receiver.

Hybrid Ring or Rate-race Circuits

A hybrid ring consists of an annular line of proper electrical length to sustain standing waves to which four arms are connected at proper intervals by means of series or parallel junctions.

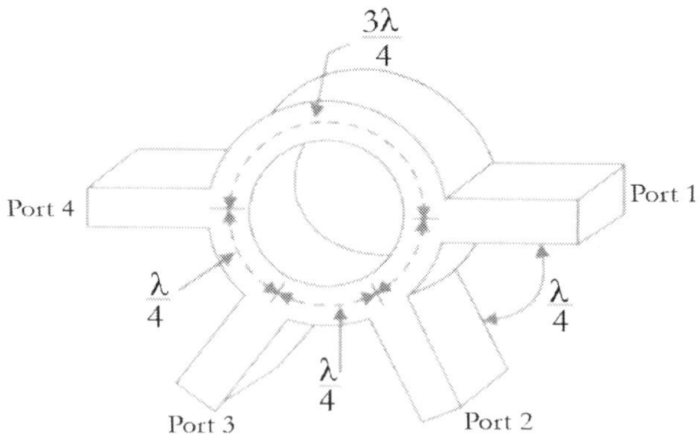

S-matrix of hybrid ring is given by

$$S = \begin{bmatrix} 0 & S_{12} & 0 & S_{14} \\ S_{12} & 0 & -S_{23} & 0 \\ 0 & S_{32} & 0 & S_{34} \\ S_{14} & 0 & S_{34} & 0 \end{bmatrix}$$

It should be noted that the phase cancellation occur only at a designated frequency for an ideal hybrid ring.

Directional Couplers

A directional coupler is a four port waveguide junction consisting of a primary waveguide 1-2 and secondary waveguide 3-4 as shown

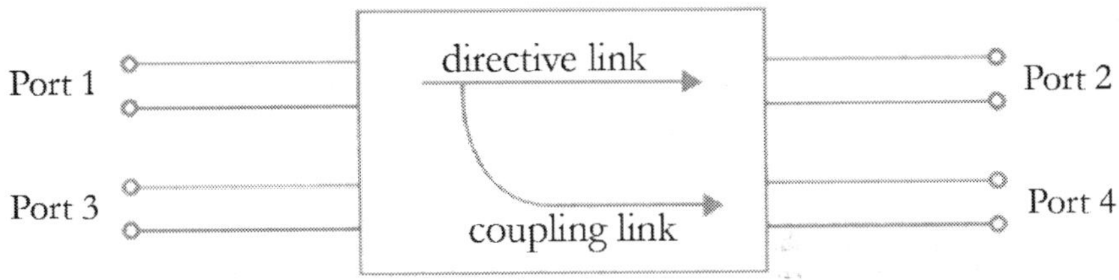

There is no transmission of power between port 1 and port 3 or between port 2 and port 4 because no coupling exists between these two pairs of ports.

S-matrix of a directional coupler is given by

$$S = \begin{bmatrix} 0 & p & 0 & jq \\ p & 0 & jq & 0 \\ 0 & jq & 0 & p \\ jq & 0 & p & 0 \end{bmatrix}$$

where,

p and q are +ve and real.

$p^2 + q^2 = 1$

The characteristic of a directional coupler can be expressed in terms of its coupling factor and its directivity and is given by

$$\text{Coupling factor (dB)} = 10 \log_{10}\left(\frac{P_1}{P_4}\right)$$

$$\text{directivity (dB)} = 10 \log_{10}\left(\frac{P_4}{P_3}\right)$$

where,

P_1 = Power input to port 1

P_3 = Power output from port 3

P_4 = Power output from port 4

Circulators and Isolators

Circulators and isolators, both are non-reciprocal transmission devices that use the property of Faraday rotation in the ferrite material. Ferrite is a family of $MeO.Fe_2O_3$ where me is a divalent iron metal.

When a piece of ferrite is affected by a dc magnetic field, the ferrite exhibits Faraday rotation. It does so because ferrite is a non-linear material.

Circulator

A circulator is a passive non-reciprocal three- or four-port device, in which a microwave or radio frequency signal entering any port is transmitted to the next port in rotation. A port in this context is a point where an external waveguide or transmission line connects to the device. It is n-port unidirectional waveguide.

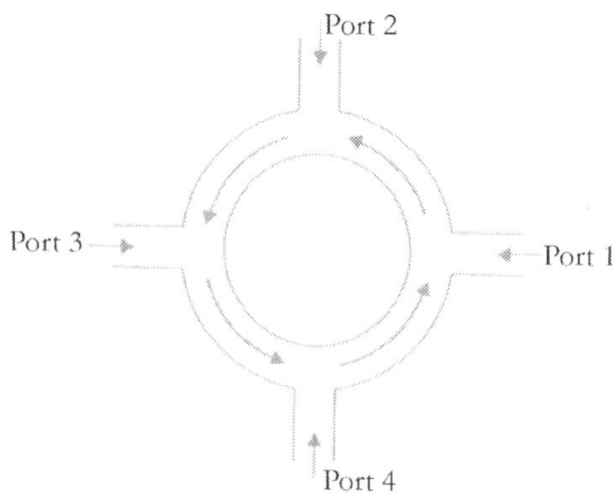

S-matrix of circulator is

$$S = \begin{bmatrix} 0 & 0 & 0 & 1 \\ 1 & 0 & 0 & 0 \\ 0 & 1 & 0 & 0 \\ 0 & 0 & 1 & 0 \end{bmatrix}$$

Propagation phase constant β^+ for the forward direction differs from the propagation phase constant β^- for the backward direction. We chose the length of ferrite slab and dc magnetic field such that

$$\left(\beta^+ - \beta^-\right).1 = \frac{\pi}{2}$$

Where

$$\beta^+ = \omega\sqrt{\mu_0 \in_0 \mu_r^+}, \; \beta^- = \omega\sqrt{\mu_0 \in_0 \mu_r^-}$$

$$\mu_r^+ = 1 + X_m^+ \text{ and } \mu_r^- = 1 + X_m^-$$

Where

μ_r^+ — Relative permittivity in clockwise direction

μ_r^- — Relative permittivity in anticlockwise direction

x_m — is susceptibility

When one port of a three-port circulator is terminated in a matched load, it can be used as an isolator, since a signal can travel in only one direction between the remaining ports. Isolator is also called aniline.

3 Microwave Transistor and Tunnel Diode

Microwave Transistor is a nonlinear device and its principle of operation is similar to that of low frequency device but requirements for dimension and other parameters are different.

The number and variety of microwave semiconductor devices, used for wireless and satellite communication and optoelectronics, have increased as new techniques, materials, and concepts have been developed and applied.

Power Frequency Limitations of Microwave Power Transistors

Voltage frequency limitation is given by

$$V_m f_T = \frac{E_m v_s}{2\pi} = \begin{cases} 2\times10^{11} & \text{for } S_i \\ 1\times10^{11} & \text{for } G_e \end{cases}$$

Where $f_T = \dfrac{1}{2\pi\tau}$ is the charge-carrier transit time cutoff frequency?

$\tau = \dfrac{L}{v}$ is the average time for charge carrier moving at an average velocity v to transverse the emitter – collector distance L.

$V_m = E_m L_{min}$ is the maximum allowable applied voltage.

E_m = maximum electric field.

v_s = maximum possible saturated drift velocity.

Current Frequency Limitation is Given By

$(I_m X_c) f_t = \frac{E_m v_s}{2\pi}$ where I_m=maximum current of the device

$X_c = \frac{1}{\omega_T C_0} = \frac{1}{2\pi f_T C_0}$ is the reactive impedance

C_0=collector base capacitance

Power Gain Frequency Limitation is Given By

$(G_m V_{th} V_m)^{1/2} f = \frac{E_m v_s}{2\pi}$ where, G_m=maximum available power gain

$V_{th} = \frac{KT}{q}$ is the thermal voltage

K=Boltzmann's constant

T=absolute temperature in Kelvin

Power Frequency Limitation is Given By

$(P_m X_c)^{1/2} f_T = \frac{E_m v_s}{2\pi}$ where, P_m=is maximum power

Maximum Available Power Gain of Transistors is Given By

Power gain included in transistor circuits, some definitions must be considered. Illustrates a signal generator E_g with an internal resistance R_g feeding into a load R_L. The total power delivered by the generator

$$P_o = \frac{E_g^{\ 2}}{R_g + R_L} \text{ the power dissipated in the load } P_L = \frac{E_L^{\ 2}}{R_L}$$

Since

$$E_L = \frac{E_g R_L}{R_g + R_L} \text{ then}$$

$$P_L = \frac{E_g^{\ 2} R_L}{\left(R_g + R_L\right)^2}$$

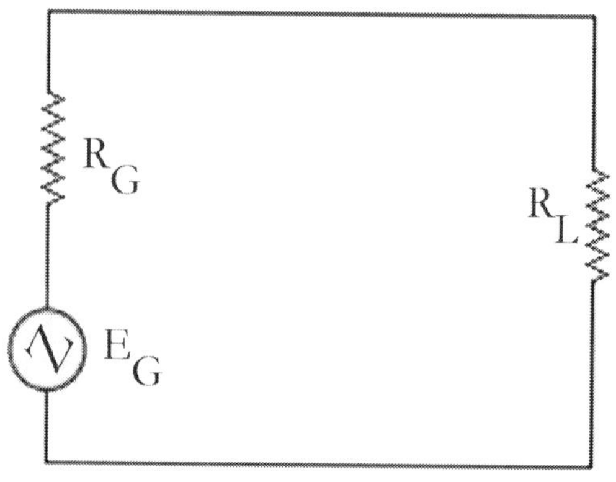

Hetrojunction Bipolar Transistor

The heterojunction bipolar transistor (HBT) is a type of bipolar junction transistor (BJT) which uses differing semiconductor materials for the emitter and base regions, creating a heterojunction. The HBT improves on the BJT in that it can handle signals of very high frequencies, up to several hundred GHz. It is commonly used in modern ultrafast circuits, mostly radio-frequency (RF) systems, and in applications requiring a high power efficiency, such as RF power amplifiers in cellular phones.

Tunnel Diode

Tunnel diode is one of the most commonly used negative conductance devices.

These diodes have a heavily doped p–n junction only some 10 nm (100 Å) wide. The heavy doping results in a broken bandgap, where conduction bandelectron states on the n-side are more or less aligned with valence band hole states on the p-side

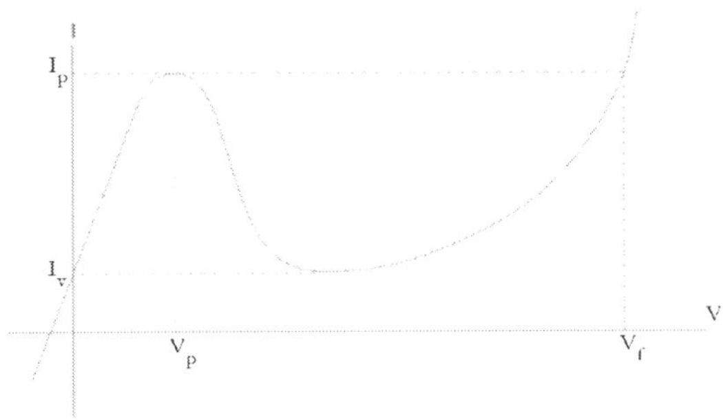

The negative resistance is used to achieve oscillation and often Ck+ function is of very high frequency frequencies.

Tunnel diode is a type of sc diode which is capable of very fast and in microwave frequency range. It was the quantum mechanical effect which is known as tunneling. It is ideal for fast oscillators and receivers for its negative slope characteristics.

Parallel Loading

The purpose of the parallel-in/ parallel-out shift register is to take in parallel data, shift it, then output it as shown below. A universal shift register is a do-everything device in addition to the parallel-in/ parallel-out function.

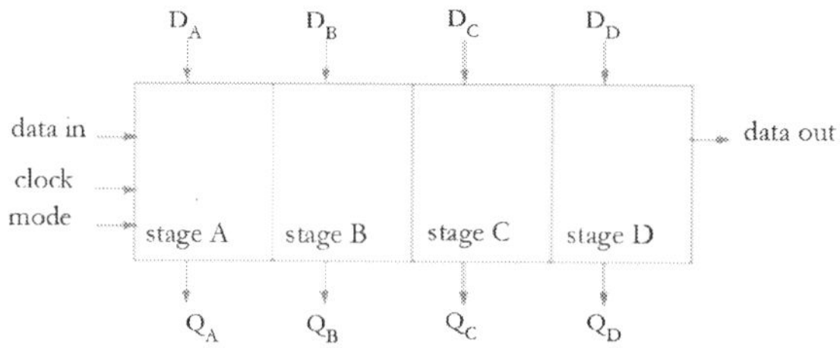

Above we apply four bit of data to a parallel-in/ parallel-out shift register at D_A D_B D_C D_D. The mode control, which may be multiple inputs, controls parallel loading vs shifting. The mode control may also control the direction of shifting in some real devices. The data will be shifted one bit position for each clock pulse. The shifted data is available at the outputs $Q_A Q_B$ Q_C Q_D.

Series Loading

For series loading equivalent circuit is given by

The power gain for series loading is given by

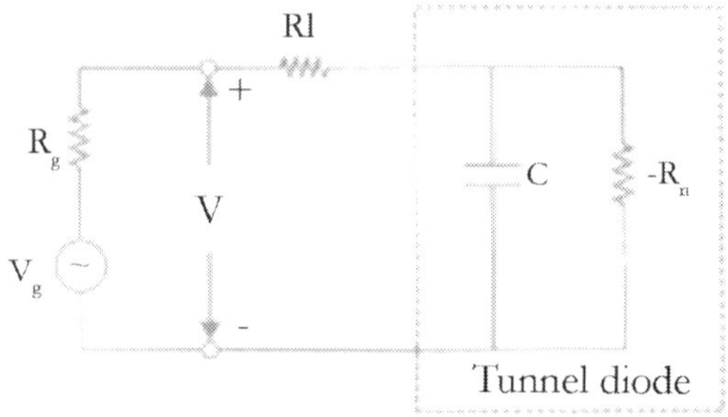

$$A = \frac{R_I}{R_I - R_n}$$

Where, A- power gain for series loading

R_l- series load

4 Microwave Field Effect Transistors

Microwave field effect transistors have several advantages over the bipolar junction transistors. Its efficiency is higher than microwave BJT. Its noise figure is low. Its operating frequency is up to X-band.

Metal Semiconductor Field Effect Transistor

The Metal-Semiconductor-Field-Effect-Transistor (MESFET) consists of a conducting channel positioned between a source and drain contact region as shown in the carrier flow from source to drain is controlled by a Schottky metal gate. The control of the channel is obtained by varying the depletion layer width underneath the metal contact which modulates the thickness of the conducting channel and thereby the current between source and drain.

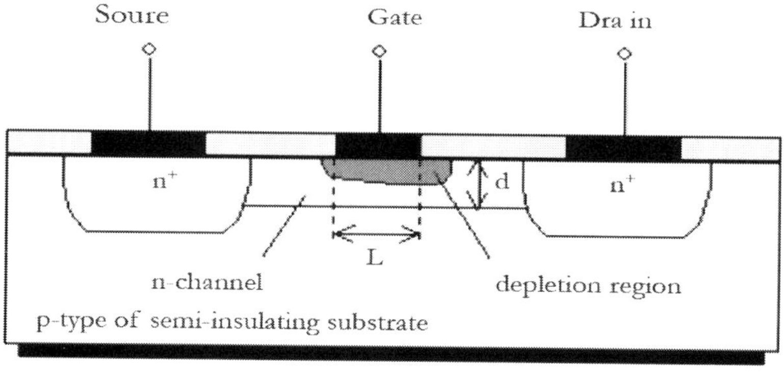

Structure of a MESFET with gate length, *L*, and channel thickness, *d*.

The higher transit frequency of the MESFET makes it particularly of interest for microwave circuits. While the advantage of the MESFET provides a superior microwave amplifier or circuit, the limitation by the diode turn-on is easily tolerated. Typically depletion-mode devices are used since they provide a larger current and larger transconductance and the circuits contain only a few transistors, so that threshold control is not a limiting factor. The buried channel also yields a better noise performance as trapping and release of carriers into and from surface states and defects is eliminated.

The use of GaAs rather than silicon MESFETs provides two more significant advantages: first, the electron mobility at room temperature is more than 5 times larger, while the peak electron velocity is about twice that of silicon. Second, it is possible to fabricate semi-insulating (SI) GaAs substrates, which eliminates the problem of absorbing microwave power in the substrate due to free carrier absorption.

Pinch-off Voltage (V_p)

Pinch off voltage for MESFETS is given by

$$V_p = \frac{qN_d a^2}{2\in_s}$$

Where, $N_d \rightarrow$ electron concentration ($/m^3$)

$a \rightarrow$ height of channel (m)

$\in_s \rightarrow$ permittivity of the material in F/m

Velocity ratio (η) is given by

$$\eta = \frac{\mu|V_p|}{v_s L}$$

Where, $v_s \rightarrow$ saturation drift velocity

L $\rightarrow$ Channel length

$\mu \rightarrow$ Low field mobility

High Electron Mobility Transistors ($HEMT_S$)

A High-electron-mobility transistor (HEMT), also known asheterostructure FET (HFET) or modulation-doped FET(MODFET), is a field-effect transistor incorporating a junction between two materials with different band gaps (i.e. aheterojunction) as the channel instead of a doped region (as is generally the case for MOSFET). A commonly used material combination is GaAs with AlGaAs, though there is wide variation, dependent on the application of the device.

The electrons generated in the thin n-type AlGaAs layer drop completely into the GaAs layer to form a depleted AlGaAs layer, because the heterojunction created by different band-gap materials forms a quantum well (a steep canyon) in the conduction band on the GaAs side where the electrons can move quickly without colliding with any impurities because the GaAs layer is undoped, and from which they cannot escape.

5 Transferred Electron Devices (TEDs)

A semiconductor device, usually a diode, that depends on internal negative resistance caused by transferred electrons in gallium arsenide or indium phosphide at high electric fields; transit time is minimized, permitting oscillation at frequencies up to several hundred megahertz.

Gunn Effect

A Gunn Effect, also known as a transferred electron device (TED), is a form of diode, a two-terminal passive semiconductor electronic component, with negative resistance, used in high-frequency electronics. Its largest use is in electronic oscillators to generate microwaves, in applications such as radar speed guns and microwave relay data link transmitters.

The production of microwave oscillations when a constant voltage in excess of a critical level is applied to opposite faces of a semiconductor.

Gunn Diode: Modes of Operation

The Gunn diode is not like a typical PN junction diode. Rather than having both p-type and n-type semiconductor, it only utilizes n-type semiconductor where electrons are the majority carriers.

The Gunn diode operation depends upon the very thin active region for its operation, it forms an ideal low power microwave RF oscillator, although it may also be used as an RF amplifier as well.

The operation of the Gunn diode can be explained in basic terms. When a voltage is placed across the device, most of the voltage appears across the inner active region. As this is particularly thin this means that the voltage gradient that exists in this region is exceedingly high.

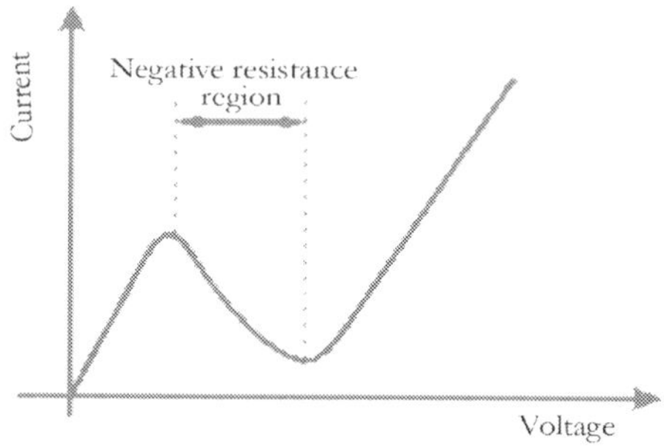

This negative resistance region means that the current flow in diode increases in the negative resistance region when the voltage falls - the inverse of the normal effect in any other positive resistance element. This phase reversal enables the Gunn diode to act as an amplifier and oscillator.

6 Avalanche Transit Time Devices

A semiconductor device with negative resistance that arises because of a phase shift between the current and the voltage at the terminals of the device as a result of the inertial properties of the avalanche multiplication of charge carriers and the finite time of their transit in the region of the p-n junction. The avalanche multiplication in the p-n junction is caused by collision ionization of atoms by the charge carriers. In contrast to other devices in this class (tunnel diodes, thyristors, and Gunn diodes), the negative resistance of avalanche-and-transit time diodes appears only at superhigh frequencies. The idea of creating this type of diode was first formulated by the American physicist W. Read in 1958.

Avalanche-and-transit time diodes are used to generate oscillations in the frequency range from 1 to 300 gigahertz. The power of the oscillations is several watts, with an efficiency of the order of 10 percent. In 1967 a new mode of operation of the diode was discovered, in which electric oscillations arise simultaneously at two frequencies: the frequency f_0, which is characteristic of the conventional mode of operation, and its sub harmonic f_0/n. This mode is distinguished by its high efficiency (up to 60 percent) and by the high level of output power at sub frequencies (up to several hundred watts).

Avalanche-and-transit time diodes can be made from structures of the p+-n-i-n+ type (the Read diode) or the p-i-n, p-n, p+-n, and p-n+ types, formed by impurity diffusion, by ion implantation, by epitaxial, or by vacuum deposition, with formation of a Schottky barrier. The semiconductor materials used in the fabrication of such diodes have a high drift rate of the charge carriers, as well as a wide forbidden gap.

Stable Amplification Mode

This mode exist when $n_0L < 10^{12} / cm^2$

In this mode device exhibits amplification at the transit time frequency rather than spontaneous oscillation.

Cd – Te Diode

It is also a crystal based diode works on same principle of transferred electron device but used for high frequency oscillations.

In – Phosphide Diode

It has three conduction band affinity between lower energy level conduction band and higher energy level conduction band is high.

Efficiency is lesser than Gunn diode.

Impatt Diodes

An IMPATT diode (IMPact ionization Avalanche Transit-Time diode) is a form of high-power semiconductor diode used in high-frequency microwave electronics devices. They have negative resistance and are used as oscillators to generate microwaves as well as amplifiers. They operate at frequencies between about 3 and 100 GHz or more. A main advantage is their high-power capability.

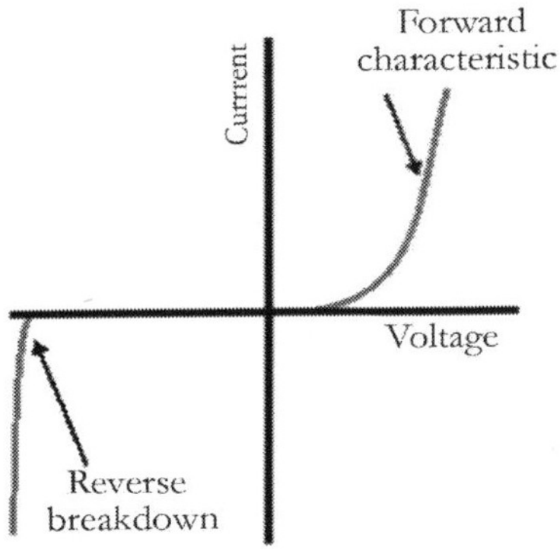

The IMPATT diode family includes many different junctions and metal semiconductor devices. The first IMPATT oscillation was obtained from a simple silicon p-n junction diode biased into a reverse avalanche break down and mounted in a microwave cavity. Because of the strong dependence of the ionization coefficient on the electric field, most of the electron–hole pairs are generated in the high field region. The generated electron immediately moves into the N region, while the generated holes drift across the P region. The time required for the hole to reach the contact constitutes the transit time delay.

Typically the device is used in a number of applications including:

- Alarms
- Radar
- Detectors using RF technology

TRAPATT Diode

The TRAPATT or TRApped, Plasma Avalanche Triggered Transit diode belongs to the same basic family as the IMPATT diode.

However the TRAPATT has a number of advantages and as a result it is used in a number of applications.

The TRAPATT diode is based around the initial concept of the IMPATT. However for the TRAPATT, the doping level between the junction and the anode.

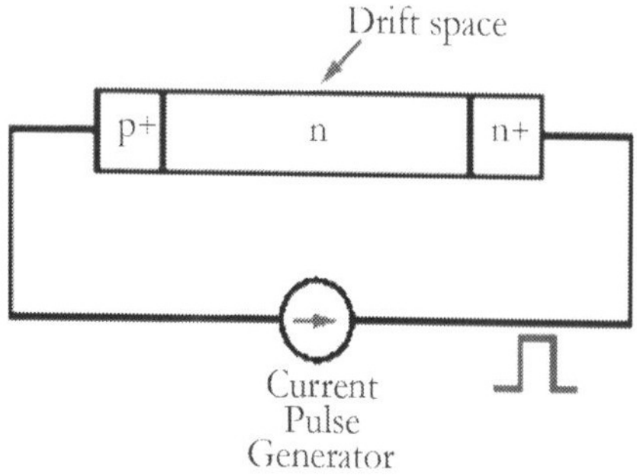

The criterion for operation in TRAPATT operation is that the avalanche front advances fasrer than the saturation velocity of the carriers.

BARITT Diodes

The BARITT diode or Barrier Injection Transit Time diode, bears many similarities to the more widely used IMPATT diode.

The BARITT is very similar, in many respects to the IMPATT, but the main difference is that the BARITT diode uses thermionic emission rather than avalanche multiplication.

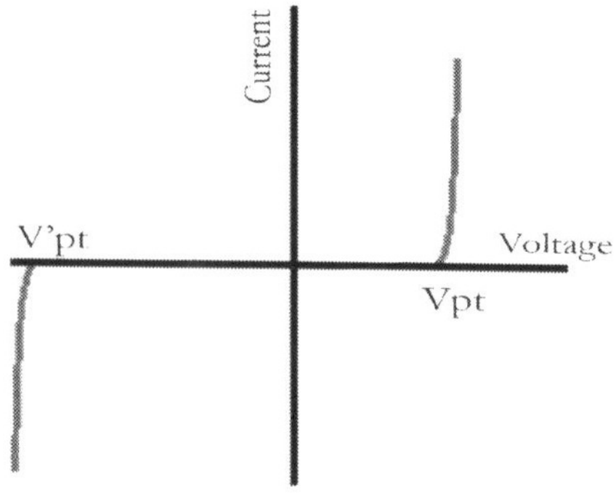

It can be seen within the diagram that the punch through voltages, Vpt are different for the two directions. This difference results from asymmetry in the two junctions and can be controlled during the manufacture stages of the diode. They can be made to be different or almost the same.

Parametric Amplifiers

A device consisting of an optically nonlinear crystal in which an optical or infrared beam draws power from a laser beam at a higher frequency and is amplified.

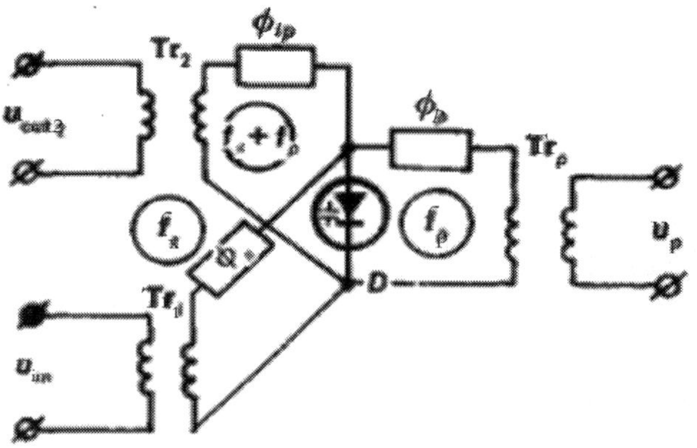

When operated at pump levels below oscillation, the parametric oscillator can amplify a signal, becoming a parametric amplifier (paramp). Varactor parametric amplifiers have been developed as low-noise amplifiers in the radio and microwave frequency range. The advantage of a parametric amplifier is that it has much lower noise than an ordinary amplifier based on a gain device like a transistor or vacuum tube. This is because in the parametric amplifier a reactance is varied instead of a (noise-producing) resistance. They have been used in very low noise radio receivers in radio telescopes and spacecraft communication antennas.

7 Microwave Linear Beam Tubes (O-Types)

Microwave tubes are lamps that produce microwaves. Microwave tubes are electron guns for generating linear beam tubes. A microwave tube generates and amplifies higher frequencies in the microwave range of frequency spectrum. When a microwave tube is energized, the electrons are emitted from the cathode and are focused on the control grid. The emitted electrons are focused by a low positive voltage. To accelerate the electron beam, a very high positive DC voltage in equal amplitude is applied at the accelerator and bunched grid. The bunched grids superimpose AC voltage on DC voltage, which generates an electrostatic field between the bunched grids. The direction of the generated electrostatic field is governed by the frequency present in the microwave tube cavity. The continuous change in an electrostatic field accelerates and de-accelerates the electron beam. There are many types of microwave tubes. Examples include Klystrons, two-cavity Klystron, and crossed-field amplifiers. Two-cavity Klystron or Klystrons are velocity-modulated tubes. Two-cavity Klystrons are widely used as communication equipment and radar. Klystrons require a strong electrostatic field for efficient emission of an electron beam. Crossed-field amplifiers are vacuum electron devices and are generally used in master oscillator power amplifiers. Other microwave tubes are commonly available.

KLYSTRON

A klystron is a specialized linear-beam vacuum tube, which is used as an amplifier for high radio frequencies, from UHF up into the microwave range. Low-power klystrons are used as oscillators in terrestrial microwave relay communications links, while high-power klystrons are used as output tubes in UHF television transmitters, satellite communication, and radar transmitters, and to generate the drive power for modern particle accelerators.

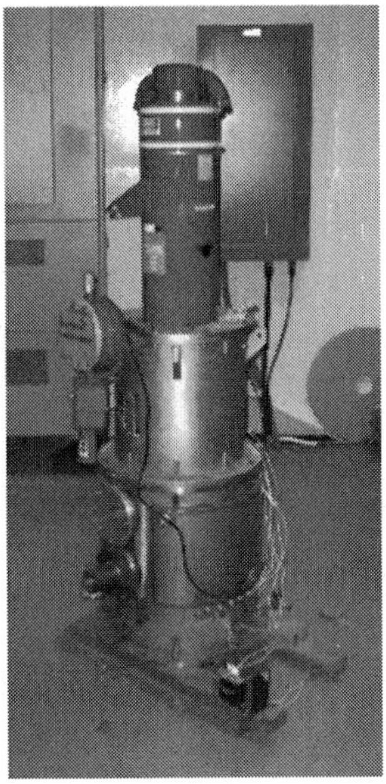

In the klystron, an electron beam interacts with the radio waves as it passes through resonant cavities, metal boxes along the length of the tube. The electron beam first passes through a cavity to which the input signal is applied. The energy of the electron beam amplifies the signal, and the amplified signal is taken from a cavity at the other end of the tube. The output signal can be coupled back into the input cavity to make an electronic oscillator to generate radio waves.

Two-cavity klystron

In the two-cavity klystron, there are two microwave cavity resonators, the "catcher" and the "buncher". When used as an amplifier, the weak microwave signal to be amplified is applied to the buncher cavity through a coaxial cable or waveguide, and the amplified signal is extracted from the catcher cavity.

At one end of the tube is the hot cathode heated by a filament which produces electrons. The electrons are attracted and pass through an anode cylinder at a high positive potential; these act as an electron gun to produce a high velocity stream of electrons.

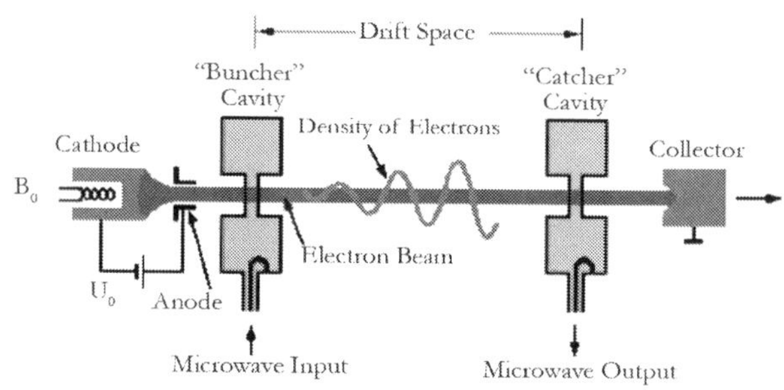

The beam first passes through the "buncher" cavity resonator, through grids attached to each side. The buncher grids have an oscillating AC potential across them, produced by standing wave oscillations within the cavity, excited by the input signal at the cavity's resonant frequency applied by a coaxial cable or waveguide. The direction of the field between the grids changes twice per cycle of the input signal. Electrons entering when the entrance grid is negative and the exit grid is positive encounter an electric field in the same direction as their motion, and are accelerated by the field.

Velocity Modulation Process

Velocity of e^- due to high dc voltage v_0 is given by

$$v_0 = \sqrt{\frac{2qV_0}{m}} = 0.593\times10^6\sqrt{V_0}$$

Let us assume that microwave signal applied is $V_s=V_1$

The average transit time through buncher gap distance d is given by

$$\tau \cong \frac{d}{v_0} = (t_1 - t_0)$$

Where, $t_0 \rightarrow$ time at which e- enters in buncher gap

$t_1 \rightarrow$ time at which e- exit from buncher gap

Average microwave voltage in the buncher gap is given by

$$\langle V_s \rangle = V_{av} = V_1 \frac{\sin\left(\frac{\omega.d}{2v_0}\right)}{\left(\frac{\omega.d}{2v_0}\right)}.\sin\left(\omega t_0 + \frac{\omega.d}{2v_0}\right)$$

$$V_{av} = V_1 \frac{\sin\left(\frac{\theta_g}{2}\right)}{\left(\frac{\theta_g}{2}\right)}\sin\left(\omega t_0 + \frac{\theta_g}{2}\right)$$

Bunching Process

Maximum and minimum velocity of e-s exit from buncher cavity is

$$v_{max} = v_0\left[1+\frac{\beta_i V_1}{2V_0}\right]; v_{min} = v_0\left[1-\frac{\beta_i V_1}{2V_0}\right]$$

Optimum bunching will take place at a distance ΔL from buncher cavity and is given by

$$\Delta L = \frac{v_0 \pi V_0}{\omega \beta_i V_1}$$

Time taken by e^- to reach second cavity after exiting from buncher cavity is given by

$$T = (t_2 - t_1) = \frac{L}{v(t_1)} = T_0\left[1 - \frac{\beta_i V_1}{2V_0}\sin\left(\omega t_1 - \frac{\theta_g}{2}\right)\right]$$

Where

$t_1 \rightarrow$ time at which e^- beam exit the buncher cavity

$t_2 \rightarrow$ time at which e- beam enter into second cavity

$L \rightarrow$ distance between two cavities

$T_0 \rightarrow \frac{L}{v_0}$ is called dc transit time

Output Power

Magnitude of fundamental component of induced current at catcher cavity is

$$I_{2ind} = \beta_0 2 I_0 J_1(X)$$

Where, $\beta_0 \rightarrow$ beam coupling coefficient of catcher cavity if buncher and catcher cavity are identical then $\beta_1 = \beta_0$.

Equivalent circuit at o/p is

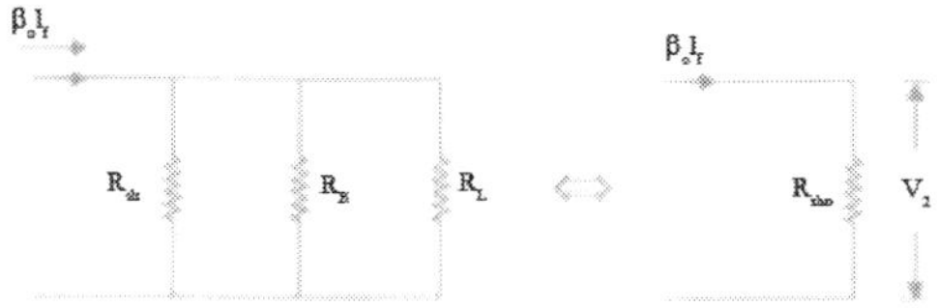

Where,

R_{sh} $\rightarrow$ wall resistance of catcher cavity

R_B $\rightarrow$ beam loading resistance

R_L $\rightarrow$ external load resistance

R_{sho} $\rightarrow$ effective shunt resistance

o/p power delivered to catcher cavity is given by

$$P_{out} = \frac{(\beta_0 I_f)^2}{2} R_{sho} = \frac{\beta_0 I_f V_2}{2}$$

Efficiency of two cavity Klystron is given by

$$\eta = \frac{P_{out}}{P_{in}} = \frac{\beta_0 I_f V_2}{2 I_0 V_0}$$

Input voltage V_1 can be expressed in terms of bunching parameter as given by

$$V_1 = \frac{2V_2}{\beta_0 \theta_0} X$$

Voltage gain of a klystron amplifier is defined as

$$A_v = \frac{|V_2|}{|V_1|} = \frac{\beta_0 I_f R_{sho}}{V_1} = \frac{\beta_0^2 \theta_0}{R_0} \frac{J_1(X)}{X} R_{sho}$$

Assuming $\beta_0 = \beta_1$

Where

$$R_0 = \frac{V_0}{I_0} \rightarrow$$ dc beam resistance

Or

$$A_v = G_m R_{sho}$$

Where,

$$G_m = \frac{2\beta_0 I_0 J_1(X)}{V_1}$$ equivalent mutual inductance of Klystron.

Reflex Klystron

The reflex klystron operates in a different mode for each additional cycle that the electrons remain in the repeller field. Mode 1 is obtained when the repeller voltage produces an electron transit time of 3/4 cycle. Additional modes follow in sequence. Mode 2 has an electron transit time of 1 3/4 cycles; mode 3 has an electron transit time of 2 3/4 cycles; etc. The physical design of the tube limits the number of modes possible in practical applications. A range of four modes of operation are normally available. The actual mode used (1 3/4 cycles through 4 3/4 cycles, 2 3/4 cycles through 6 3/4 cycles, etc.) depends upon the application. The choice of mode is determined by the difference in power available from each mode and the band of frequencies over which the circuit can be tuned.

The constant-speed electrons must remain in the reflecting field space for a minimum time of 3/4 cycle of the grid field for maximum energy transfer. The period of time the electrons remain in the repeller field is determined by the amount of negative repeller voltage. The reflex klystron will continue to oscillate if the electrons remain in the repeller field longer than 3/4 cycle (as long as the electrons return to the grid gap when the field is of the proper polarity to decelerate the electrons).

Velocity Modulation

The e- entering the cavity gap from the cathode at time t_0 it is assumed to have uniform velocity

$$v_0 = 0.593 \times 10^6 \sqrt{V_0}$$

Some e^- leaves the cavity gap at $t=t_1$ with velocity

$$v_0(t_1) = v_0\left[1 + \frac{\beta_i V_1}{2V_0}\sin\left(\omega t_1 - \frac{\theta_g}{2}\right)\right]$$

Same electron beam is forced back to the cavity at $t=t_2$ by retarding electric field E, which is given by

$$E = \frac{V_r + V_0 + V_1 \sin \omega t}{L}$$

Where,

$V_r \rightarrow$ retarding voltage or repeller voltage

$L \rightarrow$ distance between repeller and cavity

Power o/p and Efficiency

For maximum energy transfer the round trip transit angle must be

$$\omega(t_2 - t_1) = \omega T' \cong \omega T_0 = \left(n - \frac{1}{4}\right) 2\pi = N2\pi$$

Assuming $V_1 << V_0$ and n is any positive integer for cycle number.

$$N = \left(n - \frac{1}{4}\right) \text{ is the number of modes.}$$

Induced current magnitude (Fundamental component) is given by

$$I_2 = 2I_0 \beta_i J_1(X')$$

Where,

$$I_0 \rightarrow \text{dc current}$$

The ratio of signal voltage V_1 to dc voltage V_0 is given by

$$\frac{V_1}{V_0} = \frac{2X'}{\beta_i (2\pi n - \pi/2)}$$

A.C. power is given by

$$P_{ac} = \frac{V_1 V_2}{2} = \frac{2V_0 I_0 X' J_1(X')}{(2\pi n - \pi/2)}$$

Efficiency of reflex Klystron is given by

$$\eta = \frac{2X' J_1(X')}{(2\pi n - \pi/2)} = \frac{P_{ac}}{P_{dc}}$$

η is maximum when X'=2.408 and at this X'.

$X' J_1(X') = 1.25$

For a given beam voltage V_O, the relationship between the repeller voltage and cycle number n required for oscillation is given by

$$\frac{V_0}{(V_r + V_0)^2} = \frac{(2n\pi - \pi/2)^2}{8\omega^2 L^2} \frac{q}{m}$$

Power o/p can be expressed in the terms of repeller voltage V_r as

$$P_{ac} = \frac{V_0 I_0 X' J_1(X')(V_r + V_0)}{\omega L} \sqrt{\frac{q}{2mV_0}}$$

HELIX Travelling Wave Tubes (TWTs)

A traveling-wave tube (TWT) is a specialized vacuum tube that is used in electronics to amplify radio frequency (RF) signals in the microwave range.

Helix TWT - in which the radio waves interact with the electron beam while traveling down a wire helix which surrounds the beam. These have wide bandwidth, but output power is limited to a few hundred watts.

Modulated velocity in case of Helical travelling wave tube is given by

$$v(t) = v_0 \left[1 + \frac{qE_0}{m\omega V_0} \cos \omega t \right]$$

where,

E_0 - magnitude of applied signal (electric field).

Gain parameter C of travelling wave tube is given by

$$C = \left(\frac{I_0 Z_0}{4V_0} \right)^{1/3}$$

where,

I_0 is dc current or beam current

V_0 is beam voltage or dc voltage

Z_0 is characteristic impedance of Helix

o/p power gain Ap in decibels is given by

$$A_p(dB) = -9.54 + 4.73NC$$

where, N is circuit length

The values of propagation constants γ are given by

$$\gamma_1 = -\beta_e C \frac{\sqrt{3}}{2} + j\beta_e \left(1 + \frac{C}{2} \right);$$

$$\gamma_2 = \beta_e C \frac{\sqrt{3}}{2} + j\beta_e \left(1 + \frac{C}{2} \right)$$

$$\gamma_3 = j\beta_e (1 - C);$$

$$\gamma_4 = -j\beta_e \left(1 - \frac{C^3}{4} \right)$$

where

$$\beta_e = \frac{\omega}{v_0}$$

8 Microwave Crossed Field Tubes (M-type)

Crossed-field tubes get their name from the fact that the dc electric field and the magnetic field are crossed at right angles. One of the first crossed-field tubes used in early radars was the magnetron, and it was the only one available for quite a few years. Crossed-field tubes are also known as M-type devices, since they deal with propagation of waves in a magnetic field.

Cylindrical Magnetron

This type of cylindrical magnetron is a reliable solution for high power applications. The solid design, proven reliability and complete application range makes the end blocks an ideal solution for large area coaters.

- Hull cut off magnetic equation is given by

$$B_{oc} = \frac{\left(\dfrac{8V_0 m}{q}\right)^{1/2}}{b\left(1-\dfrac{a^2}{b^2}\right)}$$

Where,

V_0 is dc voltage

a is radius of cathode cylinder

b is radius from the center of cathode to the edge of the anode.

- Cutoff voltage is given by

$$V_{oc} = \frac{q}{8m} B_0^2 b^2 \left(1 - \frac{a^2}{b^2}\right)^2 \quad \text{Hull cut off voltage equation}$$

B_0 is applied magnetic flux.

If applied voltage V_0 will be less than V_{OC} for a given B_0, the electron will not reach the anode.

- Cyclotron angular frequency is given by

$$\omega_c = \frac{q}{m} B_o$$

where

ω is cyclotron angular frequency

m mass of e-

B_O is applied magnetic flux

- Time period of one complete revolution is given by

$$T = \frac{2\pi}{\omega_c} = \frac{2\pi m}{qB_o}$$

- If there are N reentrant cavities in the anode, the phase shift between two adjacent cavities is given

$$\phi_n = \frac{2\pi p}{N}$$

where

p is an integer

- If L is the mean separation between cavities, the phase constant of fundamental mode field is given by

$$\beta_0 = \frac{2\pi p}{NL}$$

where

β_0 Phase constant

9 Microwave Measurement

For frequency measurement we use cylindrical cavity resonator of adjustable dimensions. Inside cavity there is either parallel resonance circuit or series resonance circuit. If we are using parallel resonance circuit then for unknown frequency at i/p there will be no o/p. If we are using series resonance circuit then for unknown frequency at i/p there will be maximum o/p.

Bolometer

A bolometer is a device for measuring the power of incident electromagnetic radiation via the heating of a material with a temperature-dependent electrical resistance.

VSMR Meter

Works with slotted waveguide. It is also called current detecting device. VSWR is given by

$$\text{VSWR} = \frac{V_{max}}{V_{min}} = \sqrt{\frac{I_{max}}{I_{min}}}$$

where

V_{max}, V_{min} is max. and min. voltage.

I_{max}, I_{min} is max. and min. current.

If d_1 is position of V_{max} and d_2 is position of next V_{max} then

$$\frac{\lambda_g}{2} = (d_1 - d_2)$$

and

$$\frac{1}{\lambda^2} = \frac{1}{\lambda_c^2} + \frac{1}{\lambda_g^2}$$

where

λ_g is guide wave length

λ_c is cutoff wavelength

λ is wavelength of microwave signal

VSWR meter can also be used to measure frequency indirectly.

Microwave Strip Lines

Strip line is a transverse electromagnetic (TEM) transmission line medium. Microwave strip lines support quasi TEM wave. Characteristic impedance of microwave strip lines is given by

$$Z_0 = \frac{87}{\sqrt{\in_r + 1.41}} \text{In}\left[\frac{5.98h}{0.8w + t}\right]\Omega$$

$\in_r$ is relative permittivity of material of strip line

h is height from the micro-strip line to ground in miles

w is width of microwave strip line in miles

t is thickness of microwave strip line in miles

Index

Symbols

A

B

C

D

E

F

Q

R

S

T

U

V

W

Z